Seismic Wave Propagation in Stratified Media

Seismic Wave Propagation in Stratified Media

BRIAN L. N. KENNETT

THE AUSTRALIAN NATIONAL UNIVERSITY

E PRESS

E PRESS

Published by ANU E Press
The Australian National University
Canberra ACT 0200, Australia
Email: anuepress@anu.edu.au
This title is also available online at: http://epress.anu.edu.au/seismic_citation.html

National Library of Australia
Cataloguing-in-Publication entry

National Library of Australia Cataloguing-in-Publication entry

Author:	Kennett, B. L. N. (Brian Leslie Norman), 1948-
Title:	Seismic wave propagation in stratified media / Brian Kennett.
Edition:	New ed.
ISBN:	9781921536724 (pbk.) 9781921536731 (pdf.)
Subjects:	Seismology. Seismic waves. Seismic tomography.
Dewey Number:	551.220287

Cover design by ANU E Press.

Table of Contents

Preface

The primary sources of data in seismology are seismic records at the Earth's surface of natural or man-made events. The object of modern seismic analysis is to extract as much information as possible from these surface records about the nature of the seismic parameter distribution with the Earth and the source which generated the waves.

One of the most important techniques which has been developed in recent years is the construction of theoretical seismograms, as an aid to structural and source studies. In order to model such seismograms we must take into account the generation of seismic waves by the source, the passage of these waves through the Earth, and their subsequent detection and recording at the receiver.

In this book I have endeavoured to present a unified account of seismic waves in stratified media. The emphasis is on the propagation of seismic waves in realistic earth models, and the way in which this can be understood in terms of the reflection and transmission properties of portions of the stratification. With this approach I have tried to show the interrelation between the major methods used for the calculation of theoretical seismograms, and to indicate the circumstances in which they are most useful. The theoretical techniques developed in this book are applicable to a wide range of problems with distance scales which vary from a few kilometres in geophysical prospecting, to many thousands of kilometres for seismic phases returned from the Earth's core. These applications are illustrated by using examples taken from reflection and refraction seismic work, as well as earthquake studies.

I have assumed an acquaintance with the basics of elastodynamics see, e.g., Hudson (1980), and the elements of geometrical ray theory. I have not repeated material which is available in many other sources. Thus, there is no discussion of the classical Lamb's problem for a uniform half space, but I present a physically based description of the more general problem of the excitation of seismic waves in a stratified half space.

Very low frequency seismic wave problems involve the properties of the whole Earth; these are best studied in terms of the free oscillations of the Earth and are not treated here. A description of this approach may be found in the book by Lapwood & Usami (1981).

This book is a revised version of an essay which shared the Adams' Prize, in the University of Cambridge, for 1979–1980. The material has grown out of lectures for graduate students given at the University of California, San Diego, and for the Mathematical Tripos Part III at the University of Cambridge. The manuscript was prepared whilst I was a Visiting Fellow at the Cooperative Institute for Research in the Environmental Sciences, University of Colorado, Boulder, and I am very grateful for the generous provision of facilities during my stay.

I would like to thank my research students N.J. Kerry, T.J. Clarke and M.R. Illingworth, who have done much to shape my ideas on seismic wave propagation, for their help and criticism. I have tried to illustrate many of the seismic wave phenomena with actual seismograms and I extend my thanks to the many people who have helped me in my search for such examples. I am grateful to the Royal Astronomical Society for permission to reproduce a number of diagrams from the Geophysical Journal, and to the National Center for Earthquake Research, U.S. Geological Survey for the provision of figures.

Finally I would like to thank my wife, Heather, without whom this book would never have been finished.

B.L.N. Kennett

This ANU E-Press Edition has been produced by converting the original `troff` source to LaTeX. In the process some minor changes to notation have been made to aid clarity. An appendix is introduced to provide a definition of the notation. The index has also been reworked to reflect the changed page layout.

Chapter 1

Introduction

1.1 Seismic signals

The surface of the Earth is in constant slight movement and the motion at any point arises from both local effects, e.g., man-made disturbances or wind-induced rocking of trees, and from vibrations arising from afar, such as microseisms generated by the effect of distant ocean storms. If we look at the records from an observatory seismometer in its carefully constructed vault, or from a geophone whose spike is simply driven into the ground, we find that these largely consist of such seismic 'noise'. From time to time the irregular pattern of the records is interrupted by a disturbance which rises above the background noise with a well defined wavetrain (figure 1.1). This feature arises from the excitation of seismic waves, away from the receiver, by some natural or artificial source.

Earthquakes are the most common natural generators of seismic waves, and in the period range 0.001 Hz to 4 Hz their effect may be detected at considerable ranges from the source (e.g., with a surface displacement of around 10^{-8} m at 9000 km for a surface wave magnitude of 4). Indeed for the largest earthquakes we can observe waves that have circled the globe a number of times.

Most artificial sources such as chemical explosions and surface vibrators or weight-dropping devices have a much shorter range over which they give detectable arrivals. This distance is about 2 km for a single surface vibrator and may be as large as 1000 km for a charge of several tons of TNT. Only large nuclear explosions rival earthquakes in generating seismic waves which are observable over a considerable portion of the Earth's surface.

The nature of the seismic noise spectrum has had a profound influence on the nature of the instruments which have been emplaced to record earthquake signals and this in turn has affected the way in which seismic wave theory has developed. The power spectral density for the velocity of the Earth's surface as a function of frequency is shown in figure 1.2, based on a study at the Gräfenburg array (Harjes, 1981). The range of noise conditions for a reasonable quality station is indicated by the shaded region and the solid line indicates a typical smoothed velocity spectrum for the noise. Superimposed on the noise results are typical

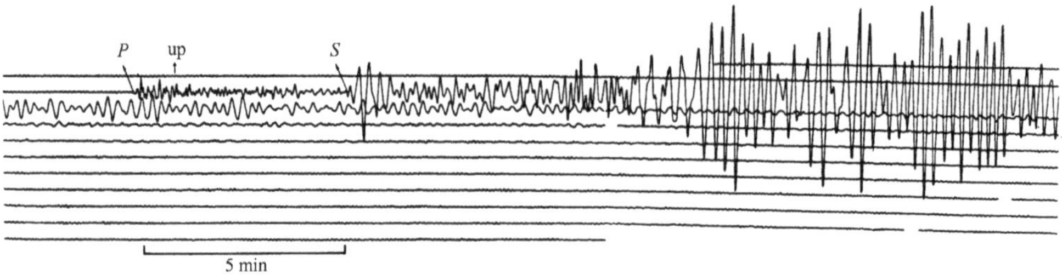

Figure 1.1. Eskdalemuir Seismogram for the Dardenelles earthquake 1912, Δ = 26.5°, Galitzin seismograph, vertical component.

amplitude spectra for the *P* wave, *S* wave and fundamental mode Rayleigh wave (*R*) for a magnitude 7 earthquake. For smaller events the spectral peaks shift to higher frequencies. The major noise peaks associated with the microseisms at 0.07-0.15 Hz cause considerable difficulties on analogue recording systems, since it is difficult to achieve adequate dynamic range to cope with the whole range of signals. For a photographic recording system there is an unavoidable limit to the smallest signal which can be discerned due to the width of the light beam and large signals simply disappear off scale. To avoid swamping the records of medium size events with microseismic noise the commonest procedure is to operate two separate instruments with characteristics designed to exploit the relatively low noise conditions on the two sides of the noise peak. This is the procedure followed in the World Wide Seismograph Network (WWSSN) which installed separate long-period and short-period instruments at over 100 sites around the world in the 1960s. The records from the long-period seismometers are dominated by the fundamental modes of surface waves although some body waves are present. The short-period records show principally body waves. A similar arrangement has been made for the digital recording channels of the SRO network which was designed to enhance the WWSSN system. The response curves for these instruments are illustrated in figure 1.2b on the same frequency scale as the power spectra in figure 1.2a. We see that these responses have a very rapid fall off in the neighbourhood of the noise peak so that even with digital techniques it is hard to recover information in this region.

For midcontinental stations the microseismic noise levels are much reduced and it then becomes possible to follow the original approach of Galitzin and use a single instrument over the frequency band of interest, as for example, in the Kirnos instrument used in the Soviet Union. The actual sensor for the SRO system is also a broad-band seismometer but the recording channels conform to the scheme originally devised for analogue purposes.

With digital recording it is now possible to use such a 'broad band' instrument even in noisier areas and to filter the data after recording if it is necessary to suppress the microseisms. The displacement response of the Wielandt system used

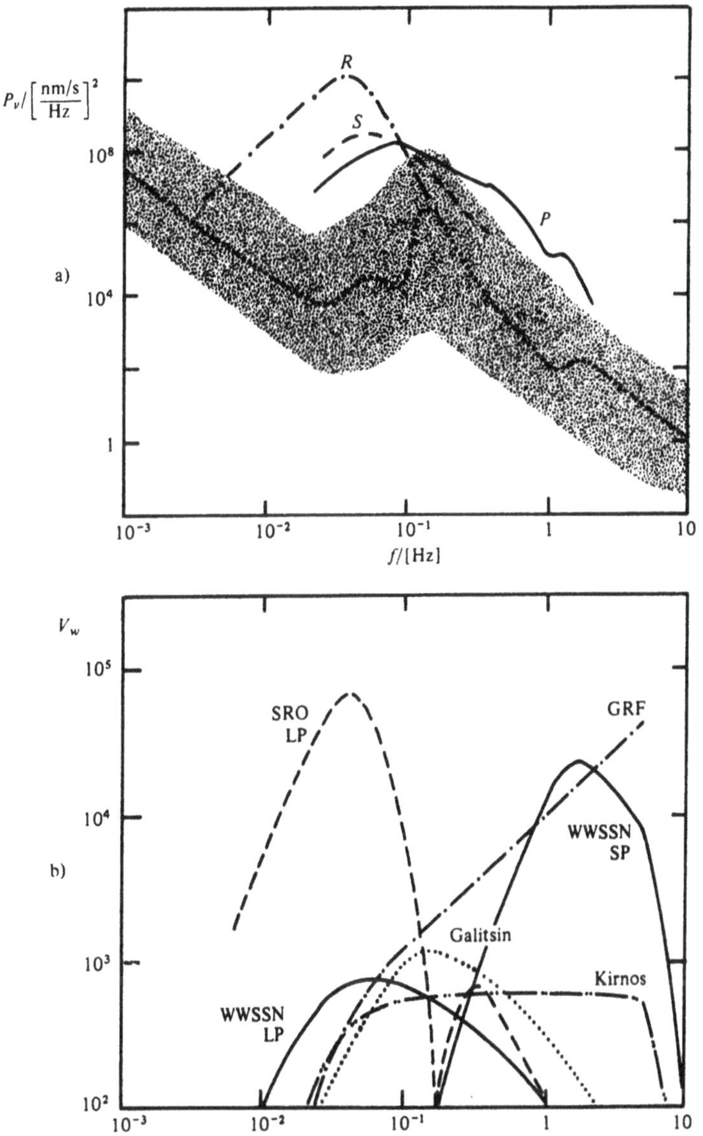

Figure 1.2. a) Power spectral density for velocity at the earth's surface and the range of noise conditions (shaded). The corresponding spectra for *P*, *S* and Rayleigh waves for a magnitude 7 earthquake are also shown. b) Instrumental response curves for major seismometer systems.

at the Grafenburg array is indicated in figure 1.2b (GRF); in velocity the response is flat from 0.05-15.0 Hz. With digital techniques it is possible to simulate the response of the narrower band WWSSN and SRO systems (Seidl, 1980) but the full information is still available.

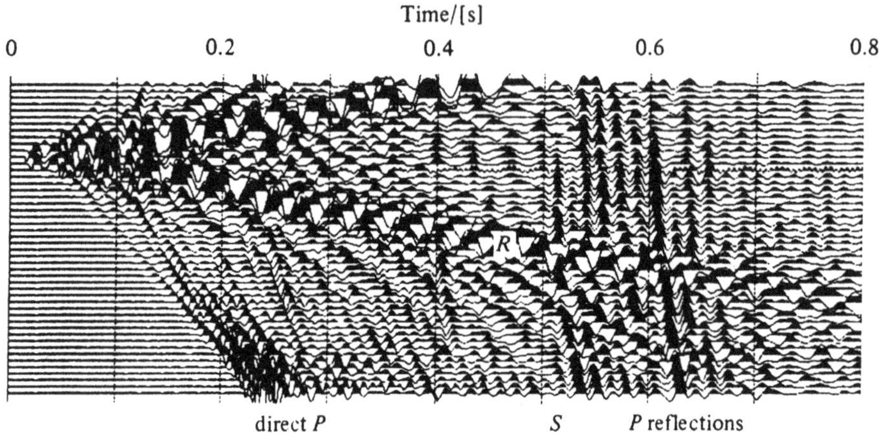

Figure 1.3. Single shot spread for shallow reflection work showing *P* refractions and reflections and prominent Rayleigh waves (*R*).

In near earthquake studies and in most work with artificial sources the recording bandwidth is sufficient to include all the major wave phenomena. Here, however, attention may be concentrated on just one part of the records, e.g., the first arriving energy. In prospecting work the recording configuration with groups of geophones may be designed to suppress the slowest surface waves (*ground roll* - Telford et al., 1976). When high frequency information is sought in reflection work, single geophones or very tight clusters may be employed to avoid problems with lateral variations in near-surface properties. In this case monitor records show very clearly the onset of compressional *P* wave energy and the Rayleigh wave energy giving rise to the ground-roll (figure 1.3). The direct *S* wave is not seen very clearly on vertical component geophones, but can just be discerned on figure 1.3. At larger offsets *S* waves reflected by the near-surface layering separate from the ground-roll, and in figure 1.3 there is also some indication of higher mode surface waves. Normally the ground-roll and *S* waves would be suppressed by some form of velocity filtering, but they do contain useful information about the shallow structure which can complement the *P* wave information.

The advent of broad-band recording blurs the separation of seismic signals into body waves and surface waves. Both are present on the same records and indeed we see features that cannot be readily assigned to either class. The development of seismic wave theory has tended to mirror this separation into body wave and surface wave studies. We are now, however, able to adopt a more broadly based approach and extract the full range of wave propagation effects from a unified treatment which we shall develop in subsequent chapters.

1.2 Seismogram analysis

The primary sources of information in seismology are the records of seismic events obtained at the Earth's surface. This means that we have a comparatively limited sampling of the entire seismic wavefield in the Earth. We would, however, like not only to deduce the detailed nature of the source which generated the seismic waves but also to try to determine the elastic parameter distribution within the Earth.

During the years up to 1940, the analysis of the arrival times of seismic pulses using ray theoretical methods led to the construction of models for the *P* wavespeed distribution within the Earth (Jeffreys & Bullen 1940). The analysis technique, and the smoothing applied to the observational data led to a continuously varying wavespeed profile with radius. This was only interrupted by the boundary between the mantle of the Earth and the fluid core and between the core and inner core. The construction of such models stimulated the development of methods to handle wave propagation in realistic media.

At this time the detailed character of the source was ignored and only its location was of relevance to the travel-time studies. Subsequently the sense of the initial *P* motions on a suite of seismograms from stations surrounding the earthquake were used to assign a simple faulting model to an earthquake (fault-plane solutions - see, e.g., Sykes, 1967) and propagation characteristics were largely ignored. It was, however, noted that allowance must be made for earthquakes with epicentres in the crust rather than in the mantle.

To improve on these very useful simple descriptions of the Earth's structure and the nature of earthquake sources, one must make further use of the information contained in the original seismograms. Over the last decade, methods based on the calculation of theoretical seismograms have been developed to aid in both source and structural studies.

In order to model the nature of the wavetrains recorded by a seismometer we have to take account of the entire process whereby the seismic energy reaches the recording site. This may be separated into three major elements. Firstly, the generation of the waves by the source, secondly, the passage of the waves through the Earth to the vicinity of the receiver and finally the detection and recording characteristics of the receiver itself.

The character of the propagation effects depends on the nature of the elastic parameter distribution within the Earth and the scale of the paths of interest and can display a wide variety of phenomena. Although we shall be principally concerned with these propagation problems, we shall need to keep in mind the effect of both source and receiver on the nature of the seismograms.

As we have seen above, the nature of the recording system can have a significant effect on the nature of the wavetrains observed by any particular seismometer. The construction of detailed models for seismic sources is rather difficult, particularly when the rupturing processes in faulting are included, since the reaction of the medium itself cannot be ignored. These aspects are discussed in detail in Volume II of Aki & Richards (1980). We shall be concerned with simple source descriptions, in particular point sources specified by a moment tensor or a specified fault model, without attempting to describe the mechanics of faulting.

1.3 Seismic waves

Although it is customary to treat seismic waves as if they satisfy the equations of linear isotropic elasticity this is an approximation and we should be aware of its limitations.

The level of stress within the Earth, predominantly due to gravitation, reaches values of the order of 10^{11} Pa. The elastic moduli at depth are of this same order, in fact in the lower mantle the stress is a little less than half the shear modulus and one-fifth of the bulk modulus. If, therefore, we start from a reference state of a non-gravitating earth we have strains of order unity, even in the absence of seismic waves, and we could certainly not use linear theory. We therefore have to adopt an incremental treatment about the gravitationally prestressed state.

As seismic waves pass through the Earth they lose energy by the geometrical effect of the enlargement of the wavefront and by the intrinsic absorption of the Earth. In most circumstances the loss due to scattering and absorption is relatively small so that we are able to treat this attenuation of the seismic energy as a small perturbation on the propagation process.

1.3.1 The effect of prestress

In the equilibrium state of the Earth, in the absence of seismic activity, the gradient of the stress tensor $\boldsymbol{\sigma}^0$ matches the gravitational accelerations derived from a potential ψ^0,

$$\frac{\partial \sigma^0_{ij}}{\partial x_i} + \rho^0 \frac{\partial \psi^0}{\partial x_j} = 0. \tag{1.0}$$

where ρ^0 is the equilibrium density and we have used the convention of summation over repeated suffices. This initial stress field will be predominantly hydrostatic. For perfect isostatic compensation at some level (by Airy or Pratt mechanisms) there would be no deviatoric component at greater depths.

Except in the immediate vicinity of a seismic source the strain levels associated with seismic waves are small. We therefore suppose incremental changes of displacement ($\mathbf{u}$) and stress (σ_{ij}) from the equilibrium state behave as for an elastic medium, and so these quantities satisfy the equation of motion

$$\frac{\partial \sigma_{ij}}{\partial x_i} + \rho f_j = \rho \frac{\partial^2 u_j}{\partial t^2}. \tag{1.1}$$

The body force term f_j includes the effect of self-gravitation and in particular the perturbation in the gravitational potential consequent on the displacement.

In order to express (1.1) in terms of the displacement alone we need a constitutive relation between the stress and strain increments away from the reference state. The usual assumption is that this incremental relation is that for linear elasticity and so we model σ_{ij} by a stress field τ_{ij} generated from the displacement $\mathbf{u}$,

$$\tau_{ij} = c_{ijkl}\partial_l u_k, \tag{1.2}$$

with $\partial_l = \partial/\partial x_l$. The tensor of incremental adiabatic elastic moduli has the symmetries

$$c_{ijkl} = c_{jikl} = c_{ijlk} = c_{klij}. \tag{1.3}$$

If however there is a significant level of stress in the reference state the relation (1.2) would not be appropriate and a more suitable form (Dahlen, 1972) is provided by

$$\tau_{ij} = d_{ijkl}\partial_k u_l - u_k\partial_k\sigma^0_{ij}. \tag{1.4}$$

The second term arises because it is most convenient to adopt a Lagrangian viewpoint for the deformation of the solid material. The constants d_{ijkl} depend on the initial stress

$$d_{ijkl} = c_{ijkl} + \tfrac{2}{3}\left(\delta_{ij}\sigma^0_{kl} - \delta_{kl}\sigma^0_{ij} + \delta_{il}\sigma^0_{jk} - \delta_{jk}\sigma^0_{il} + \delta_{jl}\sigma^0_{ik} - \delta_{ik}\sigma^0_{jl}\right), \tag{1.5}$$

and the tensor c_{ijkl} possesses the symmetries (1.3). For a hydrostatic initial stress state d_{ijkl} reduces to c_{ijkl}. The slight influence of the second term in (1.4), is frequently neglected. The main gradient of the stress tensor σ^0_{ij} is normally that with depth and, for the hydrostatic component, is about 40 Pa/m in the Earth's mantle. In this region the elements of c_{ijkl} are of order 10^{10} Pa and if we consider a disturbance with a wavelength 200 km (i.e. a frequency around 0.05 Hz), the term $c_{ijkl}\partial_k u_l$ will be about 10^4 times the correction $u_k\partial_k\sigma^0_{ij}$, and this ratio will increase with increasing frequency. For teleseismic studies the correction is therefore negligible.

Deviatoric components of the initial stress are likely to be most significant in the outer portions of the Earth, where spatial variability of the elastic constants is also important. The initial stress state may therefore have significant spatial variation on scales comparable to seismic wavelengths, and so the correction $u_k\partial_k\sigma^0_{ij}$ in (1.4) will be of greater significance than at depth.

1.3.2 Material anisotropy

A constitutive relation such as (1.2) expresses the macroscopic characteristics of the material within the Earth. On a fine scale we will have a relatively chaotic assemblage of crystal grains with anisotropic elastic moduli. However, the overall properties of a cube with the dimensions of a typical seismic wavelength (a few kilometres in the mantle) will generally be nearly isotropic. In consequence the elastic constant tensor may often be approximated in terms of only the bulk modulus κ and shear modulus μ

$$c_{ijkl} = (\kappa - \tfrac{2}{3}\mu)\delta_{ij}\delta_{kl} + \mu(\delta_{ik}\delta_{jl} + \delta_{il}\delta_{jk}). \tag{1.6}$$

Significant large-scale anisotropy has only been established in a limited number of circumstances; for example, at the top of the upper mantle under the oceans there is about five per cent anisotropy in *P* wavespeed (Raitt, 1969). Such anisotropy is likely to arise when there is preferential alignment of crystal grains, associated with some prevailing tectonic stress.

In the period before a major earthquake, significant prestrain can be built up in the epicentral region, which will be relieved by the earthquake itself. The presence of such a strain modifies the local constitutive relation as in (1.4), where σ^0_{ij} is to be taken now as the stress associated with the prestrain (Walton 1974), and this will give rise to apparent anisotropy for propagation through the region. The presence of non-hydrostatic stress will have a significant effect on the crack distribution in the crust. At low ambient stress and low pore pressure within the rocks, systems of open cracks may be differentially closed. At high ambient stress, systems of closed cracks may be opened if the pore pressure and non-hydrostatic stress are large enough, and new cracks may also be formed. Such *dilatancy* effects lead to aligned crack systems over a fair size area, and this will give apparent anisotropy to seismic wave propagation through the region. Recent evidence suggests such effects are observable in favourable circumstances (Crampin et al., 1980).

For near-surface rocks, patterns of cracking and jointing can also give rise to anisotropic variation in wavespeed.

Transverse anisotropy, where the vertical and horizontal wavespeeds differ, can be simulated by very fine bedding in sedimentary sequences below the scale of seismic disturbances. Evidence from well logs suggests that this effect can be important in some prospecting situations for compressional wave propagation (Levin, 1979).

Transverse isotropy has also been postulated for the outer part of the upper mantle above 250 km depth, in an attempt to reconcile the observed dispersion of Love and Rayleigh waves at moderate periods (Dziewonski & Anderson, 1981). Here the differences in horizontal and vertical wavespeed are needed principally for shear waves.

We shall subsequently mostly study propagation in isotropic materials. We will, however, wish to consider coupling between *P* and *SV* waves through the nature of the seismic wavespeed distribution and the methods which we shall use are directly extendable to full anisotropy.

1.3.3 Attenuation

We have a very complex rheology for the mineral assemblages in the crust and mantle. Over geological time scales they can sustain flow, and in the fold belts of mountain systems we can see considerable deformation without fracture. However, on the relatively short time scales appropriate to seismic wave propagation (0.01s-1000s) we cannot expect to see the influence of the long-term rheology, and the behaviour will be nearly elastic. The small incremental strains associated with seismic disturbances suggest that departures from our constitutive relations (1.2) or (1.4) should obey some linear law.

Any such anelastic processes will lead to the dissipation of seismic energy as a wave propagates through the Earth. Among phenomena which may be of importance are crystal defects, grain boundary processes and some thermoelastic effects (see, e.g., Jackson & Anderson, 1970). Anderson & Minster (1979) have suggested that the dislocation microstructure of mantle materials can account for long-term steady-state creep and for seismic wave attenuation. In this model the glide of dislocations within grains leads to attenuation, whilst climb and defect annihilation processes in the grain boundaries account for the long term rheology.

The anelastic behaviour may be included in our constitutive laws by introducing the assumption that the stress at a point depends on the time history of strain, so that the material has a 'memory' (Boltzman, 1876). The theory of such linear viscoelasticity has been reviewed by Hudson (1980) and he shows that the appropriate modification of the isotropic constitutive law is

$$\begin{aligned}\tau_{ij} &= \lambda_0\delta_{ij}\partial_k u_k + \mu_0(\partial_i u_j + \partial_j u_i)\\ &\quad + \int_0^t ds \begin{pmatrix} \dot{R}_\lambda(t-s)\delta_{ij}\partial_k u_k(s) \\ +\dot{R}_\mu(t-s)[\partial_i u_j(s) + \partial_j u_i(s)] \end{pmatrix}.\end{aligned} \tag{1.7}$$

Here $\lambda_0 = \kappa_0 - \frac{2}{3}\mu_0$ and μ_0 are the instantaneous elastic moduli which define the local wavespeeds and R_λ, R_μ are the relaxation functions specifying the dependence on the previous strain states.

We now take the Fourier transform of (1.7) with respect to time, and then the stress and displacement at frequency ω are related by

$$\bar{\tau}_{ij}(\omega) = [\lambda_0 + \lambda_1(\omega)]\delta_{ij}\partial_k\bar{u}_k(\omega) + [\mu_0 + \mu_1(\omega)](\partial_i\bar{u}_j(\omega) + \partial_j\bar{u}_i(\omega)), \tag{1.8}$$

where λ_1 and μ_1 are the transforms of the relaxation terms

$$\lambda_1(\omega) = \int_0^\infty dt\dot{R}_\lambda(t)e^{i\omega t}, \qquad \mu_1(\omega) = \int_0^\infty dt\dot{R}_\mu(t)e^{i\omega t} \tag{1.9}$$

So that if we are indeed in the regime of linear departures from elastic behaviour, the stress-strain relation at frequency ω is as in an elastic medium but now with complex moduli.

A convenient measure of the rate of energy dissipation is provided by the loss factor $Q^{-1}(\omega)$, which may be defined as

$$Q^{-1}(\omega) = -\Delta E(\omega)/(2\pi E_0(\omega)). \tag{1.10}$$

Here $\Delta E(\omega)$ is the energy loss in a cycle at frequency ω and E_0 is the 'elastic' energy stored in the oscillation. Thus E_0 is the sum of the strain and kinetic energy calculated with just the instantaneous elastic moduli. The energy dissipation δE is just associated with the imaginary part of the elastic moduli. For purely dilatational disturbances

$$Q_\kappa^{-1}(\omega) = -\mathrm{Im}\{\kappa_1(\omega)\}/\kappa_0, \tag{1.11}$$

and for purely deviatoric effects

$$Q_\mu^{-1}(\omega) = -\mathrm{Im}\{\mu_1(\omega)\}/\mu_0. \tag{1.12}$$

For the Earth it appears that loss in pure dilatation is much less significant than loss in shear, and so $Q_\kappa^{-1} \ll Q_\mu^{-1}$.

Since the relaxation contributions to (1.7) depend only on the past history of the strain, $\dot{R}_\mu(t)$ vanishes for $t < 0$, so that the transform $\mu_1(\omega)$ must be analytic in the upper half plane ($\mathrm{Im}\,\omega \geq 0$). In consequence the real and imaginary parts of $\mu_1(\omega)$ are the Hilbert transforms of each other (see, e.g., Titchmarsh, 1937) i.e.

$$\mathrm{Re}\{\mu_1(\omega)\} = \frac{1}{\pi} P \int_{-\infty}^{\infty} d\omega' \frac{\mathrm{Im}\{\mu_1(\omega')\}}{\omega' - \omega}, \tag{1.13}$$

where P denotes the Cauchy principal value. We cannot therefore have dissipative effects without some frequency dependent modification of the elastic moduli. This property is associated with any causal dissipative mechanism, and the analogous result to (1.13) in electromagnetic work is known as the Kramers-Krönig relations.

From our definition (1.12) of Q_μ^{-1} we can rewrite the relation (1.13) in a way which shows the dependence of $\mathrm{Re}\{\mu_1(\omega)\}$ on the behaviour of the loss factor with frequency,

$$\mathrm{Re}\{\mu_1(\omega)\} = -\frac{2\mu_0}{\pi} P \int_0^{\infty} d\omega' \frac{\omega' Q_\mu^{-1}(\omega')}{\omega'^2 - \omega^2}. \tag{1.14}$$

When we wish to use observational information for the loss factor $Q_\mu^{-1}(\omega)$ we are faced with the difficulty that this only covers a limited range of frequencies, but the detailed form of $\mathrm{Re}\{\mu_1(\omega)\}$ depends on the extrapolation of $Q_\mu^{-1}(\omega)$ to both high and low frequencies.

The distribution of Q_μ^{-1} with depth in the earth is still imperfectly known, because of the difficulties in isolating all the factors which effect the amplitude of a recorded seismic wave. However, most models show a moderate loss factor in the crust ($Q_\mu^{-1} \sim 0.004$) with an increase in the uppermost mantle ($Q_\mu^{-1} \sim 0.01$) and then a decrease to crustal values, or lower, in the mantle below 1000 km. Over the frequency band 0.001-10 Hz the intrinsic loss factor Q_μ^{-1} appears to be essentially constant, but in order for there to be a physically realisable loss mechanism, Q_μ^{-1} must depend on frequency outside this band. A number of different forms have been suggested (Azimi et al., 1968; Liu, Anderson & Kanamori, 1976; Jeffreys, 1958) but provided Q_μ^{-1} is not too large ($Q_\mu^{-1} < 0.01$) these lead to the approximate relation

$$\mathrm{Re}\{\mu_1(\omega)\} = 2\mu_0 \ln(\omega a) Q_\mu^{-1}/\pi, \tag{1.15}$$

in terms of some time constant a. A similar development may be made for the complex bulk modulus $\kappa_0 + \kappa_1(\omega)$ in terms of the loss factor Q_κ^{-1}.

For a locally uniform region, at a frequency ω, substitution of the stress-strain relation (1.8) into the equations of motion shows that, as in a perfectly elastic medium, two sets of plane waves exist. The *S* waves have a complex wavespeed $\bar{\beta}$ given by

$$\bar{\beta}^2(\omega) = [\mu_0 + \mu_1(\omega)]/\rho \tag{1.16}$$

influenced only by shear relaxation processes. In terms of the wavespeed $\beta_0 = (\mu_0/\rho)^{1/2}$ calculated for the instantaneous modulus, (1.16) may be rewritten as

$$\bar{\beta}^2(\omega) = \beta_0^2\left(1 + \mathrm{Re}\{\mu_1(\omega)\}/\mu_0 - i\,\mathrm{sgn}(\omega) Q_\mu^{-1}(\omega)\right), \tag{1.17}$$

where we have used the definition of Q_μ^{-1} in equation (1.12). Even if Q_μ^{-1} is frequency independent in the seismic band, our previous discussion shows that $\bar{\beta}$ will have weak frequency dispersion through $\mathrm{Re}\{\mu_1(\omega)\}$.

For a small loss factor ($Q_\mu^{-1} \ll 1$) the ratio of complex velocities at two different frequencies ω_1 and ω_2 will from (1.15) be approximately

$$\frac{\bar{\beta}(\omega_1)}{\bar{\beta}(\omega_2)} = 1 + \frac{Q_\mu^{-1}}{\pi} \ln\left(\frac{\omega_1}{\omega_2}\right) - i\,\mathrm{sgn}(\omega)\tfrac{1}{2}Q_\mu^{-1}. \tag{1.18}$$

We can thus overcome the problem of the unknown constant a by agreeing to fix a reference frequency (most commonly 1 Hz) and then

$$\bar{\beta}(\omega) \approx \beta_1\left[1 + \pi^{-1} Q_\mu^{-1} \ln(\omega/2\pi) - i\,\mathrm{sgn}(\omega)\tfrac{1}{2}Q_\mu^{-1}\right], \tag{1.19}$$

where β_1 is the velocity at 1 Hz. The presence of the frequency dependent terms in (1.17, 1.19) arises from the requirement that all dissipative processes will be causal. In consequence there will be no seismic energy arriving with wavespeed faster than that in the reference elastic medium (β_0). For an initially sharp pulse, propagation through the lossy medium leads to an assymetric pulse shape with a fairly sharp onset, illustrated in figure 1.4a.

When $Q_\mu(\omega)$ has some significant frequency dependence, we will still obtain a similar structure to (1.19) although the nature of the frequency dependence $\mathrm{Re}\,\bar{\beta}(\omega)\}$ will vary. Smith & Dahlen (1981) have shown that, as suggested originally by Jeffreys (1958), a weak frequency variation in loss factor, $Q_\mu^{-1} \propto \omega^{-\gamma}$ with $\gamma \approx 0.1$, will fit the observed period (435.2 days) and damping of the

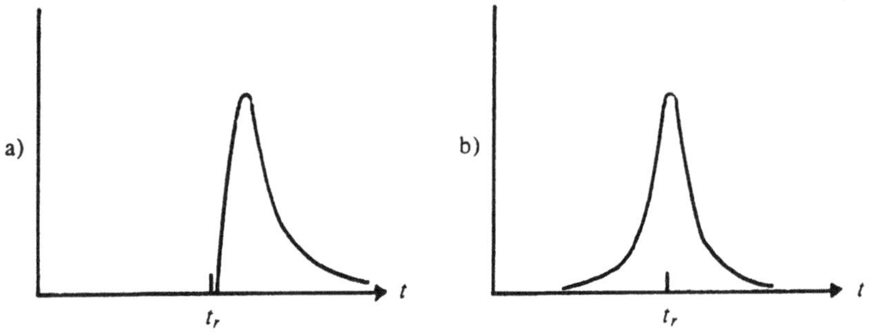

Figure 1.4. a) Pulse after passage through a medium with causal Q_μ^{-1} and associated velocity dispersion; t_r arrival time in reference medium. b) Pulse broadening due to scattering.

Chandler wobble, as well as the results in the seismic band. The value of γ is dependent on the reference loss model and is primarily influenced by the properties of the lower mantle. Lundquist & Cormier (1980) have suggested that the loss factor in the upper mantle may vary significantly for frequencies between 1 and 10 Hz, and relate this to relaxation time scales for absorption processes. For shallow propagation at high frequencies (10-60 Hz) O'Brien & Lucas (1971) have shown that the constant Q^{-1} model gives a good explanation of observed amplitude loss in prospecting situations.

For *P* waves the situation is a little more complicated since the anelastic effects in pure dilatation and shear are both involved. The complex wavespeed $\bar{\alpha}$ is given by

$$\begin{aligned}\bar{\alpha}^2(\omega) &= [\kappa_0 + \tfrac{4}{3}\mu_0 + \kappa_1(\omega) + \tfrac{4}{3}\mu_1(\omega)]/\rho, \\ &= \alpha_0\{1 + A(\omega) - i\,\mathrm{sgn}(\omega) Q_A^{-1}(\omega)\}, \end{aligned} \tag{1.20}$$

where

$$\alpha_0 = [(\kappa_0 + \tfrac{4}{3}\mu_0)/\rho]^{1/2} \tag{1.21}$$

and we have introduced the loss factor for the *P* waves

$$Q_A^{-1} = -\mathrm{Im}\{\kappa_1 + \tfrac{4}{3}\mu_1\}/(\kappa_0 + \tfrac{4}{3}\mu_0). \tag{1.22}$$

If loss in dilatation is very small compared with that in shear (i.e. $Q_\kappa^{-1} \ll Q_\mu^{-1}$)

$$Q_A^{-1} \approx \tfrac{4}{3}(\beta_0^2/\alpha_0^2) Q_\mu^{-1}, \tag{1.23}$$

as suggested by Anderson, Ben-Menahem & Archambeau (1965). The real dispersive correction to the wave speed $A(\omega)$ will have a rather complex form in general but under the conditions leading to (1.23), we will have a similar form to (1.15)

$$A(\omega) = 2(\kappa_0 + \tfrac{4}{3}\mu_0) Q_A^{-1} \ln(\omega a)/\pi. \tag{1.24}$$

We may therefore once again get a form for the complex wavespeed in terms of the wave speed at 1 Hz (α_1),

$$\bar{\alpha}(\omega) = \alpha_1 \left[1 + \pi^{-1} Q_A^{-1} \ln(\omega/2\pi) - i\,\mathrm{sgn}(\omega) \tfrac{1}{2} Q_A^{-1}\right]. \tag{1.25}$$

In the frequency domain, calculations with these complex velocities turn out to be little more complicated than in the perfectly elastic case.

In addition to the dissipation of elastic energy by anelastic processes, the apparent amplitude of a seismic wave can be diminished by scattering which redistributes the elastic energy. As we have noted above, our choice of elastic moduli defines a reference medium whose properties smooth over local irregularities in the properties of the material. The fluctuations of the true material about the reference will lead to scattering of the seismic energy out of the primary wave which will be cumulative along the propagation path, and the apparent velocity of transmission of the scattered energy will vary from that in the reference. Since locally the material may be faster or slower than the assigned wavespeed, the effect of scattering is to give a pulse shape which is broadened and diminished in amplitude relative to that in the reference medium, with an emergent onset before the reference travel time (figure 1.4b). At a frequency ω we may once again describe the effect of the scattering by a loss factor ${}_sQ^{-1}(\omega)$ and the changing character of the scattering process leads to a strong frequency dependence. As the wavelength diminishes, the effect of local irregularities becomes more pronounced and so ${}_sQ^{-1}$ tends to increase until the wavelength is of the same order as the size of the scattering region.

This scattering mechanism becomes important in areas of heterogeneity and its influence seems largely to be confined to the lithosphere. There are also considerable regional variations, with earthquake zones showing the most significant effects (Aki, 1981).

For each wave type the overall rate of seismic attenuation Q^{-1}, which is the quantity which would be derived from observations, will be the sum of the loss factors from intrinsic anelasticity and scattering. Thus for *S* waves

$$Q_\beta^{-1}(\omega) = Q_\mu^{-1}(\omega) + {}_sQ_\beta^{-1}(\omega). \tag{1.26}$$

For *P* waves,

$$Q_\alpha^{-1}(\omega) = Q_A^{-1}(\omega) + {}_sQ_\alpha^{-1}(\omega); \tag{1.27}$$

since the scattering component here is arising from a totally distinct mechanism to the dissipation there is no reason to suppose that ${}_sQ_\alpha^{-1}$, ${}_sQ_\beta^{-1}$ are related in a similar way to (1.23).

Recent observational results (Aki, 1981) suggest that the contributions Q_μ^{-1} and ${}_sQ_\beta^{-1}$ are separable via their different frequency behaviour (figure 1.5). The intrinsic absorption Q_μ^{-1} is quite small and nearly frequency independent over the seismic band and then superimposed on this with characteristics which vary from

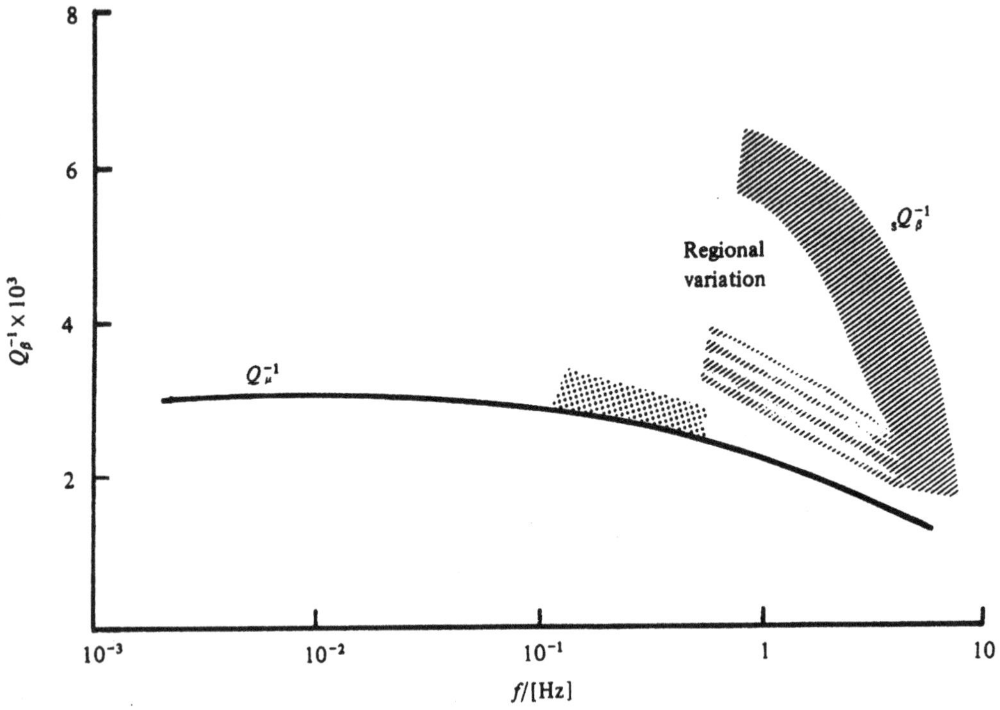

Figure 1.5. Frequency separation of intrinsic loss factor Q_μ^{-1} from scattering contribution $_sQ_\beta^{-1}$.

region to region is the more rapidly varying scattering loss. Other studies also indicate such an increase in the loss factors Q^{-1} as the frequency rises towards a few Hz. The results of figure 1.5 could be fitted with some postulated dependence of Q^{-1} on frequency. However, the wavespeed dispersion estimated by (1.14) from such relations would be very misleading. It is only for the anelastic portion Q_μ^{-1}, Q_A^{-1} that we have dispersive wavespeed terms. The scattering contribution $_sQ_\beta^{-1}$, $_sQ_\alpha^{-1}$ does not have the same restriction to a local 'memory' effect and there is no consequent dispersion.

1.4 Heterogeneity

As yet we have no means, besides the pure numerical, to consider seismic wave propagation in a completely general medium with arbitrary variations in even isotropic elastic properties. Even numerical methods are limited by storage and time requirements to a restricted range of propagation. At high frequencies we may make a ray theoretical development and this approach is described in some detail in the book by Červený, Molotkov & Pšenčík (1978). For intermediate and low frequencies ray theory results are hardly adequate.

We will therefore adopt, for most of this work, a stratified model of the elastic parameter distribution within the Earth in which the only dependence is on depth (or radius). For this model we can use a range of methods to calculate the propagational characteristics of the seismic waves. But we must bear in mind that this, like our assumption of isotropy, is only an approximation of limited validity.

The considerable variability in the near-surface portion of the Earth means that, at best, a horizontally stratified model has local meaning. It may be appropriate for undisturbed sediments but certainly not in the regions disturbed by the intrusion of diapiric salt domes which are of considerable commercial importance because of their oil potential. Stratified models have been used with considerable success in examining the details of the seismic properties in the oceanic crust (see, e.g., Helmberger, 1977; Kennett, 1977; Spudich & Orcutt, 1980). But for the continents the lateral heterogeneity seems to be rather higher and stratified models are mostly used to describe the broad features of the crustal properties.

The most significant lateral variations in properties which are excluded from the stratified model are the transition zone from continent to ocean and the effect of a subduction zone. The latter is of particular importance since it is the region in which intermediate and deep focus earthquakes occur. When, therefore, we try to simulate the propagation from such events by using a source embedded in a stratified model, we must take care in the specification of the source. For higher frequency propagation (around 1 Hz) the effect of the downgoing slab can be quite important at teleseismic distance.

On a local scale, heterogeneity in the material properties gives rise to scattering of seismic energy. If the fluctuations in elastic parameters are quite small the main effect will be an attenuation of a seismic pulse and can be described by the loss factors ${}_sQ_{\alpha}^{-1}$, ${}_sQ_{\beta}^{-1}$. Larger fluctuations in properties give rise to significant features on seismic records, in addition to those predicted from the averaged model. Thus, following the main *P* and *S* arrivals from local earthquakes is the coda, an elongated train of waves with exponential decay of amplitude which appears to arise from back-scattering from velocity variations in the crust and upper mantle (Aki & Chouet, 1975). A horizontally stratified structure with a strong crustal waveguide will also give rise to a similar style of coda, associated with multiple reverberations in the waveguide.

In addition to significant irregularities in the seismic properties of the outer regions of the Earth, there are a number of indications of heterogeneity in a region about 200 km thick above the core-mantle boundary. The presence of scattering from these regions is most apparent in those cases where scattered energy arrives in a quiet portion of the records as precursors to a large phase. Such occurrences are usually associated with stationary, but non-minimum-time, ray paths. Arrivals which have been interpreted as *P* wave scattering in the near-surface region occur before the *PP* phase for epicentral distances of about 100° (King, Haddon & Husebye, 1975). Energy arriving before *PKIKP* between 120° and 142° has also been attributed to scattering from *PKP* near, or at, the core-mantle boundary (Haddon & Cleary, 1974; Husebye, King & Haddon, 1976). It is possible that the increases in loss factors near the core-mantle boundary in recent Q^{-1} models can be attributed to scattering loss rather than enhanced intrinsic absorption.

1.5 Stratified models

In the previous sections we have discussed the characteristics of seismic wave propagation within the Earth and the extent to which we are able to match this behaviour with a relatively simple description of the constitutive relation and the spatial variation of material properties.

As a simple, but reasonably realistic, model for studying the effect of the Earth's structure on seismic wave propagation we shall consider stratified media composed of isotropic, nearly elastic, material. The weak attenuation will be included in the frequency domain by working with complex wavespeeds.

To simplify the configuration, whilst retaining the physical features of interest, we shall start by studying a horizontally stratified half space. On a local scale this is often a good approximation, but as waves penetrate deeper into the Earth the effects of sphericity become more important.

There is no exact transformation which takes the seismic properties in a stratified sphere into those in an equivalent half space for *P-SV* waves (Chapman, 1973) although this can be achieved for *SH* waves (Biswas & Knopoff, 1970). However, by a suitable 'earth-flattening' transformation we can map the wavespeed profile with radius R in a sphere into a new wavespeed distribution with depth z in a half space so that transit times from source to receiver are preserved. Thus in the flattened model we take

$$z = r_e \ln(r_e/R), \tag{1.28}$$

and the wavespeeds after flattening α_f, β_f are

$$\alpha_f(z) = \alpha(R)(r_e/R), \qquad \beta_f(z) = \beta(R)(r_e/R), \tag{1.29}$$

where r_e is the radius of the Earth. This wavespeed mapping needs to be supplemented by an approximate density transformation, e.g.,

$$\rho_f(z) = \rho(R)(R/r_e), \tag{1.30}$$

which leads to the same reflection coefficients at normal incidence in the spherical and flattened models.

In the outer regions of the Earth the distortion introduced by flattening is not too severe and the 'flattened earth' can provide useful quantitative results. The increased velocity gradients in the flattened model compensate for the crowding effect of sphericity as the radius diminishes.

For studies of the deep interior from core-mantle boundary towards the centre it is desirable to work directly with a spherical model, even though comparison studies have shown the flattening approximation to give quite good results (Choy et al. 1980). The spherical model is of course essential for very long-period phenomena which involve a substantial fraction of the Earth. For the spherical case we are able to carry over the calculation methods developed for the half space to give a unified treatment of the whole range of propagation problems.

1.6 Preview

Although our goal is to understand the way in which the features on observed seismograms are related to the properties of the source and seismic structure of the Earth, we need to establish a variety of mathematical and physical tools which will help us in this task. These will be developed over the next few chapters.

We start in Chapter 2 by representing the seismic displacement within a stratified medium as a superposition of cylindrical wave elements modulated by angular terms depedent on source excitation. For each of these wave elements, characterised by frequency ω, horizontal waveslowness p, and angular order m, we are able to follow the development of the associated displacements and tractions with depth z by means of sets of coupled first order differential equations in z. For an isotropic medium these equations separate into two sets:

i) *P-SV*, coupling compressional and shear wave propagation in a vertical plane,
ii) *SH*, shear waves with motion confined to a horizontal plane.

Such coupled equations are well suited to the solution of initial value problems, and in this context we introduce the propagator matrix which acts as a transfer matrix between the stress and displacement elements at two levels in the stratification.

In Chapter 3 we will discuss the construction of stress-displacement fields. For a uniform medium we show how the seismic displacements can be expressed in terms of upward and downward travelling waves and use this result to derive the corresponding propagator. When the seismic wavespeeds vary smoothly with depth an approximate stress-displacement field can be found for which the asymptotic behaviour is like upgoing or downgoing waves.

Chapter 4 is devoted to the excitation of seismic waves by seismic sources. A physical source is represented by an equivalent force system within our model of a stratified medium. For finite size sources, the low frequency radiation may be modelled by a seismic moment tensor, which gives the relative weighting of force doublet contributions. For large source regions an extended multipole expansion is needed if an equivalent point source is used. For a point source the seismic wave excitation enters into the stress-displacement vector picture as a discontinuity across the source plane.

The reflection and transmission of seismic waves in stratified regions are introduced in Chapter 5. For coupled *P-SV* wave propagation we introduce reflection and transmission matrices, whose entries are the reflection and transmission coefficients between the different wave types. These matrices enable us to develop systematic techniques for handling conversion between wave types and can also be related to the propagator matrix for a region.

In Chapter 6 we show that the overall reflection response of a zone of the stratification can be built up by an addition rule, from the reflection and transmission properties of its subregions. This addition rule enables us to produce effective computational methods to construct reflection coefficients for a stack of uniform layers or for piecewise smooth media, based on the stress-displacement field representations introduced in Chapter 3.

Chapter 7 brings together the discussion of the excitation and reflection of seismic waves to construct the full response of a stratified medium to excitation by a source. A number of different representations exist for the full response which exhibit different facets of the propagation process. By working with the reflection properties of the stratification we are able to make a clear physical interpretation of the contributions to the response. Once we have constructed the surface displacements in the transform domain we may generate theoretical seismograms by direct integration over the cylindrical wave representation; different algorithms will be used depending on whether the integration over frequency or slowness is performed first.

Complete theoretical seismograms including all body waves and surface effects are expensive to calculate. They are most useful when the time separation between the different seismic phases are small. Once the different types of seismic wave contributions are well separated in time it becomes worthwhile to develop approximate techniques designed to model the particular portions of the seismic record which are of interest.

In Chapter 8 we consider the nature of seismic records as a function of distance from the source and frequency content, so that we can use these results as a guide to the appropriate approximations developed in Chapters 9–11. We consider reflection seismograms as recorded in typical prospecting work and the refraction technique used for lower resolution work to greater depths. We then turn our attention to the records obtained at seismographic stations and look at the evolution of the seismic field with epicentral distance.

With the aid of the reflection matrix approach it is fairly easy to construct systematic approximations to the response which give a good representation of certain parts of the seismic wave field. In Chapter 9 we show how to make use of partial expansions of reflector operators to examine the effect of the free surface and the near-surface zone with low wavespeeds. These approximations in the frequency-slowness are then combined with suitable integration schemes to show how the 'reflectivity' and 'full-wave' techniques are related to the full response. We also show how simple approximate calculations may be made for teleseismic P and S waveforms.

In Chapter 10 we carry the expansion of the frequency-slowness response much farther and represent the displacement field as a sum of generalized ray contributions with a specific form of frequency and slowness dependence. This functional dependence can be exploited to produce the time and space response for each generalized ray in uniform layer models (Cagniard's method), and asymptotically for piecewise smooth models using a method introduced by Chapman.

The last chapter is devoted to a discussion of seismic surface waves and the other contributions to the seismic response arising from pole singularities in the representation of seismic displacements via reflection matrices. These poles arise at a combination of frequency and slowness such that a single stress-displacement vector can satisfy both the free surface condition and the radiation condition at the base of the stratification. Once the modal dispersion curves as a function of frequency and slowness have been found we can calculate theoretical seismograms by superposition of modal contributions. With many mode branches we get, in addition to a surface wave train with low group velocity, faster pulses with a body wave character.

Chapter 2

Coupled Equations for Seismic Waves

The incremental displacement **u** induced by the passage of a seismic wave is governed by the equation of motion

$$\rho(\mathbf{x})\partial_{tt}u_i(\mathbf{x}) = \partial_j\tau_{ij}(\mathbf{x}), \tag{2.1}$$

in the absence of sources. The behaviour of the material enters via the constitutive relation connecting the incremental stress and strain. When the approximation of local isotropic behaviour is appropriate, the displacement satisfies

$$\rho(\mathbf{x})\partial_{tt}u_i(\mathbf{x}) = \partial_j\left[\lambda(\mathbf{x})\partial_k u_k(\mathbf{x})\delta_{ij} + \mu(\mathbf{x})(\partial_i u_j(\mathbf{x}) + \partial_j u_i(\mathbf{x}))\right], \tag{2.2}$$

which leads to rather complex behaviour for arbitrary spatial variation of the elastic moduli λ, μ. As discussed in Section 1.3.3 we may accommodate slight dissipation within the medium by allowing the moduli λ, μ to be complex functions of frequency within the seismic band.

Even in a uniform medium where λ and μ are constants the three components of displacement are coupled but can be represented in terms of two simple classes of disturbance. These are firstly, compressional (*P*) waves for which the dilatation $(\partial_k u_k)$ satisfies

$$\rho\partial_{tt}(\partial_k u_k) = (\lambda + 2\mu)\partial_{jj}(\partial_k u_k), \tag{2.3}$$

which we see to be a wave equation with associated *P* wavespeed

$$\alpha = [(\lambda + 2\mu)/\rho]^{1/2}. \tag{2.4}$$

In the second class, shear waves, the dilatation vanishes so that

$$\rho\partial_{tt}u_i = \mu(\partial_{ij}u_j + \partial_{jj}u_i), \tag{2.5}$$

and we may reduce (2.5) to the form of a wave equation by applying the curl operator to give

$$\rho\partial_{tt}(\mathrm{curl}\,\mathbf{u}) = \mu\partial_{jj}(\mathrm{curl}\,\mathbf{u}), \tag{2.6}$$

and so we have an *S* wavespeed

$$\beta = [\mu/\rho]^{1/2}. \tag{2.7}$$

In a uniform medium the P and S waves can exist separately, and the total displacement will be a superposition of contributions from these two wave types.

For general spatial variation of $\lambda(\mathbf{x})$, $\mu(\mathbf{x})$ and $\rho(\mathbf{x})$ we cannot make such a separation of the wavefield into P and S waves. If the spatial gradients in elastic properties are slight, then waves which travel at the local P and S wavespeeds, defined via (2.5), (2.7), can exist but these are coupled to each other by the gradients of $\lambda(\mathbf{x})$, $\mu(\mathbf{x})$ and $\rho(\mathbf{x})$

2.1 Depth dependent properties

The complexity of the propagation problem is reduced somewhat if the elastic properties depend only on depth. For such a horizontally stratified medium we are able to set up coupled equations involving displacement and traction elements in which the dependence of the wavefield on depth is emphasised.

We will adopt a cylindrical set of coordinates (r,ϕ,z) with the vertical axis perpendicular to the stratification. The displacement $\mathbf{u}$ may be represented in terms of its components

$$\mathbf{u}(r,\phi,z,t) = u_r\mathbf{e}_r + u_\phi\mathbf{e}_\phi + u_z\mathbf{e}_z, \tag{2.8}$$

using the orthogonal unit vectors $\mathbf{e}_r$, $\mathbf{e}_\phi$, $\mathbf{e}_z$. Since we are now working with a spatially varying coordinate system, the gradient of the stress tensor, which appears in the equation of motion (2.1), is not quite as simple as in the cartesian case. Explicitly we have

$$\begin{aligned}
&\partial_z\tau_{rz} + \partial_r\tau_{rr} + r^{-1}\partial_\phi\tau_{r\phi} + r^{-1}(\tau_{rr} - \tau_{\phi\phi}) = \rho\partial_{tt}u_r - \rho f_r,\\
&\partial_z\tau_{\phi z} + \partial_r\tau_{r\phi} + r^{-1}\partial_\phi\tau_{\phi\phi} + 2r^{-1}\tau_{r\phi} = \rho\partial_{tt}u_\phi - \rho f_\phi,\\
&\partial_z\tau_{zz} + \partial_r\tau_{rz} + r^{-1}\partial_\phi\tau_{\phi z} + r^{-1}\tau_{rz} = \rho\partial_{tt}u_z - \rho f_z,
\end{aligned} \tag{2.9}$$

in the presence of a body force per unit mass $\mathbf{f}$. The relation between stress and strain follows the usual functional form, so that for isotropy

$$\tau_{ij} = \lambda\delta_{ij}e_{kk} + 2\mu e_{ij}, \tag{2.10}$$

in terms of the components of the strain tensor e_{ij}. A further consequence of the curvilinear coordinate system is that these strain elements now allow for the distortion of the reference grid as well as displacement gradients. The stress elements are related to the displacements by

$$\begin{aligned}
\tau_{rr} &= (\lambda + 2\mu)\partial_r u_r + \lambda(\partial_z u_z + r^{-1}\partial_\phi u_\phi + r^{-1}u_r),\\
\tau_{\phi\phi} &= (\lambda + 2\mu)r^{-1}(\partial_\phi u_\phi + u_r) + \lambda(\partial_z u_z + \partial_r u_r),\\
\tau_{zz} &= (\lambda + 2\mu)\partial_z u_z + \lambda(\partial_r u_r + r^{-1}\partial_\phi u_\phi + r^{-1}u_r),\\
\tau_{rz} &= \mu(\partial_z u_r + \partial_r u_z),\\
\tau_{\phi z} &= \mu(r^{-1}\partial_\phi u_z + \partial_z u_\phi),\\
\tau_{r\phi} &= \mu(\partial_r u_\phi - r^{-1}u_\phi + r^{-1}\partial_\phi u_r).
\end{aligned} \tag{2.11}$$

In the equations of motion, derivatives with respect to z appear only on the components of the traction across a horizontal plane

$$\mathbf{t}(r, \phi, z, t) = \tau_{rz}\mathbf{e}_r + \tau_{\phi z}\mathbf{e}_\phi + \tau_{zz}\mathbf{e}_z, \tag{2.12}$$

and in the stress-strain relations (2.11) z derivatives appear on the displacement components. The traction $\mathbf{t}$ and displacement $\mathbf{u}$ are both continuous across any plane $z = const$ within the stratification, under the assumption of welded contact between any dissimilar materials. We therefore want to rearrange the equations of motion and stress-strain relations so that derivatives with respect to z appear only on the left hand side of the equations.

The additional gradient contributions arising from the cylindrical coordinates complicate the behaviour for the horizontal elements $u_r, u_\phi, \tau_{rz}, \tau_{\phi z}$. However, a simple form may be found if we introduce the new elements (cf., Hudson, 1969b)

$$\begin{aligned} u_V &= r^{-1}[\partial_r(ru_r) + \partial_\phi u_\phi], \\ \tau_{Vz} &= r^{-1}[\partial_r(r\tau_{rz}) + \partial_\phi \tau_{\phi z}], \end{aligned} \tag{2.13}$$

and

$$\begin{aligned} u_H &= r^{-1}[\partial_r(ru_\phi) - \partial_\phi u_r], \\ \tau_{Hz} &= r^{-1}[\partial_r(r\tau_{\phi z}) - \partial_\phi \tau_{rz}], \end{aligned} \tag{2.14}$$

with similar definitions for f_V, f_H. We also introduce the horizontal Laplacian ∇_1^2 such that

$$\nabla_1^2\psi = r^{-1}\partial_r(r\partial_r\psi) + r^{-2}\partial_{\phi\phi}\psi. \tag{2.15}$$

In terms of these quantities we can rearrange the equations to a form where we have managed to isolate the z derivatives. This leads to six coupled equations which separate into two sets.

The first set is

$$\begin{aligned} \partial_z u_z &= -\lambda(\lambda + 2\mu)^{-1}u_V + (\lambda + 2\mu)^{-1}\tau_{zz}, \\ \partial_z u_V &= -\nabla_1^2 u_z + \mu^{-1}\tau_{Vz}, \\ \partial_z \tau_{zz} &= \rho\partial_{tt}u_z - \tau_{Vz} - \rho f_z, \\ \partial_z \tau_{Vz} &= (\rho\partial_{tt} - \rho\nu\nabla_1^2)u_V - \lambda(\lambda + 2\mu)^{-1}\nabla_1^2\tau_{zz} - \rho f_V, \end{aligned} \tag{2.16}$$

where we have introduced the composite modulus

$$\rho\nu = (\lambda + 2\mu) - \lambda^2/(\lambda + 2\mu) = 4\mu(\lambda + \mu)/(\lambda + 2\mu). \tag{2.17}$$

These equations couple *P* waves with local wavespeed α to *SV* shear waves, involving vertical displacement, with wavespeed β. The second set comprises shear disturbances entirely confined to a horizontal plane (*SH*) with the same wavespeed β :

$$\begin{aligned} \partial_z u_H &= \mu^{-1}\tau_{Hz}, \\ \partial_z \tau_{Hz} &= (\rho\partial_{tt} - \mu\nabla_1^2)u_H - \rho f_H. \end{aligned} \tag{2.18}$$

This decomposition into coupled *P-SV* and *SH* systems also occurs for a transversely isotropic medium with a vertical symmetry axis (Takeuchi & Saito, 1972) and this case is discussed in the appendix to Chapter 3. For a generally anisotropic medium it is still possible to arrange the elastic equations in a form where derivatives with respect to z appear only on the left hand side of the equations (Woodhouse, 1974) but now there is no decoupling.

The sets of coupled equations (2.16) and (2.18) still involve partial derivatives with respect to the horizontal coordinates and time, and include all effects of vertical gradients in the elastic parameters. Since the elastic properties do not depend on horizontal position, we may use transforms over time and the horizontal coordinates to reduce (2.16) and (2.18) to a set of ordinary differential equations in the depth variable z. We take a Fourier transform with respect to time and, for the horizontal coordinates, a Hankel transform of order m over radial distance from the origin and a finite Fourier transform over the angular variable:

$$\begin{aligned}\mathcal{F}_m[\psi] &= \hat{\psi}(k, m, \omega)\\ &= \frac{1}{2\pi}\int_{-\infty}^{\infty} dt e^{i\omega t}\int_0^{\infty} dr\, r J_m(kr)\int_{-\pi}^{\pi} d\phi e^{-im\phi}\psi(r, \phi, t),\end{aligned} \tag{2.19}$$

for which

$$\mathcal{F}_m[\nabla_1^2\psi] = -k^2\mathcal{F}_m[\psi]. \tag{2.20}$$

For each azimuthal order m we introduce a set of variables related to the transforms of the displacement and stress variables appearing in (2.16), (2.18) by

$$\begin{aligned} U &= \hat{u}_z, & P &= \omega^{-1}\hat{\tau}_{zz},\\ V &= -k^{-1}\hat{u}_V, & S &= -(\omega k)^{-1}\hat{\tau}_{Vz},\\ W &= -k^{-1}\hat{u}_H, & T &= -(\omega k)^{-1}\hat{\tau}_{Hz}.\end{aligned} \tag{2.21}$$

The scaling factors are designed to give a set of variables in each group with equal dimensionality; the scaling via horizontal wavenumber k arises from the horizontal differentation in (2.13), (2.14) and the frequency scaling for stress simplifies the form of subsequent equations. These new displacement and stress quantities are related by

$$\begin{aligned}\omega P &= \rho\alpha^2\partial_z U - k\rho(\alpha^2 - 2\beta^2)V,\\ \omega S &= \rho\beta^2(\partial_z V + kU),\\ \omega T &= \rho\beta^2\partial_z W,\end{aligned} \tag{2.22}$$

in terms of the *P* and *S* wavespeeds α , β .

The body force terms must also be transformed and we apply a comparable scaling to that for the stress variables

$$\begin{aligned}F_z &= \rho\omega^{-1}\hat{f}_z,\\ F_V &= -\rho(\omega k)^{-1}\hat{f}_V,\\ F_H &= -\rho(\omega k)^{-1}\hat{f}_H.\end{aligned} \tag{2.23}$$

When we apply the Fourier-Hankel transform operator (2.19) to the equation sets (2.16) and (2.18) we obtain coupled sets of ordinary differential equations for the new displacement and stress quantities $U(k, m, z, \omega)$, $P(k, m, z, \omega)$ etc. These transformed equations take a very convenient form if we work in terms of the horizontal slowness $p = k/\omega$, with units of reciprocal wavespeed, rather than the horizontal wavenumber k. Thus for *P-SV* waves we have

$$\frac{\partial}{\partial z}\begin{bmatrix} U \\ V \\ P \\ S \end{bmatrix} = \omega \begin{bmatrix} 0 & p(1-2\beta^2/\alpha^2) & (\rho\alpha^2)^{-1} & 0 \\ -p & 0 & 0 & (\rho\beta^2)^{-1} \\ -\rho & 0 & 0 & p \\ 0 & \rho[\nu p^2 - 1] & -p(1-2\beta^2/\alpha^2) & 0 \end{bmatrix} \begin{bmatrix} U \\ V \\ P \\ S \end{bmatrix} - \begin{bmatrix} 0 \\ 0 \\ F_z \\ F_V \end{bmatrix},$$

where $\nu = 4\beta^2(1-\beta^2/\alpha^2)$, (2.24)

and for *SH* waves

$$\frac{\partial}{\partial z}\begin{bmatrix} W \\ T \end{bmatrix} = \omega \begin{bmatrix} 0 & (\rho\beta^2)^{-1} \\ \rho[\beta^2 p^2 - 1] & 0 \end{bmatrix} \begin{bmatrix} W \\ T \end{bmatrix} - \begin{bmatrix} 0 \\ F_H \end{bmatrix}. \tag{2.25}$$

For an isotropic medium the coefficients appearing in (2.24), (2.25) are independent of the azimuthal order m and the azimuthal dependence of $U(k, m, z, \omega)$ etc. will arise solely from the nature of the force system $\mathbf{F}$. The elements of the coupling matrices involve only the elastic parameters at the depth z and not their vertical derivatives. This desirable property was first pointed out by Alterman, Jarosch & Pekeris (1959) in an analogous development for a sphere, and this makes (2.24), (2.25) well suited to numerical solution since the errors involved in interpolating the elastic parameters are minimised.

Each of the sets of coupled equations (2.24) and (2.25) can be written in the form

$$\partial_z \mathbf{b}(k, m, z, \omega) = \omega \mathbf{A}(p, z)\mathbf{b}(k, m, z, \omega) + \mathbf{F}(k, m, z, \omega), \tag{2.26}$$

in terms of a column vector $\mathbf{b}$ whose entries are the displacement and stress quantities. For *P-SV* waves

$$\mathbf{b}_P(k, m, z, \omega) = [U,\ V,\ P,\ S]^T, \tag{2.27}$$

where ^{T} denotes the transpose of the row vector. For *SH* waves

$$\mathbf{b}_H(k, m, z, \omega) = [W,\ T]^T. \tag{2.28}$$

When we wish to look at the general structure of the results we will write a general stress-displacement vector $\mathbf{b}$ in the form

$$\mathbf{b}(k, m, z, \omega) = [\mathbf{w},\ \mathbf{t}]^T, \tag{2.29}$$

and we will specialise, when appropriate, to the *P-SV* and *SH* systems.

The displacement $\mathbf{u}$ and traction $\mathbf{t}$ across any horizontal plane are continuous, and since only horizontal derivatives enter into the definitions of $u_V, u_H, \tau_{Vz}, \tau_{Hz}$ (2.13), (2.14) these will also be continuous across a horizontal plane. The transform operator $\mathcal{F}_m$ preserves these continuity properties, and thus the stress-displacement vector $\mathbf{b}$ will be continuous across any plane of discontinuity in material properties as well as all other planes $z = const$.

In the depth intervals where the elastic properties are continuous we may solve (2.26) to construct the stress-displacement vector $\mathbf{b}$, and then we are able to use the continuity of $\mathbf{b}$ to carry the solution across the level of any jump in the elastic parameters.

2.1.1 Coupled second order equations and propagation invariants

Although most recent studies of seismic waves in stratified media have made use of the sets of first order differential equations we have introduced in equations (2.24) and (2.25), there is an alternative formulation in terms of coupled second order equations (Keilis-Borok, Neigauz & Shkadinskaya, 1965). This representation gives further insight into the character of the displacement which will later be useful when we consider the excitation of seismic waves by a source.

For *SH* waves, the two first order equations (2.25) are equivalent to the single equation

$$\partial_z(\rho\beta^2\partial_z W) - \rho\omega^2(\beta^2 p^2 - 1)W = -\omega F_H. \tag{2.30}$$

In the *P-SV* wave case we have two coupled second order equations which are conveniently expressed in terms of the displacement vector $\boldsymbol{w} = [U, V]^T$,

$$\partial_z[\mathbf{A}\partial_z \boldsymbol{w} + \omega p \mathbf{B}\boldsymbol{w}] - \omega p \mathbf{B}^T\partial_z \boldsymbol{w} - \rho\omega^2(p^2\mathbf{C} - \mathrm{I})\boldsymbol{w} = -\omega \mathsf{F}, \tag{2.31}$$

where I is the identity matrix and the other 2×2 matrices $\mathbf{A}, \mathbf{B}, \mathbf{C}$ are given by

$$\mathbf{A} = \rho\begin{bmatrix}\alpha^2 & 0\\ 0 & \beta^2\end{bmatrix}, \quad \mathbf{B} = \rho\begin{bmatrix}0 & (2\beta^2 - \alpha^2)\\ \beta^2 & 0\end{bmatrix}, \quad \mathbf{C} = \begin{bmatrix}\beta^2 & 0\\ 0 & \alpha^2\end{bmatrix}. \tag{2.32}$$

From (2.22) we may recognise the traction contibution to (2.32) as

$$\omega \mathbf{t} = \mathbf{A}\partial_z \boldsymbol{w} + \omega p \mathbf{B}\boldsymbol{w}. \tag{2.33}$$

Both (2.30) and (2.31) have the form

$$\partial_z(\omega \mathbf{t}) + \mathbf{K}\boldsymbol{w} = -\omega \mathsf{F}, \tag{2.34}$$

in terms of an operator $\mathbf{K}$, and are self adjoint. We may make use of this property to establish propagation invariants for the *P-SV* and *SH* wave systems. For frequency ω and slowness p, consider two distinct displacement fields $\boldsymbol{w}_1, \boldsymbol{w}_2$, which satisfy different boundary conditions, then the structure of the operator $\mathbf{K}$ is such that

$$\partial_z[\boldsymbol{w}_1^T \mathbf{t}_2 - \mathbf{t}_1^T \boldsymbol{w}_2] = \boldsymbol{w}_2^T \mathsf{F}_1 - \boldsymbol{w}_1^T \mathsf{F}_2, \tag{2.35}$$

as may be verified by direct evaluation. In the absence of sources the quantity

$$<\boldsymbol{w}_1, \boldsymbol{w}_2> = \boldsymbol{w}_1^T \mathbf{t}_2 - \mathbf{t}_1^T \boldsymbol{w}_2, \tag{2.36}$$

is therefore independent of depth and thus a propagation invariant. $<w_1, w_2>$ is a weighted Wronskian for the coupled equations: explicitly, for *P-SV* waves

$$<w_1, w_2> = \mathsf{U}_1\mathsf{P}_2 + \mathsf{V}_1\mathsf{S}_2 - \mathsf{P}_1\mathsf{U}_2 - \mathsf{S}_1\mathsf{V}_2, \tag{2.37}$$

and for *SH* waves we have a comparable form

$$<w_1, w_2> = \mathsf{W}_1\mathsf{T}_2 - \mathsf{T}_1\mathsf{W}_2. \tag{2.38}$$

These invariants may also be established from the coupled first order equations. For *P-SV* waves the coefficient matrix **A** appearing in (2.26) has the symmetry properties:

(a) for a dissipative medium when α, β are complex

$$\mathbf{NA} + \mathbf{A}^{\mathsf{T}}\mathbf{N} = \mathbf{0}, \tag{2.39}$$

(b) for a perfectly elastic medium

$$\mathbf{NA} + \mathbf{A}^{\mathsf{T}*}\mathbf{N} = \mathbf{0}, \tag{2.40}$$

where a star indicates a complex conjugate, and **N** is a block off-diagonal matrix

$$\mathbf{N} = \begin{bmatrix} 0 & \mathbf{I} \\ -\mathbf{I} & 0 \end{bmatrix}. \tag{2.41}$$

If we construct the quantity $\mathbf{b}_1^{\mathsf{T}}\mathbf{N}\mathbf{b}_1$, then, in the absence of sources, from (2.26) we find

$$\partial_z(\mathbf{b}_1^{\mathsf{T}}\mathbf{N}\mathbf{b}_1) = \omega\mathbf{b}_1^{\mathsf{T}}[\mathbf{A}^{\mathsf{T}}\mathbf{N} + \mathbf{NA}]\mathbf{b}_2 = 0; \tag{2.42}$$

by (2.39), and performing the matrix multiplication,

$$\mathbf{b}_1^{\mathsf{T}}\mathbf{N}\mathbf{b}_1 = <w_1, w_2> = \mathsf{U}_1\mathsf{P}_2 + \mathsf{V}_1\mathsf{S}_2 - \mathsf{P}_1\mathsf{U}_2 + \mathsf{S}_1\mathsf{V}_2. \tag{2.43}$$

For *SH* waves we have similar behaviour,

(a) in the presence of dissipation

$$\mathbf{nA} + \mathbf{A}^{\mathsf{T}}\mathbf{n} = \mathbf{0}, \tag{2.44}$$

(b) for a perfectly elastic medium

$$\mathbf{nA} + \mathbf{A}^{\mathsf{T}*}\mathbf{n} = \mathbf{0}, \tag{2.45}$$

where

$$\mathbf{n} = \begin{bmatrix} 0 & 1 \\ -1 & 0 \end{bmatrix}. \tag{2.46}$$

Once again $\mathbf{b}_1^{\mathsf{T}}\mathbf{n}\mathbf{b}_2 = <w_1, w_2>$.

For a perfectly elastic medium, we may make use of the two symmetries (2.40), (2.45) to establish a further propagation invariant $\mathbf{b}_1^{\mathsf{T}*}\mathbf{N}\mathbf{b}_2$. This invariant property is a consequence of the conservation of energy in a lossless medium. When $\mathbf{b}_1$

and $\mathbf{b}_2$ are the same this invariant is simply a multiple of the energy flux crossing planes $z = const$. It is, however, rather more difficult to give any simple physical interpretation of $<\mathbf{w}_1, \mathbf{w}_2>$ in the presence of dissipation.

We have hitherto considered the circumstance in which both stress-displacement vectors have the same frequency and slowness. We shall later wish to have available the results for different frequencies and slownesses, and so taking note of the structure of (2.34) we use the combination

$$[\mathbf{w}_1^T(\omega_1, p_1)\omega_2\mathbf{t}_2(\omega_2, p_2) - \omega_1\mathbf{t}_1^T(\omega_1, p_1)\mathbf{w}_1(\omega_2, p_2)]. \tag{2.47}$$

For *P-SV* waves

$$\begin{aligned}\partial_z[U_1\omega_2P_2 + V_1\omega_2S_2 - \omega_1P_1U_2 - \omega_1S_1V_2]\\ = \rho(\omega_1^2 - \omega_2^2)[U_1U_2 + V_1V_2] + 4\rho(1-\beta^2/\alpha^2)\beta^2[\omega_2^2p_2^2 - \omega_1^2p_1^2]V_1V_2\\ +(\omega_2p_2 - \omega_1p_1)\{[U_1\omega_2S_2 + U_2\omega_1S_1]\\ -(1-2\beta^2/\alpha^2)[V_1\omega_2P_2 + V_2\omega_1P_1]\},\end{aligned} \tag{2.48}$$

and for *SH* waves

$$\begin{aligned}\partial_z[W_1\omega_2T_2 - \omega_1T_1W_2]\\ = \rho(\omega_1^2 - \omega_2^2)W_1W_2 + \rho\beta^2(\omega_2^2p_2^2 - \omega_1^2p_1^2)W_1W_2.\end{aligned} \tag{2.49}$$

The propagation invariants play an important role in the description of the seismic wavefield, even in the presence of sources, and we will frequently need to use the quantity $<\mathbf{w}_1, \mathbf{w}_2>$.

2.1.2 Recovery of spatial displacement from b vector

The displacement $\mathbf{u}$ within the stratification can be recovered from the transformed quantities U, V, W which appear as elements of the stress-displacement vectors $\mathbf{b}$ by inverting the Fourier-Hankel transform (2.19).

The simplest case is that for vertical displacement since here a direct transformation was made, thus in terms of the quantities $U(k, m, z, \omega)$

$$u_z(r, \phi, z, t) = \frac{1}{2\pi}\int_{-\infty}^{\infty} d\omega e^{-i\omega t}\int_0^{\infty} dk\, k \sum_m U(m, z)J_m(kr)e^{im\phi}. \tag{2.50}$$

This representation of the displacement u_z may be regarded as a superposition of cylindrical waves whose order dictates the nature of their azimuthal modulation. At each frequency and angular order the radial contribution is obtained by superposing all horizontal wavenumbers k from 0 to infinity. This corresponds to including all propagating waves at the level z within the stratification, from vertically travelling to purely horizontal, for all wave types as well as the whole spectrum of evanescent waves out to infinite wavenumber: a Sommerfeld-Weyl integral. At any particular distance r the relative contributions of the wavenumbers are imposed by the radial phase functions $J_m(kr)$.

For the horizontal components we must recall that the transform operation was applied to the composite quantities u_V, u_H. When we recover u_r, u_ϕ the horizontal derivatives in (2.13), (2.14) are reflected by derivatives of the Bessel functions and angular terms in the component expressions. The coupling of the horizontal displacements in u_V, u_H means that the displacements u_r and u_ϕ involve both $V(k, m, z, \omega)$ and $W(k, m, z, \omega)$:

$$u_r(r, \phi, z, t) = \frac{1}{2\pi}\int_{-\infty}^{\infty} d\omega e^{-i\omega t}\int_0^{\infty} dk\, k \times \sum_m \left[V(m, z)\frac{\partial J_m(kr)}{\partial(kr)} + W(m, z)\frac{im}{kr}J_m(kr)\right] e^{im\phi}, \quad (2.51)$$

$$u_\phi(r, \phi, z, t) = \frac{1}{2\pi}\int_{-\infty}^{\infty} d\omega e^{-i\omega t}\int_0^{\infty} dk\, k \times \sum_m \left[V(m, z)\frac{im}{kr}J_m(kr) - W(m, z)\frac{\partial J_m(kr)}{\partial(kr)}\right] e^{im\phi}. \quad (2.52)$$

The summation over angular order m will in principle cover the entire range from minus infinity to infinity, but the actual range of non-zero transform elements will depend on the nature of the source exciting the disturbance.

The expressions (2.50)-(2.52) depend on the quantity

$$Y_k^m(r, \phi) = J_m(kr)e^{im\phi}, \quad (2.53)$$

and its horizontal gradient $\nabla_1 Y_k^m$, where ∇_1 is the operator

$$\nabla_1 = \mathbf{e}_r\partial_r + \mathbf{e}_\phi r^{-1}\partial_\phi. \quad (2.54)$$

This dependence may be emphasised by rewriting the expressions for the three components of displacement as a vector surface-harmonic expansion (Takeuchi & Saito, 1972)

$$\mathbf{u}(r, \phi, z, t) = \frac{1}{2\pi}\int_{-\infty}^{\infty} d\omega e^{-i\omega t}\int_0^{\infty} dk\, k \sum_m [U\mathbf{R}_k^m + V\mathbf{S}_k^m + W\mathbf{T}_k^m], \quad (2.55)$$

where the vector harmonics are

$$\mathbf{R}_k^m = \mathbf{e}_z Y_k^m, \quad \mathbf{S}_k^m = k^{-1}\nabla_1 Y_k^m, \quad \mathbf{T}_k^m = -\mathbf{e}_z \wedge \mathbf{S}_k^m. \quad (2.56)$$

These vector harmonics are orthogonal to each other so that, for example,

$$\int_0^{\infty} dr\, r\int_{-\pi}^{\pi} d\phi\, \mathbf{R}_k^m.[\mathbf{S}_k^\mu]^* = 0 \quad (2.57)$$

where the star denotes a complex conjugate. For an individual harmonic we have the orthonormality property

$$\int_0^{\infty} dr\, r\int_{-\pi}^{\pi} d\phi\, \mathbf{R}_k^m.\, [\mathbf{R}_\kappa^\mu]^* = (k\kappa)^{-1/2}\, 2\pi\delta_{m\mu}\delta(k-\kappa) \quad (2.58)$$

with similar results for $\mathbf{S}_k^m$, $\mathbf{T}_k^m$. These properties enable us to simplify the calculation of source excitation coefficients. The corresponding harmonic expansion for the traction $\mathbf{t}$ has ωP, ωS, ωT in place of U, V, W in (2.55). The harmonic $\mathbf{T}_k^m$ lies wholly within a horizontal plane and, as we have already seen, for an isotropic medium this part of the displacement and traction separates from the rest to give the *SH* portion of the seismic field.

In the course of this book we will concentrate on three-dimensional problems. It is, however, interesting to note that if we consider a two-dimensional situation, where all stresses and displacements are independent of the cartesian coordinate y and take a Fourier transform over time and horizontal position x

$$\bar{\psi}(k,\omega) = \frac{1}{2\pi}\int_{-\infty}^{\infty} dt\, e^{i\omega t}\int_{-\infty}^{\infty} dx\, e^{-ikx}\psi(x,t), \tag{2.59}$$

then the sets of equations (2.24), (2.25) are recovered if we work in cartesian components and set

$$\begin{aligned} U &= i\bar{u}_z, & P &= i\bar{\tau}_{zz}, \\ V &= \bar{u}_x, & S &= \bar{\tau}_{xz}, \\ W &= \bar{u}_y, & T &= \bar{\tau}_{yz}. \end{aligned} \tag{2.60}$$

This treatment gives a plane wave decomposition rather than the cylindrical wave decomposition implied by (2.50)-(2.55).

In a spherically stratified model we may make an expansion of the seismic displacement in a spherical coordinate system (R, Δ, ϕ) in terms of vector tesseral harmonics on a sphere (cf., 2.55). The transform variables are now frequency ω, angular order l and azimuthal order m. We have a purely radial harmonic $\mathbf{R}_l^m = \mathbf{e}_R Y_l^m(\Delta, \phi)$ and as in (2.48), a second orthogonal harmonic $\mathbf{S}_l^m$ is generated by the action of the gradient operator on a spherical shell. The displacement elements U, V associated with $\mathbf{R}_l^m$, $\mathbf{S}_l^m$ are coupled and represent the *P-SV* wave part of the field. A stress-displacement vector $\mathbf{b}(l, m, R, \omega)$ can be constructed from U, V and their associated stress variables P, S. This vector satisfies a set of first order differential equations with respect to radius R with the same structure as (2.26) (see, e.g., Woodhouse 1978). As in the horizontally stratified case these equations are independent of the angular order m. The remaining vector harmonic $\mathbf{T}_l^m$ is orthogonal to both $\mathbf{R}_l^m$ and $\mathbf{S}_l^m$ and its displacement element W represents the *SH* wave portion of the seismic field.

The angular order l only takes discrete values, but we can think in terms of an angular slowness $\wp = (l + \frac{1}{2})/\omega$. At the surface $R = r_e$, the horizontal slowness $p = \wp/r_e$. The phase term $Y_l^m(\Delta, \phi)$ appearing in all the vector harmonics has angular dependence $P_l^m(\cos\Delta)e^{im\phi}$, and so we are faced with a summation in l over a sequence of terms depending on associated Legendre functions. With the aid of the Poisson sum formula the sum over l can be converted into an integral over the variable $\upsilon = l + \frac{1}{2}$, together with a sum which Gilbert (1976) has shown can be associated with successive orbits of waves around the Earth. If we consider just the first orbit this gives a representation

$$\bar{u}_R = \int_0^\infty d\upsilon \sum_m \bar{U}(\upsilon, m, r_e, \omega) P^m_{\upsilon-1/2}(\cos\Delta) e^{im\phi} \tag{2.61}$$

for the radial displacement. At high frequencies,

$$P^m_{\upsilon-1/2}(\cos\Delta) e^{im\phi} \sim \left(\frac{\Delta}{\sin\Delta}\right)^{1/2} J_m(\omega p r) e^{im\phi}, \tag{2.62}$$

in terms of the horizontal range $r = r_e\Delta$. Thus, asymptotically, we recover the expressions (2.50)–(2.52) for the surface displacements, with a scaling factor $(\Delta/\sin\Delta)^{1/2}$ to compensate for sphericity.

In the limit $\omega \to \infty$, $\wp = \upsilon/\omega$ fixed, the first order differential equations for the displacement vector **b** reduce to those for horizontal stratification (2.24) and (2.25).

2.2 Fundamental and propagator matrices

In the previous section we have introduced the coupled differential equations (2.26) for the stress-displacement vector **b**. When we come to consider the excitation of the seismic wavefield we will want to solve these equations subject to the source effects and boundary conditions imposed by the nature of the stratification. At a free surface we will require the traction elements P, S, T to vanish. If the stratification is terminated by a half space then we will impose a radiation condition that the field in the half space should consist only of outgoing propagating waves or evanescent waves which are exponentially damped as one penetrates into the half space.

In order to use such boundary conditions with the coupled sets of first order equations, we have to be able to relate the stress-displacement vectors at different levels within the stratification. We do this by introducing matrices whose columns consist of stress-displacement vectors satisfying particular boundary conditions.

As we wish to look at the evolution of the stress-displacement field with depth, we fix the angular order m, horizontal wavenumber k and frequency ω, and use the shortened form $\mathbf{b}(z)$ to mean $\mathbf{b}(k, m, z, \omega)$. In the absence of any forcing term a stress-displacement vector $\mathbf{b}(z)$ satisfies the equation

$$\partial_z \mathbf{b}(z) = \omega \mathbf{A}(p, z) \mathbf{b}(z). \tag{2.63}$$

If, therefore, we construct a square matrix $\mathbf{B}(z)$ whose columns are independent stress-displacement vectors satisfying different initial conditions, then $\mathbf{B}(z)$ is governed by the matrix equation

$$\partial_z \mathbf{B}(z) = \omega \mathbf{A}(p, z) \mathbf{B}(z). \tag{2.64}$$

Normally the columns of such a *fundamental* matrix would be chosen to have some common characteristics. Thus, for example, they may be the **b** vectors corresponding to upward and downward travelling waves at some level.

For *SH* waves the fundamental matrix **B** is constructed from two **b** vectors

$$\mathbf{B}_H = [\mathbf{b}_1; \mathbf{b}_2], \tag{2.65}$$

but for *P-SV* waves we have four **b** vectors present. Normally these would divide into two groups, within which the properties are similar, e.g., downgoing *P* waves and downgoing *SV* waves,

$$\mathbf{B}_P = [\mathbf{b}_{11}, \mathbf{b}_{12}; \mathbf{b}_{21}, \mathbf{b}_{22}]. \tag{2.66}$$

In each case we can partition the fundamental **B** matrix to display the displacement and stress elements, thus we write

$$\mathbf{B} = \begin{bmatrix} \mathsf{W}_1 & \mathsf{W}_2 \\ \mathsf{T}_1 & \mathsf{T}_2 \end{bmatrix}, \tag{2.67}$$

in terms of displacement matrices W and their associated traction matrices T. For the *P-SV* system W will be a 2×2 matrix whose columns can be thought of as independent solutions of the second order system (2.31) and the traction matrix is generated as in (2.33), $\omega\mathsf{T} = \mathsf{A}\partial_z\mathsf{W} + \omega p\mathsf{B}\mathsf{W}$. In the *SH* case W and T are just the displacement and traction elements W, T. Any particular stress-displacement vector can be created by taking a linear combination of the columns of **B**, thus $\mathbf{b} = \mathbf{Bc}$ in terms of some constant vector **c**.

We may establish a general form for the inverse of a fundamental matrix by extending our treatment of propagation invariants. For the displacement matrices W_1 and W_2 we introduce the matrix

$$<\mathsf{W}_1, \mathsf{W}_2> = \mathsf{W}_1^T\mathsf{T}_2 - \mathsf{T}_1^T\mathsf{W}_2, \tag{2.68}$$

and then the ij th entry in $<\mathsf{W}_1, \mathsf{W}_2>$ is the expression $<\mathsf{w}_{1i}, \mathsf{w}_{2j}>$ (2.36) constructed from the ith column of W_1 and the jth column of W_2. For *SH* waves there is no distinction between (2.68) and (2.36), but for *P-SV* waves $<\mathsf{W}_1, \mathsf{W}_2>$ is a 2×2 matrix. Since each of the entries in (2.68) is independent of depth so is $<\mathsf{W}_1, \mathsf{W}_2>$ and we have a matrix propagation invariant. From the definition (2.68)

$$<\mathsf{W}_1, \mathsf{W}_2>^T = -<\mathsf{W}_2, \mathsf{W}_1>. \tag{2.69}$$

When the displacement vectors in W_1 satisfy a *common* boundary condition that a linear combination of displacement and traction vanishes at some level z_0 i.e.

$$\mathsf{C}\mathsf{w}(z_0) + \mathsf{D}\mathsf{t}(z_0) = 0, \tag{2.70}$$

for some matrices C, D, then

$$<\mathsf{w}_{1i}, \mathsf{w}_{1j}> = 0, \qquad \text{for all } i, j, \tag{2.71}$$

(these class of boundary conditions include free surface and radiation conditions). In this case

$$<\mathbf{W}_1, \mathbf{W}_1> = 0. \tag{2.72}$$

We will assume that (2.72) holds for both $\mathbf{W}_1$ and $\mathbf{W}_2$ and then the form of (2.68) suggests that we should try to construct the inverse of the fundamental matrix (2.67) from the transposes of the displacement and traction matrices. The matrix product

$$\begin{bmatrix} \mathsf{T}_2^T & -\mathsf{W}_2^T \\ -\mathsf{T}_1^T & \mathsf{W}_1^T \end{bmatrix} \begin{bmatrix} \mathsf{W}_1 & \mathsf{W}_2 \\ \mathsf{T}_1 & \mathsf{T}_2 \end{bmatrix} = \begin{bmatrix} <\mathsf{W}_1, \mathsf{W}_2>^T & 0 \\ 0 & <\mathsf{W}_1, \mathsf{W}_2> \end{bmatrix}, \tag{2.73}$$

when we use (2.72). The inverse of the fundamental matrix has, therefore, the partitioned form

$$\mathbf{B}^{-1} = \begin{bmatrix} <\mathsf{W}_1, \mathsf{W}_2>^{-T}\mathsf{T}_2^T & -<\mathsf{W}_1, \mathsf{W}_2>^{-T}\mathsf{W}_2^T \\ <\mathsf{W}_1, \mathsf{W}_2>^{-1}\mathsf{T}_1^T & -<\mathsf{W}_1, \mathsf{W}_2>^{-1}\mathsf{W}_1^T \end{bmatrix}, \tag{2.74}$$

where we have used the superscript $^{-T}$ to indicate the inverse of a transpose. For many cases of interest $\mathbf{W}_1$ and $\mathbf{W}_2$ can be chosen so that $<\mathbf{W}_1, \mathbf{W}_2>$ has a simple form, often just a multiple of the unit matrix, in which case (2.74) simplifies by the extraction of a common factor.

2.2.1 The propagator matrix

From any fundamental matrix we may construct a propagator matrix $\mathbf{P}(z, z_0)$ (Gilbert & Backus, 1966) for a portion of the medium as

$$\mathbf{P}(z, z_0) = \mathbf{B}(z)\mathbf{B}^{-1}(z_0). \tag{2.75}$$

The propagator is a fundamental matrix satisfying the constraint

$$\mathbf{P}(z_0, z_0) = \mathbf{I}, \tag{2.76}$$

where $\mathbf{I}$ is the unit matrix of appropriate dimensionality. In principle, at least, the propagator can always be constructed by solving an initial value problem for (2.64) with the starting condition $\mathbf{B}(z_0) = \mathbf{I}$.

In terms of this propagator matrix the solution of (2.63) with the stress-displacement vector specified at some level z_0 is

$$\mathbf{b}(z) = \mathbf{P}(z, z_0)\mathbf{b}(z_0). \tag{2.77}$$

The character of the propagator can be seen by writing (2.77) in partitioned form

$$\begin{bmatrix} \mathsf{w}(z) \\ \mathsf{t}(z) \end{bmatrix} = \begin{bmatrix} \mathsf{P}_{WW} & \mathsf{P}_{WT} \\ \mathsf{P}_{TW} & \mathsf{P}_{TT} \end{bmatrix} \begin{bmatrix} \mathsf{w}(z_0) \\ \mathsf{t}(z_0) \end{bmatrix}. \tag{2.78}$$

Thus the displacement elements at z depend on both the displacement and stress elements at z_0. The partitions of the propagator correspond to the displacement and traction matrices introduced in (2.68). We may use the general expression for the inverse of a fundamental matrix (2.74) to find a partitioned form for $\mathbf{P}^{-1}(z, z_0)$. Since $\mathbf{P}(z_0, z_0) = \mathbf{I}$,

$$<\mathsf{P}_{WW}, \mathsf{P}_{WT}> = \mathrm{I}, \tag{2.79}$$

and thus

$$\mathbf{P}^{-1}(z, z_0) = \begin{bmatrix} \mathsf{P}_{TT}^{T} & -\mathsf{P}_{WT}^{T} \\ -\mathsf{P}_{TW}^{T} & \mathsf{P}_{WW}^{T} \end{bmatrix}, \tag{2.80}$$

a result which has also been pointed out by Chapman & Woodhouse (1981).

The continuity of the $\mathbf{b}$ vector at discontinuities in the elastic parameters means that a propagator matrix, defined in (2.77) as a transfer function for the stress and displacement between two levels, can be constructed for an arbitrary structure with depth. $\mathbf{P}(z, z_0)$ as a function of z will be continuous across all planes $z = const$.

We can illustrate the role of the propagator by a simple scalar example: consider the equation

$$\partial_z y = \omega a y. \tag{2.81}$$

With a constant, the solution of the initial value problem with y specified at z_0 is

$$y(z) = \exp[\omega(z - z_0)a] y(z_0). \tag{2.82}$$

The exponential here acts as a transfer operator for the value of y. In a similar way if the coefficient matrix $\mathbf{A}$ is constant, so that we are considering a portion of the medium with uniform properties, the solution of (2.63) is

$$\mathbf{b}(z) = \exp[\omega(z - z_0)\mathbf{A}]\mathbf{b}(z_0), \tag{2.83}$$

where the matrix exponential may be defined by its series expansion.

The propagator in (2.77) represents a generalisation of the initial value solution (2.83) to the case where $\mathbf{A}$ is not necessarily constant. In the context of general matrix theory, the propagator is often referred to as a 'matricant' (see, e.g., Frazer, Duncan & Collar, 1938). The propagator may be constructed for arbitrary $\mathbf{A}$ by a recursive scheme, for a finite interval $a \leq z, z_0 \leq b$,

$$\mathbf{P}_{j+1} = \mathbf{I} + \omega \int_{z_0}^{z} d\zeta\, \mathbf{A}(\zeta)\mathbf{P}_j(\zeta, z_0), \tag{2.84}$$

with $\mathbf{P}_0(z, z_0) = \mathbf{I}$. This procedure is derived from the equivalent Volterra integral equation to (2.64) and will converge uniformly as $j \to \infty$, provided that all the elements in $\mathbf{A}$ are bounded. This will always be true for solids, and so the propagator is given by

$$\mathbf{P}(z, z_0) = \mathbf{I} + \omega \int_{z_0}^{z} d\zeta\, \mathbf{A}(\zeta) + \omega^2 \int_{z_0}^{z} d\zeta\, \mathbf{A}(\zeta) \int_{z_0}^{\zeta} d\eta\, \mathbf{A}(\eta) + \dots . \tag{2.85}$$

When $\mathbf{A}$ is constant this series is just the expansion of an exponential, as expected from (2.83).

For the seismic case, the propagator always has unit determinant and so we have no problems with singular behaviour. In general,

$$\det \mathbf{P}(z, z_0) = \exp \left[\int_{z_0}^{z} d\zeta \, \mathrm{tr}\, \mathbf{A}(\zeta) \right], \tag{2.86}$$

and tr $\mathbf{A}$, the sum of the diagonal elements, is zero for the coefficient matrices in (2.26), and thus

$$\det \mathbf{P}(z, z_0) = 1. \tag{2.87}$$

We have so far considered the propagator from the level z_0 but when we are building up the seismic response we need the relation between propagators from different starting levels. This can be found by recognising that $\mathbf{P}(z, z_0)$ and $\mathbf{P}(z, \zeta)$ are both fundamental matrices and so each column in $\mathbf{P}(z, z_0)$ can be expressed as a linear combination of the columns of $\mathbf{P}(z, \zeta)$,

$$\mathbf{P}(z, z_0) = \mathbf{P}(z, \zeta)\mathbf{C}, \tag{2.88}$$

for some constant matrix $\mathbf{C}$. If we set $z = \zeta$, $\mathbf{P}(\zeta, \zeta)$ is just the unit matrix and so $\mathbf{C} = \mathbf{P}(\zeta, z_0)$. The propagator matrices thus satisfy a chain rule

$$\mathbf{P}(z, z_0) = \mathbf{P}(z, \zeta)\mathbf{P}(\zeta, z_0), \tag{2.89}$$

which also leads to an interesting relation between the propagator and its inverse (cf., 2.80)

$$\mathbf{P}^{-1}(z, z_0) = \mathbf{P}(z_0, z). \tag{2.90}$$

The chain rule may be extended to an arbitrary number of intermediate levels. Consider a portion of the stratification between z_n and z_0 and divide this into n parts with dividers at $z_1 \leq z_2 \leq \ldots\ldots \leq z_{n-1}$; the overall propagator may be obtained by successive use of (2.89) and is just a continued product of the propagators for the subdivisions

$$\mathbf{P}(z_n, z_0) = \prod_{j=1}^{n} \mathbf{P}(z_j, z_{j-1}). \tag{2.91}$$

If the interval $z_j - z_{j-1}$ is sufficiently small, we may take $\mathbf{A}$ to be essentially constant over this depth range and so, by the mean value theorem,

$$\mathbf{A}(z) \approx \mathbf{A}(\zeta_j), \qquad z_{j-1} \leq \zeta_j, z \leq z_j. \tag{2.92}$$

Thus with many fine intervals the overall propagator may be represented by a product of matrix exponential terms,

$$\mathbf{P}(z_n, z_0) = \prod_{j-1}^{n} \exp[\omega(z_j - z_{j-1})\mathbf{A}(\zeta_j)]. \tag{2.93}$$

The approximation of taking $\mathbf{A}$ constant over the interval (z_j, z_{j-1}) is equivalent to assuming that the elastic properties are uniform in this region i.e. we have a homogeneous layer. The representation (2.93) is thus that produced by the matrix method due to Thomson (1950) and Haskell (1953), for uniform layers, and $\exp[\omega(z_j - z_{j-1})\mathbf{A}(\zeta_j)]$ is just the jth layer matrix.

2.2.2 Propagators and sources

Once we have some form of source in the stratification we must solve the inhomogeneous equation (2.26)

$$\partial_z \mathbf{b}(z) - \omega \mathbf{A}(p, z)\mathbf{b}(z) = \mathbf{F}(z), \tag{2.94}$$

subject to some initial conditions on the stress-displacement vector $\mathbf{b}$. Since the inverse of the propagator $\mathbf{P}^{-1}(z, z_0)$ satisfies

$$\partial_z \mathbf{P}^{-1}(z, z_0) = -\omega \mathbf{P}^{-1}(z, z_0)\mathbf{A}(p, z), \tag{2.95}$$

it may be recognised as an integrating factor for (2.26), and so multiplying (2.26) by this inverse propagator we have

$$\partial_z [\mathbf{P}^{-1}(z, z_0)\mathbf{b}(z)] = \mathbf{P}(z_0, z)\mathbf{F}(z). \tag{2.96}$$

On integrating with respect to z and multiplying out by $\mathbf{P}(z, z_0)$ we obtain

$$\mathbf{b}(z) = \mathbf{P}(z, z_0)\mathbf{b}(z_0) + \int_{z_0}^{z} d\zeta\, \mathbf{P}(z, \zeta)\mathbf{F}(\zeta), \tag{2.97}$$

using the propagator chain rule (2.89). The presence of the source terms modifies the previous simple form (2.77) and (2.97) displays the cumulative effect of the source terms as we move away from the reference plane $z = z_0$.

An important class of sources are confined to a plane and arise from the transformation of some point source leading to a dipolar contribution

$$\mathbf{F}(z) = \mathbf{F}_1 \delta(z - z_S) + \mathbf{F}_2 \delta'(z - z_S). \tag{2.98}$$

With this form for $\mathbf{F}(z)$, the integral contribution to (2.97) becomes

$$\int_{z_0}^{z} d\zeta\, \mathbf{P}(z, \zeta)\mathbf{F}(\zeta) = H(z - z_S)\mathbf{P}(z, z_S)\mathbf{S}(z_S), \tag{2.99}$$

in terms of the Heaviside step function $H(z)$, where the vector $\mathbf{S}(z_S)$ is given by

$$\mathbf{S}(z_S) = \mathbf{P}(z_S, z_S)\mathbf{F}_1 - \partial_\zeta \mathbf{P}(z_S, \zeta)|_{\zeta = z_S} \mathbf{F}_2. \tag{2.100}$$

We recall that $\mathbf{P}(z_S, \zeta) = \mathbf{P}^{-1}(\zeta, z_S)$ and so from (2.95)

$$\mathbf{S}(z_S) = \mathbf{F}_1 + \omega \mathbf{A}(p, z_S)\mathbf{F}_2. \tag{2.101}$$

With such a source confined to a plane, the effect of the forcing term appears explicitly only for depths z below the level of the source z_S. Across the source plane itself the stress-displacement vector suffers a discontinuity at the source level with a jump

$$\mathbf{b}(z_S+) - \mathbf{b}(z_S-) = \mathbf{S}(z_S) = \mathbf{F}_1 + \omega \mathbf{A}(p, z_S)\mathbf{F}_2. \quad (2.102)$$

Above the source the seismic field is governed by the initial conditions on $\mathbf{b}$ at z_0, which will normally involve the source indirectly via the boundary conditions on the stress-displacement field.

Chapter 3
Stress-Displacement Fields

In Chapter 2 we have shown how the governing equations for seismic wave propagation can be represented as coupled sets of first order equations in terms of the stress-displacement vector **b**. We now turn our attention to the construction of stress-displacement fields in stratified media.

We start by considering a uniform medium for which we can make an unambiguous decomposition of the wavefield into up and downgoing parts. We then treat the case where the seismic properties vary smoothly with depth. Extensions of the approach used for the uniform medium run into problems at the turning points of *P* or *S* waves. These difficulties can be avoided by working with uniform approximations based on Airy functions, which behave asymptotically like up and downgoing waves.

3.1 A uniform medium

An important special case of a 'stratified' medium is a uniform medium, for which we can split up a seismic disturbance into its *P* and *S* wave contributions. This separation is preserved under the Fourier-Hankel transformation (2.19) and the cylindrical waves for each wave type can be further characterised as up or downgoing by the character of their dependence on the z coordinate.

We will now show how to relate the stress-displacement vector **b** to the up and downgoing waves in a uniform medium, and then use this relation to illustrate the fundamental and propagator matrices introduced in Section 2.2.

For a cylindrical wave with frequency ω, slowness p and angular order m, we introduce a transformation which connects the stress-displacement vector **b** to a new vector **v**

$$\mathbf{b} = \mathbf{D}\mathbf{v}, \tag{3.1}$$

and try to choose the matrix **D** to give a simple form for the evolution of **v** with z. In a source-free region **v** must satisfy

$$\partial_z(\mathbf{D}\mathbf{v}) = \omega\mathbf{A}(p, z)\mathbf{D}\mathbf{v}, \tag{3.2}$$

and so

$$\partial_z \mathbf{v} = [\omega \mathbf{D}^{-1}\mathbf{A}\mathbf{D} - \mathbf{D}^{-1}\partial_z \mathbf{D}]\mathbf{v}. \tag{3.3}$$

If we choose $\mathbf{D}$ to be the local eigenvector matrix for $\mathbf{A}(p, z)$, the first element on the right hand side of (3.3) reduces to diagonal form,

$$\omega \mathbf{D}^{-1}\mathbf{A}\mathbf{D} = i\omega\boldsymbol{\Lambda}, \tag{3.4}$$

where $i\boldsymbol{\Lambda}$ is a diagonal matrix whose entries are the eigenvalues of $\mathbf{A}$. From the explicit forms of the coefficient matrices in (2.24) and (2.25) we find that for *P-SV* waves,

$$\boldsymbol{\Lambda}_P = \mathrm{diag}[-q_\alpha, -q_\beta, q_\alpha, q_\beta], \tag{3.5}$$

and for *SH* waves

$$\boldsymbol{\Lambda}_H = \mathrm{diag}[-q_\beta, q_\beta]; \tag{3.6}$$

where

$$q_\alpha = (\alpha^{-2} - p^2)^{1/2}, \qquad q_\beta = (\beta^{-2} - p^2)^{1/2}, \tag{3.7}$$

are the vertical slownesses for *P* and *S* waves for a horizontal slowness p. The choice of branch cuts for the radicals q_α, q_β will normally be

$$\mathrm{Im}\,\omega q_\alpha \geq 0, \qquad \mathrm{Im}\,\omega q_\beta \geq 0, \tag{3.8}$$

the frequency factor enters from (3.4).

In a uniform medium the coefficient matrix $\mathbf{A}$ is constant and so the eigenvector matrix $\mathbf{D}$ is independent of z, with the result that $\mathbf{D}^{-1}\partial_z\mathbf{D}$ vanishes. The vector $\mathbf{v}$ is then governed by the differential equation

$$\partial_z \mathbf{v} = i\omega\boldsymbol{\Lambda}\mathbf{v}, \tag{3.9}$$

with a solution

$$\mathbf{v}(z) = \exp[i\omega(z - z_0)\boldsymbol{\Lambda}]\mathbf{v}(z_0) = \mathbf{Q}(z, z_0)\mathbf{v}(z_0), \tag{3.10}$$

in terms of a 'wave-propagator' $\mathbf{Q}$, which depends on the difference between the current depth z and the reference level z_0. The exponential of a diagonal matrix is a further diagonal matrix with exponential entries and so for *P-SV* waves,

$$\mathbf{Q}_P(h, 0) = \mathrm{diag}[e^{-i\omega q_\alpha h}, e^{-i\omega q_\beta h}, e^{i\omega q_\alpha h}, e^{i\omega q_\beta h}] \tag{3.11}$$

and for *SH* waves,

$$\mathbf{Q}_H(h, 0) = \mathrm{diag}[e^{-i\omega q_\beta h}, e^{i\omega q_\beta h}]. \tag{3.12}$$

With our convention that z increases with increasing depth, these exponentials correspond to the phase increments that we would expect for the propagation of upward and travelling *P* and *S* waves through a vertical distance h. For example, suppose that we have a plane *S* wave travelling downward at an angle j to the z axis, then

$$p = \sin j/\beta, \qquad q_\beta = \cos j/\beta, \tag{3.13}$$

and the phase difference we would expect to be introduced in traversing a depth interval h is

$$\exp[i\omega h \cos j/\beta] = \exp[i\omega q_\beta h]; \tag{3.14}$$

for upgoing waves we would have the inverse of (3.14).

From (3.10) the wavevector $\mathbf{v}$ at z is just a phase shifted version of its value at z_0 and we may identify the elements of $\mathbf{v}$ with up or downgoing *P* and *S* waves by (3.11), (3.12). For *P-SV* waves we set

$$\mathbf{v}_P = [P_U, S_U, P_D, S_D]^T, \tag{3.15}$$

where P, S are associated with *P* and *SV* propagation and the suffices U, D represent up and downgoing waves; for *SH* waves we denote the elements by H, so that

$$\mathbf{v}_H = [H_U, H_D]^T. \tag{3.16}$$

We may summarise the behaviour of the wavevector $\mathbf{v}$ by introducing partitions corresponding to up and downgoing waves

$$\mathbf{v} = [\mathbf{v}_U, \mathbf{v}_D]^T. \tag{3.17}$$

When the horizontal slowness becomes larger than the inverse wavespeeds α^{-1}, β^{-1} the corresponding radicals q_α, q_β become complex. With our choice of radical, in a perfectly elastic medium with $p > \beta^{-1}$,

$$\exp[i\omega q_\beta z] = \exp[-\omega|q_\beta|z], \tag{3.18}$$

and so downgoing waves $\mathbf{v}_D$ in the propagating regime $(p < \beta^{-1})$ map to evanescent waves which decay with depth. This property extends to a dissipative medium but is not as easily illustrated. In a similar way the upgoing waves $\mathbf{v}_U$ map to evanescent waves which increase exponentially with increasing depth z.

From the initial value solution for the wavevector $\mathbf{v}$ (3.10) we can construct the initial value solution for the stress-displacement $\mathbf{b}$, in the form

$$\mathbf{b}(z) = \mathbf{D}\exp[i\omega(z - z_0)\mathbf{\Lambda}]\mathbf{D}^{-1}\mathbf{b}(z_0), \tag{3.19}$$

and so from (2.77) we may recognise the propagator for the uniform medium as

$$\mathbf{P}(z, z_0) = \exp[\omega(z - z_0)\mathbf{A}] = \mathbf{D}\exp[i\omega(z - z_0)\mathbf{\Lambda}]\mathbf{D}^{-1}. \tag{3.20}$$

We have been able to simplify the calculation of the matrix exponential by the use of the similarity transformation provided by $\mathbf{D}$. From the representation of the propagator matrix in terms of a fundamental stress-displacement matrix $\mathbf{B}$, (2.64), we can recognise a fundamental matrix for the uniform medium

$$\mathbf{B}(z) = \mathbf{D}\exp[i\omega(z - z_{ref})\mathbf{\Lambda}], \tag{3.21}$$

where z_{ref} is the reference level for the phase of the *P* and *S* wave elements. The eigenvector matrix **D** may now be seen to be this fundamental matrix evaluated at the reference level z_{ref}, and thus its columns may be identified as 'elementary' stress-displacement vectors corresponding to the different wavetypes. For *P-SV* waves

$$\mathbf{D}_P = [\epsilon_\alpha \mathbf{b}^P_U, \epsilon_\beta \mathbf{b}^S_U; \epsilon_\alpha \mathbf{b}^P_D, \epsilon_\beta \mathbf{b}^S_D], \tag{3.22}$$

where

$$\begin{aligned}\mathbf{b}^P_{U,D} &= [\mp i q_\alpha, p, \rho(2\beta^2 p^2 - 1), \mp 2i\rho\beta^2 p q_\alpha]^T,\\ \mathbf{b}^S_{U,D} &= [p, \mp i q_\beta, \mp 2i\rho\beta^2 p q_\beta, \rho(2\beta^2 p^2 - 1)]^T,\end{aligned} \tag{3.23}$$

and we take the upper sign for the upgoing elements and the lower for downgoing elements. For *SH* waves

$$\mathbf{D}_H = [\epsilon_H \mathbf{b}^H_U; \epsilon_H \mathbf{b}^H_D], \tag{3.24}$$

with

$$\mathbf{b}^H_{U,D} = [\beta^{-1}, \mp i\rho\beta q_\beta]^T. \tag{3.25}$$

We have chosen the scaling to give comparable dimensionality to corresponding elements of $\mathbf{b}^P$, $\mathbf{b}^S$, $\mathbf{b}^H$; the *SH* waveslowness β^{-1} appears in (3.25) in a similar role to the horizontal slowness in (3.23). We have a free choice of the scaling parameters ϵ_α, ϵ_β, ϵ_H and we would like the quantities P_U, S_U, H_U, etc. to have comparable meanings. It is convenient to normalise these **b** vectors so that, in a perfectly elastic medium, each of them carries the same energy flux in the *z* direction for a propagating wave.

The energy flux crossing a plane $z = const$ is given by an area integral of the scalar product of the velocity and the traction on the plane

$$\mathcal{E} = 2\pi \int_0^r dr\, r \sum_j \dot{w}_j \tau_{jz}. \tag{3.26}$$

At a frequency ω we may represent the areal flux, averaged over a cycle in time, as

$$\langle \mathcal{E} \rangle = 2\pi \int_0^r dr\, r \frac{-i\omega}{4} [\bar{\mathbf{w}}.\bar{\mathbf{t}}^* - \bar{\mathbf{w}}^*.\bar{\mathbf{t}}], \tag{3.27}$$

where the overbars denote a Fourier transform with respect to time. We may now use the vector harmonic expansion for displacement and tractions (2.55) and the orthonormality properties of the vector harmonics (2.57-(2.58)) to evaluate (3.27) as

$$\langle \mathcal{E} \rangle = \frac{1}{2\pi} \int_0^\infty dk\, k \sum_m \frac{-i\omega^2}{4} [UP^* + VS^* + WT^* - U^*P - V^*S - W^*T]. \tag{3.28}$$

For an individual cylindrical wave we can therefore construct measures of the associated energy flux : for *P-SV* waves we take

$$\Upsilon_P(\mathbf{b}) = i[UP^* + VS^* - U^*P - V^*S], \tag{3.29}$$

which is just $i\mathbf{b}^{T*}\mathbf{N}\mathbf{b}$ (cf. 2.37), and for *SH* waves

$$\Upsilon_H(\mathbf{b}) = i[WT^* - W^*T]. \tag{3.30}$$

Let us now consider the vector $\epsilon_\alpha \mathbf{b}_D^P$ for a downgoing propagating *P* wave in a perfectly elastic medium, i.e. q_α real, then

$$\Upsilon_P = |\epsilon_\alpha|^2 2\rho q_\alpha. \tag{3.31}$$

A convenient normalisation is to take

$$\epsilon_\alpha = (2\rho q_\alpha)^{-1/2}, \tag{3.32}$$

and the actual energy flux associated with $\epsilon_\alpha \mathbf{b}_D^P$ is $\omega^2/4$. We may make a corresponding choice for the normalisations for both the *SV* and *SH* elements by choosing

$$\epsilon_\beta = \epsilon_H = (2\rho q_\beta)^{-1/2}. \tag{3.33}$$

Thus for propagating *P* and *S* waves

$$\begin{aligned}\Upsilon(\epsilon_\alpha \mathbf{b}_D^P) = \Upsilon(\epsilon_\beta \mathbf{b}_D^S) = \Upsilon(\epsilon_\beta \mathbf{b}_D^H) = 1,\\ \Upsilon(\epsilon_\alpha \mathbf{b}_U^P) = \Upsilon(\epsilon_\beta \mathbf{b}_U^S) = \Upsilon(\epsilon_\beta \mathbf{b}_U^H) = -1.\end{aligned} \tag{3.34}$$

Although we have constructed ϵ_α, ϵ_β for a perfectly elastic medium, we will use the normalisations (3.32-3.33) in both the propagating and evanescent regimes.

For evanescent waves in a perfectly elastic medium, q_α and q_β will be pure imaginary and so ϵ_α and ϵ_β become complex and then, for example,

$$\Upsilon(\epsilon_\beta \mathbf{b}_D^S) = 0 \tag{3.35}$$

confirming that evanescent waves carry no energy flux in the *z* direction.

We have already noted that the eigenvector matrix $\mathbf{D}$ is a special case of a fundamental matrix, and we may display its role as a transformation by writing $\mathbf{D}$ in partitioned form

$$\mathbf{D} = \begin{bmatrix} \mathfrak{m}_U & \mathfrak{m}_D \\ \mathfrak{n}_U & \mathfrak{n}_D \end{bmatrix}. \tag{3.36}$$

The partition $\mathfrak{m}_U$ transforms the upgoing elements $\mathbf{v}_U$ of the wavevector into displacements and $\mathfrak{m}_D$ generates displacements from $\mathbf{v}_D$. The partitions $\mathfrak{n}_U$, $\mathfrak{n}_D$ generate stresses from $\mathbf{v}_U$ and $\mathbf{v}_D$. Such a structure will occur in all cases including full anisotropy. For *P-SV* waves we have from (3.23)

$$\mathfrak{m}_{U,D} = \begin{bmatrix} \mp i q_\alpha \epsilon_\alpha & p\epsilon_\beta \\ p\epsilon_\alpha & \mp i q_\beta \epsilon_\beta \end{bmatrix}$$

$$\mathfrak{n}_{U,D} = \begin{bmatrix} \rho(2\beta^2p^2 - 1)\epsilon_\alpha & \mp 2i\rho\beta^2 p q_\beta \epsilon_\beta \\ \mp 2i\rho\beta^2 p q_\alpha \epsilon_\alpha & \rho(2\beta^2p^2 - 1)\epsilon_\beta \end{bmatrix} \tag{3.37}$$

and for *SH* waves

$$\mathfrak{m}_{U,D} = \beta^{-1}\epsilon_\beta, \quad \mathfrak{n}_{U,D} = \mp i\rho\beta q_\beta \epsilon_\beta. \tag{3.38}$$

From these partitioned forms we can construct the propagation invariants $<\mathfrak{m}_U, \mathfrak{m}_D>$, (2.68), and in each case

$$<\mathfrak{m}_U, \mathfrak{m}_D> = i\mathbf{I}. \tag{3.39}$$

This means that we have a particularly simple closed form inverse for **D** via the representation (2.74)

$$\mathbf{D}^{-1} = i\begin{bmatrix} -\mathfrak{n}_D^T & \mathfrak{m}_D^T \\ \mathfrak{n}_U^T & -\mathfrak{m}_U^T \end{bmatrix}. \tag{3.40}$$

With these expressions for the eigenvector matrix **D** and its inverse, we may now use (3.20) to construct expressions for the stress-displacement propagator in the uniform medium. For *P-SV* waves

$$\mathbf{P}_P(h,0) = \mathbf{D}_P \,\mathrm{diag}[e^{-i\omega q_\alpha h}, e^{-i\omega q_\beta h}, e^{i\omega q_\alpha h}, e^{i\omega q_\beta h}]\, \mathbf{D}_P^{-1}, \tag{3.41}$$

and so the partitions $\mathbf{P}_{WW}, \mathbf{P}_{WT}, \mathbf{P}_{TW}, \mathbf{P}_{TT}$ of the propagator are given by

$$\mathbf{P}_{WW} = \begin{bmatrix} 2\beta^2p^2C_\beta - \Gamma C_\alpha & -p[2\beta^2q_\alpha^2 S_\alpha + \Gamma S_\beta] \\ -p[\Gamma S_\alpha + 2\beta^2 q_\beta^2 S_\beta] & 2\beta^2p^2C_\alpha - \Gamma C_\beta \end{bmatrix},$$

$$\mathbf{P}_{WT} = \rho^{-1}\begin{bmatrix} q_\alpha^2 S_\alpha + p^2 S_\beta & p(C_\alpha - C_\beta) \\ p(C_\beta - C_\alpha) & p^2 S_\alpha + q_\beta^2 S_\beta \end{bmatrix},$$

$$\mathbf{P}_{TW} = -\rho\begin{bmatrix} 4\beta^4p^2q_\beta^2 S_\beta + \Gamma^2 S_\alpha & p\beta^2\Gamma(C_\beta - C_\alpha) \\ p\beta^2\Gamma(C_\alpha - C_\beta) & 4\beta^4p^2q_\alpha^2 S_\alpha + \Gamma^2 S_\beta \end{bmatrix},$$

$$\mathbf{P}_{TT} = \begin{bmatrix} 2\beta^2p^2C_\beta - \Gamma C_\alpha & -p[2\beta^2q_\beta^2 S_\beta + \Gamma S_\alpha] \\ p[\Gamma S_\beta + 2\beta^2 q_\alpha^2 S_\alpha] & 2\beta^2p^2C_\alpha - \Gamma C_\beta \end{bmatrix}, \tag{3.42}$$

where

$$C_\alpha = \cos\omega q_\alpha h, \quad C_\beta = \cos\omega q_\beta h,$$
$$S_\alpha = q_\alpha^{-1}\sin\omega q_\alpha h, \quad S_\beta = q_\beta^{-1}\sin\omega q_\beta h,$$

and

$$\Gamma = 2\beta^2p^2 - 1. \tag{3.43}$$

The *SH* wave propagator is rather simpler

$$\mathbf{P}_H(h,0) = \begin{bmatrix} C_\beta & (\rho\beta^2)^{-1}S_\beta \\ -\rho\beta^2q_\beta^2 S_\beta & C_\beta \end{bmatrix}; \tag{3.44}$$

in fact this expression may be easily constructed by summing the matrix exponential series (2.85). The inverses of the propagators may be found from (2.80) and for (3.42) and (3.44) we may verify the inverse propagator relation (2.90).

These uniform layer propagators are identical to the layer matrices of Haskell (1953) although they have been derived via a different route. We have followed Dunkin (1965) and diagonalised **A** via the eigenvector matrix **D**, but other choices are possible and lead to the same result. For example Hudson (1969a) describes a transformation to variables $\mathsf{P}_\mathsf{U} \pm \mathsf{P}_\mathsf{D}$ (in our notation) and this is closely related to the original treatment of Haskell. Hudson is able to calculate the exponential of his transformed matrix by summing the series (2.85), since direct exponentiation is only convenient for a diagonal matrix.

In our construction of the stress-displacement propagator via (3.20) we have split the wavefield in the uniform medium into its component parts via $\mathbf{D}^{-1}$. We have then added in the phase increments for the separate up and downgoing *P* and *S* wave contributions for a depth interval h and finally reconstituted displacements and stresses via the matrix **D**.

With the aid of the expressions (3.36) and (3.40) for **D** and its inverse, we can represent the uniform layer propagator as a sum of upgoing and downgoing contributions

$$\mathbf{P}(\mathsf{h},0) = \mathrm{i}\begin{bmatrix} -\mathsf{m}_\mathsf{U}\mathsf{E}_\mathsf{U}\mathsf{n}_\mathsf{D}^T & \mathsf{m}_\mathsf{U}\mathsf{E}_\mathsf{U}\mathsf{m}_\mathsf{D}^T \\ -\mathsf{n}_\mathsf{U}\mathsf{E}_\mathsf{U}\mathsf{n}_\mathsf{D}^T & \mathsf{n}_\mathsf{U}\mathsf{E}_\mathsf{U}\mathsf{m}_\mathsf{D}^T \end{bmatrix} + \mathrm{i}\begin{bmatrix} \mathsf{m}_\mathsf{D}\mathsf{E}_\mathsf{D}\mathsf{n}_\mathsf{U}^T & -\mathsf{m}_\mathsf{D}\mathsf{E}_\mathsf{D}\mathsf{m}_\mathsf{U}^T \\ \mathsf{n}_\mathsf{D}\mathsf{E}_\mathsf{D}\mathsf{n}_\mathsf{U}^T & -\mathsf{n}_\mathsf{D}\mathsf{E}_\mathsf{D}\mathsf{m}_\mathsf{U}^T \end{bmatrix}, \tag{3.45}$$

where the diagonal matrix E_D is the phase income for downgoing waves, e.g., for *P-SV* waves

$$\mathsf{E}_\mathsf{D} = \mathrm{diag}[e^{\mathrm{i}\omega q_\alpha \mathsf{h}}, e^{\mathrm{i}\omega q_\beta \mathsf{h}}], \tag{3.46}$$

and $\mathsf{E}_\mathsf{U} = \mathsf{E}_\mathsf{D}^{-1}$.

3.2 A smoothly varying medium

In many cases of interest we need to consider the propagation of seismic waves in continuously stratified regions; as for example, the wavespeed gradients due to compaction in a sedimentary sequence or the velocity profile in the Earth's mantle. Although we may simulate a gradient by a fine cascade of uniform layers, we would like to make a direct construction of a fundamental stress-displacement matrix **B** so that we may define the propagator matrix and establish the reflection properties of the medium.

As an extension of the treatment for a uniform medium we consider a local eigenvector transformation (3.1) at each level z so that $\mathbf{b}(z) = \mathbf{D}(z)\mathbf{v}(z)$. The wavevector **v** will then be governed by the evolution equation

$$\partial_z \mathbf{v} = [\mathrm{i}\omega\boldsymbol{\Lambda} - \mathbf{D}^{-1}\partial_z\mathbf{D}]\mathbf{v}, \tag{3.47}$$

and the presence of the matrix $\boldsymbol{\Delta} = -\mathbf{D}^{-1}\partial_z\mathbf{D}$ will introduce coupling between the elements of **v** which we would characterise as up or downgoing in a uniform medium. The normalisation of the columns of **D** to constant energy flux in the z direction means that the diagonal elements of $\boldsymbol{\Delta}$ vanish. A wave quantity such as $\mathrm{P_U}$ is therefore modified by loss to, or gain from, other components, arising from the nature of the variation of the elastic properties with depth. The coupling matrix $\boldsymbol{\Delta}$ depends on the gradients of the vertical slownesses q_α, q_β and the elastic parameters (Chapman, 1974a).

For *SH* waves the coupling matrix is symmetric with only off-diagonal elements

$$\boldsymbol{\Delta}_H = -\mathbf{D}_H^{-1}\partial_z\mathbf{D}_H = \begin{bmatrix} 0 & \gamma_H \\ \gamma_H & 0 \end{bmatrix} \tag{3.48}$$

where

$$\gamma_H = \tfrac{1}{2}\partial_z q_\beta / q_\beta + \tfrac{1}{2}\partial_z \mu/\mu, \tag{3.49}$$

in terms of the shear modulus $\mu = \rho\beta^2$. The coefficient γ_H determines the transfer between $\mathrm{H_U}$ and $\mathrm{H_D}$. For *P-SV* waves

$$\boldsymbol{\Delta}_P = -\mathbf{D}_P^{-1}\partial_z\mathbf{D}_P = \begin{bmatrix} 0 & -i\gamma_T & \gamma_P & -i\gamma_R \\ -i\gamma_T & 0 & -i\gamma_R & \gamma_S \\ \gamma_P & i\gamma_R & 0 & i\gamma_T \\ i\gamma_R & \gamma_S & i\gamma_T & 0 \end{bmatrix}, \tag{3.50}$$

with

$$\begin{aligned} \gamma_T &= p(q_\alpha q_\beta)^{-1/2}[\beta^2(p^2 + q_\alpha q_\beta)\partial_z\mu/\mu - \tfrac{1}{2}\partial_z\rho/\rho], \\ \gamma_R &= p(q_\alpha q_\beta)^{-1/2}[\beta^2(p^2 - q_\alpha q_\beta)\partial_z\mu/\mu - \tfrac{1}{2}\partial_z\rho/\rho], \\ \gamma_A &= 2\beta^2 p^2 \partial_z\mu/\mu - \tfrac{1}{2}\partial_z\rho/\rho, \\ &\text{and} \\ \gamma_P &= \gamma_A + \tfrac{1}{2}\partial_z q_\alpha/q_\alpha, \\ \gamma_S &= \gamma_A + \tfrac{1}{2}\partial_z q_\beta/q_\beta. \end{aligned} \tag{3.51}$$

It is interesting to note that the *P* wavespeed appears only indirectly through the slowness q_α. The coefficient γ_T determines the rate of conversion of *P* to *S* or *vice versa* for elements of the same sense of propagation (e.g. $\mathrm{P_U}$ and $\mathrm{S_U}$); γ_R determines the rate of conversion for elements of different sense (e.g. $\mathrm{P_U}$ and $\mathrm{S_D}$). γ_P governs the rate of transfer between $\mathrm{P_U}$ and $\mathrm{P_D}$; and γ_S has a similar role for $\mathrm{S_U}$ and $\mathrm{S_D}$. If we consider a thin slab of material the elements γ are closely related to the reflection and transmission coefficients for the slab (see Section 5.5). If the elements of the coupling matrix $\boldsymbol{\Delta}$ are small compared with the diagonal terms in $\omega\boldsymbol{\Lambda}$, we may construct a good approximation to a fundamental stress-displacement matrix as

$$\mathbf{B}_0(z) = \mathbf{D}(z)\exp[i\omega\int_{z_{ref}}^{z} d\zeta\,\boldsymbol{\Lambda}(\zeta)] = \mathbf{D}(z)\mathcal{E}(z). \tag{3.52}$$

When we recall the normalisation implicit in $\mathbf{D}(z)$ we find that $\mathbf{B}_0$ corresponds to a WKBJ solution assuming independent propagation of each wave type. The phase integral allows for the variation of the vertical slownesses with depth and ϵ_α, ϵ_β maintain constant energy flux for each wave-element P_U, S_U *etc*. This WKBJ solution is just what would be predicted to the lowest order in ray theory, using the ideas of energy conservation along a ray tube and phase delay.

The approximation (3.52) makes no allowance for the presence of the coupling matrix $\mathbf{\Delta}$. In any better approximation we wish to retain the phase terms in $\mathcal{E}(z)$, arising from the $i\omega\mathbf{\Lambda}$ diagonal elements in (3.47). These diagonal elements are a factor of ω larger than the coupling terms in $\mathbf{\Delta}$ and so should dominate at high frequencies. This led Richards (1971) to suggest an asymptotic expansion in inverse powers of ω,

$$\mathbf{B}(z) = \mathbf{D}(z)[\mathbf{I} + \sum_{r=1}^{\infty} \mathbf{X}_r (i\omega)^r]\mathcal{E}(z). \tag{3.53}$$

The matrices $\mathbf{X}_r$ in this 'ray-series' expansion are determined recursively. When we substitute (3.53) into the equation (2.64) for $\mathbf{B}$ and examine the equations for each power of ω in turn, we find that $\mathbf{X}_r$ depends on $\mathbf{X}_{r-1}$ as

$$[\mathbf{X}_r\mathbf{\Lambda} - \mathbf{\Lambda}\mathbf{X}_r] = \partial_z \mathbf{X}_{r-1} + \mathbf{\Delta}\mathbf{X}_{r-1}. \tag{3.54}$$

Starting with $\mathbf{X}_0 = \mathbf{I}$, this commutator relation may be solved to find the $\mathbf{X}_r$.

An alternative approach used by Chapman (1976) and Richards & Frasier (1976) is to look for a solution in the form

$$\mathbf{B}(z) = \mathbf{D}(z)\mathcal{E}(z)\mathbf{\Psi}(z), \tag{3.55}$$

and then, since $\partial_z\mathcal{E} = i\omega\mathbf{\Lambda}\mathcal{E}$, the correction term $\mathbf{\Psi}$ satisfies

$$\partial_z\mathbf{\Psi}(z) = \mathcal{E}^{-1}(z)\mathbf{\Delta}(z)\mathcal{E}(z)\mathbf{\Psi}(z), \tag{3.56}$$

and a solution may be constructed in a similar way to our treatment of the propagator (2.85) as the series

$$\mathbf{\Psi}(z) = \mathbf{I} + \int_{z_{ref}}^{z} d\zeta \mathcal{E}^{-1}(\zeta)\mathbf{\Delta}(\zeta)\mathcal{E}(\zeta) + \dots. \tag{3.57}$$

The successive integrals in (3.57) can be identified by their phase behaviour with multiple reflections within the varying medium.

The two approximation schemes we have described are of greatest utility when only the first correction terms to $\mathbf{B}_0$ are significant. This requires that all the elements in the coupling matrix $\mathbf{\Delta}$ must be small, but even for velocity models with only slow variation with depth the elements of $\mathbf{\Delta}$ can be large. The matrix $\mathbf{\Delta}$ becomes singular when one of the vertical slownesses q_α or q_β vanishes, i.e. when $\alpha^{-1}(Z_\alpha) = p$ or $\beta^{-1}(Z_\beta) = p$. This level in a perfectly elastic medium, separates the region where q_α is real and we have propagating waves, from the evanescent region where q_α is imaginary. If the *P* wavespeed increases with depth, the turning level Z_α will correspond to the level at which an initially downgoing ray with inclination to the vertical $\sin^{-1}[\alpha(z)p]$ will be travelling horizontally.

In this ray picture the effect of wavefront curvature will be sufficient to turn the ray back up towards the surface. In the neighbourhood of this turning level any attempt to separate the wavefield into specifically downgoing and upgoing elements is imposing an artificial structure, this is reflected by the singularity in $\mathbf{\Delta}$ at the level of total reflection Z_α.

In a weakly attenuative medium there is no real turning level at which q_α vanishes, nevertheless the coefficients in $\mathbf{\Delta}$ become very large in the region where $\mathrm{Re}\, q_\alpha$ is very small, and it is still inappropriate to make a decomposition into up and downgoing waves in this zone.

Well away from a turning level, in either the propagating or evanescent regimes, the WKBJ approximation (3.52) , supplemented perhaps by a single correction term as in (3.57), will provide a good description of the propagation process in a smoothly varying medium. In particular this approach is useful for near-vertical incidence (p small) when we are interested in the waves returned by the structure.

In the following section we discuss an alternative construction for a fundamental **B** matrix in a smoothly varying medium which avoids the singular behaviour at the turning levels and leads to a uniform approximation.

3.3 Uniform approximations for a smoothly varying medium

The difficulties with the eigenvector matrix decomposition are associated with the singularities at the turning points of *P* and *S* waves. At a *P* wave turning level the coefficient γ_P linking the P_U and P_D elements has a q_α^{-1} singularity at Z_α which is the main difficulty. The integrable $q_\alpha^{-1/2}$ singularity in the coefficients γ_R, γ_T which control the conversion from *P* to *S* waves causes no major problems.

A simple example which exhibits turning point behaviour is the linear slowness profile for a scalar wave. A solution which represents both the behaviour in the propagating regime and the exponential decay below the turning level was given by Gans (1915). This solution may be written in terms of an Airy function $\mathrm{Ai}(x)$. Langer (1937) recognised that by a mapping of the argument of the Airy function a uniform asymptotic solution across a turning point can be found for a general monotonic slowness distribution.

For *SH* waves the Langer approach can be used with little modification. We look for a fundamental matrix **B**, in the form

$$\mathbf{B} = \mathbf{C}\mathbf{U}, \tag{3.58}$$

and then **U** will satisfy

$$\partial_z \mathbf{U} = [\omega \mathbf{C}^{-1}\mathbf{A}\mathbf{C} - \mathbf{C}^{-1}\partial_z \mathbf{C}]\mathbf{U}. \tag{3.59}$$

We choose the new transformation matrix $\mathbf{C}_H$ so that $\mathbf{C}_H^{-1}\mathbf{A}_H\mathbf{C}_H$ has only off-diagonal elements

$$\mathbf{H}_\beta = \mathbf{C}_H^{-1}\mathbf{A}_H\mathbf{C}_H = \begin{bmatrix} 0 & p \\ -q_\beta^2/p & 0 \end{bmatrix}, \tag{3.60}$$

and now rather than find the matrix exponential of (3.60) we will seek a matrix E_β which provides a good asymptotic representation of the solution of (3.59) at high frequencies. A suitable transformation matrix $\mathbf{C}_H$ is

$$\mathbf{C}_H = (\rho p)^{-1/2} \begin{bmatrix} \beta^{-1} & 0 \\ 0 & \rho p \beta \end{bmatrix}, \tag{3.61}$$

which provides a rescaling of the stress and displacement elements. $\mathbf{C}_H$ does not depend on the radical q_β, with the result that the new coupling matrix $-\mathbf{C}_H^{-1}\partial_z\mathbf{C}_H$ depends only on the elastic parameter gradients

$$-\mathbf{C}_H^{-1}\partial_z\mathbf{C}_H = \begin{bmatrix} \frac{1}{2}\partial_z\mu/\mu & 0 \\ 0 & \frac{1}{2}\partial_z\mu/\mu \end{bmatrix}, \tag{3.62}$$

and is well behaved at the turning level for *SH* waves.

When we consider the *P-SV* wave system we are only able to apply the Langer approach to one wave type at a time, and so we must make a transformation as in (3.58) to bring $\mathbf{C}_P^{-1}\mathbf{A}_P\mathbf{C}_P$ into block diagonal form where the entry for each wave type has the structure (3.60). Thus we seek

$$\mathbf{H} = \mathbf{C}_P^{-1}\mathbf{A}_P\mathbf{C}_P = \begin{bmatrix} \mathbf{H}_\alpha & 0 \\ 0 & \mathbf{H}_\beta \end{bmatrix}, \tag{3.63}$$

and guided by the work of Chapman (1974b) and Woodhouse (1978) we take

$$\mathbf{C}_P = (\rho p)^{-1/2} \begin{bmatrix} 0 & p & p & 0 \\ p & 0 & 0 & p \\ \rho(2\beta^2p^2-1) & 0 & 0 & 2\rho\beta^2p^2 \\ 0 & 2\rho\beta^2p^2 & \rho(2\beta^2p^2-1) & 0 \end{bmatrix} \tag{3.64}$$

As in the *SH* case we have avoided the slownesses q_α, q_β but there is no longer such a simple relation between $\mathbf{B}$ and $\mathbf{U}$. It is interesting to note that we can construct the columns of $\mathbf{C}_P$ by taking the sum and difference of the columns of $\mathbf{D}_P$ and rescaling; this suggests that we may be able to make a standing wave interpretation of $\mathbf{U}$. The *P-SV* coupling terms are given by

$$-\mathbf{C}_P^{-1}\partial_z\mathbf{C}_P = \begin{bmatrix} \gamma_A & 0 & 0 & -\gamma_C \\ 0 & -\gamma_A & -\gamma_B & 0 \\ 0 & \gamma_C & \gamma_A & 0 \\ \gamma_C & 0 & 0 & -\gamma_A \end{bmatrix}, \tag{3.65}$$

where γ_A has already been defined in (3.47) and controls the rate of change of the *P* and *S* wave coefficients. The off-diagonal terms in (3.61)

$$\gamma_B = 2\beta^2 p^2 \partial_z\mu/\mu - \partial_z\rho/\rho, \quad \gamma_C = 2\beta^2 p^2 \partial_z\mu/\mu, \tag{3.66}$$

lead to cross-coupling between *P* and *S* elements.

We now introduce a 'phase-matrix' $\hat{\mathbf{E}}_\beta$ which matches the high frequency part of (3.59). Following Woodhouse (1978) we construct $\hat{\mathbf{E}}_\beta$ from Airy function entries. The two linearly independent Airy functions $\mathrm{Ai}(x)$, $\mathrm{Bi}(x)$ are solutions of the equation

$$\partial^2 y/\partial x^2 - xy = 0 \tag{3.67}$$

and so, e.g., the derivative of Ai′ is just a multiple of Ai. Thus a matrix with entries depending on Airy functions and their derivatives will match the off-diagonal high frequency part of (3.55). In a slightly dissipative medium we take

$$\hat{\mathbf{E}}_\beta(\omega, p, z) = \pi^{1/2}\begin{bmatrix} s_\beta\omega^{1/6}r_\beta^{1/2}\mathrm{Bi}(-\omega^{2/3}\phi_\beta) & s_\beta\omega^{1/6}r_\beta^{1/2}\mathrm{Ai}(-\omega^{2/3}\phi_\beta) \\ \omega^{-1/6}r_\beta^{-1/2}\mathrm{Bi}'(-\omega^{2/3}\phi_\beta) & \omega^{-1/6}r_\beta^{-1/2}\mathrm{Ai}'(-\omega^{2/3}\phi_\beta) \end{bmatrix}, \tag{3.68}$$

with

$$s_\beta = -\partial_z\phi_\beta/|\partial_z\phi_\beta|, \quad r_\beta = p/|\partial_z\phi_\beta|. \tag{3.69}$$

The argument of the Airy functions is chosen so that the derivative of $\mathbf{E}_\beta$ can be brought into the same form as $\mathbf{H}_\beta$ (3.60), so we require

$$\phi_\beta(\partial_z\phi_\beta)^2 = q_\beta^2 = \beta^{-2} - p^2. \tag{3.70}$$

The solution for ϕ_β can be written as

$$\omega^{2/3}\phi_\beta = \mathrm{sgn}\{\mathrm{Re}\, q_\beta^2\}[\tfrac{3}{2}\omega\tau_\beta]^{2/3} \tag{3.71}$$

where

$$\begin{aligned} \omega\tau_\beta &= \int_z^{z_\beta} d\zeta\, \omega q_\beta(\zeta), \quad \mathrm{Re}\, q_\beta^2 > 0, \\ &= \int_z^{z_\beta} d\zeta\, i\omega q_\beta(\zeta), \quad \mathrm{Re}\, q_\beta^2 < 0, \end{aligned} \tag{3.72}$$

for positive frequency ω; we have here made use of our choice of branch cut (3.8) $\{\mathrm{Im}\,\omega q_\beta \geq 0\}$ for the radical q_β. We take the principal value of the $^2/_3$ power in (3.71). In a perfectly elastic medium these rather complex forms simplify to

$$\phi_\beta = \mathrm{sgn}(q_\beta^2)|\tfrac{3}{2}\tau_\beta|^{2/3}, \tag{3.73}$$

$$s_\beta = -\mathrm{sgn}(\partial_z\phi_\beta), \quad \tau_\beta = \int_z^{z_\beta} d\zeta\, |q_\beta(\zeta)|. \tag{3.74}$$

We take z_β to be some convenient reference level. In a perfectly elastic medium, when a turning point exists i.e. $q_\beta(Z_\beta) = 0$ at a real depth Z_β then ϕ_β will be regular and unique with the choice $z_\beta = Z_\beta$. For dissipative media, the location of the root $q_\beta = 0$ has to be found by analytically continuing the wavespeed profile to complex depth. If we then take this complex value for z_β the integral for τ_β will be a contour integral. For small dissipation Q_β^{-1} it is simpler to take z_β to be the depth at which $\mathrm{Re}\, q_\beta = 0$. If the slowness p is such that no turning point occurs for *S* waves in the region of interest, then any choice of z_β may be taken so that ϕ_β is non-unique. In order that the character of the phase matrix $\hat{\mathbf{E}}_\beta$ should correspond to the physical nature of the wave propagation, it is often desirable to extrapolate $\beta(z)$ so that a turning point Z_β is created. This artificial turning level is then used as the reference to reckon τ_β.

To simplify subsequent discussion we will use an abbreviated form for the elements of $\hat{\mathbf{E}}_\beta$ (Kennett & Woodhouse, 1978; Kennett & Illingworth, 1981)

$$\hat{\mathbf{E}}_\beta = \begin{bmatrix} s_\beta Bj(\omega\tau_\beta) & s_\beta Aj(\omega\tau_\beta) \\ Bk(\omega\tau_\beta) & Ak(\omega\tau_\beta) \end{bmatrix}. \tag{3.75}$$

The inverse of $\hat{\mathbf{E}}_\beta$ is readily constructed by using the result that the Wronskian of Ai and Bi is π^{-1} and has the form

$$\hat{\mathbf{E}}_\beta^{-1} = \begin{bmatrix} -s_\beta Ak(\omega\tau_\beta) & Aj(\omega\tau_\beta) \\ s_\beta Bk(\omega\tau_\beta) & -Bj(\omega\tau_\beta) \end{bmatrix}. \tag{3.76}$$

This 'phase matrix' $\hat{\mathbf{E}}_\beta$ satisfies

$$\partial_z \hat{\mathbf{E}}_\beta = [\omega \mathbf{H}_\beta + \partial_z \mathbf{\Phi}_\beta \mathbf{\Phi}_\beta^{-1}]\hat{\mathbf{E}}_\beta; \tag{3.77}$$

the diagonal matrix $\partial_z \mathbf{\Phi}_\beta \mathbf{\Phi}_\beta^{-1}$ depends on the Airy function argument ϕ_β

$$\partial_z \mathbf{\Phi}_\beta \mathbf{\Phi}_\beta^{-1} = \tfrac{1}{2}(\partial_{zz}\phi_\beta / \partial_z \phi_\beta)\mathrm{diag}[-1, 1] \tag{3.78}$$

and is well behaved even at turning points, where $\partial_{zz}\phi_\beta/\partial_z\phi_\beta = \{\partial_{zz}\beta - 3p(\partial_z\beta)^2\}/4\partial_z\beta$.

If then we construct the matrix $\mathbf{C}(p, z)\hat{\mathbf{E}}(\omega, p, z)$ as an approximation to a fundamental matrix $\mathbf{B}$, this will be effective at high frequencies when $\omega\mathbf{H}_\beta$ is the dominant term in (3.77). Comparison with (3.59)–(3.60) shows that we have failed to represent the contributions from $\mathbf{C}^{-1}\partial_z\mathbf{C}$ and $\partial_z\mathbf{\Phi}\mathbf{\Phi}^{-1}$ and we will need to modify $\mathbf{CE}$ to allow for these effects.

We will first look at the character of the high frequency approximation and then discuss ways of improving on this solution.

At high frequencies the argument of the Airy functions becomes large and we may use the asymptotic representations of the functions. Below an *S* wave turning point in a perfectly elastic medium (i.e. $q_\beta^2 < 0$), the entries of $\hat{\mathbf{E}}_\beta$ are asymptotically

$$\begin{aligned}
\mathrm{Aj}(\omega\tau_\beta) &\sim \tfrac{1}{2}p^{1/2}|q_\beta|^{-1/2}\exp(-\omega|\tau_\beta|),\\
\mathrm{Ak}(\omega\tau_\beta) &\sim -\tfrac{1}{2}p^{-1/2}|q_\beta|^{1/2}\exp(-\omega|\tau_\beta|),\\
\mathrm{Bj}(\omega\tau_\beta) &\sim p^{1/2}|q_\beta|^{-1/2}\exp(\omega|\tau_\beta|),\\
\mathrm{Bk}(\omega\tau_\beta) &\sim p^{-1/2}|q_\beta|^{1/2}\exp(\omega|\tau_\beta|).
\end{aligned} \tag{3.79}$$

In this region $\hat{\mathbf{E}}_\beta$ gives a good description of the evanescent wavefields. Above the turning point, the asymptotic behaviour of Aj and Bk is as $\cos(\omega\tau_\beta - \pi/4)$, and for Ak and Bj as $\sin(\omega\tau_\beta - \pi/4)$. These elements thus describe standing waves, but for most purposes we would prefer to consider travelling wave forms.

We can achieve this goal by constructing a new matrix $\mathbf{E}_\beta$, whose columns are a linear combination of those of $\hat{\mathbf{E}}_\beta$,

$$\mathbf{E}_\beta = \hat{\mathbf{E}}_\beta . 2^{-1/2}\begin{bmatrix} e^{-i\pi/4} & e^{i\pi/4} \\ e^{i\pi/4} & e^{-i\pi/4} \end{bmatrix}. \tag{3.80}$$

This new matrix will also satisfy an equation of the form (3.77), and so $\mathbf{C}(z)\mathbf{E}(z)$ is an equally good candidate for an approximate $\mathbf{B}$ matrix at high frequencies. In terms of the Airy function entries

$$\begin{aligned}
\mathbf{E}_\beta &= 2^{-1/2}\begin{bmatrix} s_\beta e^{i\pi/4}(\mathrm{Aj} - i\mathrm{Bj}) & s_\beta e^{-i\pi/4}(\mathrm{Aj} + i\mathrm{Bj}) \\ e^{i\pi/4}(\mathrm{Ak} - i\mathrm{Bk}) & e^{-i\pi/4}(\mathrm{Ak} + i\mathrm{Bk}) \end{bmatrix}\\
&= 2^{-1/2}\begin{bmatrix} s_\beta \mathrm{Ej} & s_\beta \mathrm{Fj} \\ \mathrm{Ek} & \mathrm{Fk} \end{bmatrix}.
\end{aligned} \tag{3.81}$$

Once again at high frequencies we may use the asymptotic forms of the Airy functions and then, well above a turning point with $q_\beta^2 > 0$:

$$\begin{aligned}
\mathrm{Ej}(\omega\tau_\beta) &\sim (q_\beta/p)^{-1/2}\exp(i\omega\tau_\beta),\\
\mathrm{Ek}(\omega\tau_\beta) &\sim -i(q_\beta/p)^{1/2}\exp(i\omega\tau_\beta),\\
\mathrm{Fj}(\omega\tau_\beta) &\sim (q_\beta/p)^{-1/2}\exp(-i\omega\tau_\beta),\\
\mathrm{Fk}(\omega\tau_\beta) &\sim i(q_\beta/p)^{1/2}\exp(-i\omega\tau_\beta).
\end{aligned} \tag{3.82}$$

With our convention that z increases downwards, τ_β (3.73) is a decreasing function of z for $z > Z_\beta$, the turning level for the *S* waves. We have adopted a time factor $\exp(-i\omega t)$ so that asymptotically Ej, Ek have the character of upgoing waves; similarly Fj, Fk have the character of downgoing waves. These interpretations for large arguments $\omega\tau_\beta$ are misleading if extrapolated too close to the turning point.

Below the turning point all the entries $\mathbf{E}_\beta$ increase exponentially with depth because of the dominance of the Bi terms, and so $\mathbf{E}_\beta$ is not useful in the evanescent regime. Once again E_β has a simple inverse which may be expressed in terms of the entries of E_β:

$$\mathsf{E}_\beta^{-1} = \mathrm{i}2^{-1/2}\begin{bmatrix} -s_\beta \mathrm{Fk} & \mathrm{Fj} \\ s_\beta \mathrm{Ek} & -\mathrm{Ej} \end{bmatrix}. \tag{3.83}$$

In the propagating regime, above a turning point we will use E_β as the basis for our fundamental **B** matrix, and in the evanescent region we will use $\hat{\mathsf{E}}_\beta$. Both of these matrices were constructed to take advantage of the uniform approximations afforded by the Airy function for an isolated turning point. A different choice of phase matrix with parabolic cylinder function entries is needed for a uniform approximation with two close turning points (Woodhouse, 1978).

For the *P-SV* wave system we have constructed $\mathbf{C}_P$ to give a high frequency block diagonal structure (3.63) and so the corresponding phase matrix **E** has a block diagonal form. Above all turning points **E** has E_α for the *P* wave contribution, and E_β for the *SV* contribution. Below both *P* and *S* turning levels $\hat{\mathbf{E}}$ is constructed from $\hat{\mathsf{E}}_\alpha$, $\hat{\mathsf{E}}_\beta$. In the intermediate zone below the *P* turning level, so that *P* waves are evanescent, but with *S* waves still propagating, we take the block diagonal form

$$\bar{\mathbf{E}} = \begin{bmatrix} \hat{\mathsf{E}}_\alpha & 0 \\ 0 & \mathsf{E}_\beta \end{bmatrix}. \tag{3.84}$$

In the high frequency limit we ignore any coupling between *P* and *SV* waves.

We now wish to improve on our high frequency approximations to the fundamental stress-displacement matrix which we will represent as

$$\mathbf{B}_0(\omega, p, z) = \mathbf{C}(p, z)\mathbf{E}(\omega, p, z), \tag{3.85}$$

in terms of some phase matrix representing the physical situation. As in our discussion of the eigenvector decomposition we want to retain the merits of **E** in representing phase terms, so we look for correction matrices which either pre- or post-multiply **E**.

3.3.1 An asymptotic expansion

If we look for a fundamental **B** matrix in the form

$$\mathbf{B}(\omega, p, z) = \mathbf{C}(p, z)\mathbf{K}(\omega, p, z)\mathbf{E}(\omega, p, z), \tag{3.86}$$

then our correction matrix **K** must satisfy

$$\partial_z \mathbf{K} = \omega[\mathbf{HK} - \mathbf{KH}] - \mathbf{C}^{-1}\partial_z\mathbf{CK} - \mathbf{K}\partial_z\mathbf{\Phi}\mathbf{\Phi}^{-1}, \tag{3.87}$$

where we have used (3.59) and the differential equation for **E** (3.77). The matrix **K** is independent of the choice for **E** and the frequency ω enters only through the commutator term, this suggests an asymptotic expansion in inverse powers of frequency,

$$\mathbf{K}(z) = \mathbf{I} + \sum_{r=1}^{\infty} \omega^{-r}\mathbf{k}_r(z). \tag{3.88}$$

On substituting (3.88) into (3.87) and equating powers of ω we obtain the recursive equations

$$[\mathbf{H}\mathbf{k}_{r+1} - \mathbf{k}_{r+1}\mathbf{H}] = -\partial_z\mathbf{k}_r + \mathbf{C}^{-1}\partial_z\mathbf{C}\mathbf{k}_R + \mathbf{k}_r\partial_z\mathbf{\Phi}\mathbf{\Phi}^{-1}, \tag{3.89}$$

for the $\mathbf{k}_r$, starting with $\mathbf{k}_0 = 0$. These equations have been solved by Woodhouse (1978) who has given detailed results for the coefficient $\mathbf{k}_1(z)$.

The fundamental stress-displacement matrix $\mathbf{B}$ has an asymptotic form, to first order,

$$\mathbf{B}_A(\omega, p, z) \sim \mathbf{C}(p, z)[\mathbf{I} + \omega^{-1}\mathbf{k}_1(p, z)]\mathbf{E}(\omega, p, z), \tag{3.90}$$

where we have shown the explicit dependence of the various terms on frequency, slowness and depth. A merit of this approach is that the correction $\mathbf{k}_1$ is independent of frequency. The propagator for a region (z_A, z_B) can be constructed from $\mathbf{B}_A$ as

$$\mathbf{P}(z_A, z_B) \sim \mathbf{C}(z_A)[\mathbf{I} + \omega^{-1}\mathbf{k}_1(z_A)]\mathbf{E}(z_A)\mathbf{E}^{-1}(z_B)[\mathbf{I} - \omega^{-1}\mathbf{k}_1(z_B)]\mathbf{C}^{-1}(z_B), \tag{3.91}$$

to first order. The representation (3.91) shows no identifiable reflections within (z_A, z_B) and this makes it difficult to gain any insight into interactions between the wavefield and parameter gradients. This asymptotic form of propagator has mostly been used in surface wave and normal mode studies (see, e.g., Kennett & Woodhouse, 1978; Kennett & Nolet, 1979).

3.3.2 Interaction series

An alternative scheme for constructing a full $\mathbf{B}$ matrix is to postmultiply the high frequency approximation by $\mathbf{L}(z)$ so that

$$\mathbf{B}(\omega, p, z) = \mathbf{C}(p, z)\mathbf{E}(\omega, p, z)\mathbf{L}(\omega, p, z). \tag{3.92}$$

The matrix $\mathbf{L}$ then satisfies

$$\partial_z\mathbf{L} = -\mathbf{E}^{-1}[\mathbf{C}^{-1}\partial_z\mathbf{C} + \partial_z\mathbf{\Phi}\mathbf{\Phi}^{-1}]\mathbf{E}\mathbf{L} = \{\mathbf{E}^{-1}\mathbf{j}\mathbf{E}\}\mathbf{L}. \tag{3.93}$$

The frequency dependence of $\mathbf{L}$ arises from the phase terms $\mathbf{E}, \mathbf{E}^{-1}$ and the form of $\mathbf{L}$ is controlled by the choice of $\mathbf{E}$. The equation for $\mathbf{L}$ is equivalent to the integral equation

$$\mathbf{L}(z; z_{ref}) = \mathbf{I} + \int_{z_{ref}}^{z} d\zeta \{\mathbf{E}^{-1}(\zeta)\mathbf{j}(\zeta)\mathbf{E}(\zeta)\}\mathbf{L}(\zeta), \tag{3.94}$$

where we would choose the lower limit of integration $z_{\rm ref}$ to correspond to the physical situation. Thus for a region containing a turning point we take $z_{\rm ref}$ at the turning level and use different forms of the phase matrix above and below this level. Otherwise we may take any convenient reference level. We may now make an iterative solution of (3.94) (Chapman 1981, Kennett & Illingworth 1981) in terms of an 'interaction series'. We construct successive estimates as

$$\mathbf{L}_r(z;z_{\rm ref}) = \mathbf{I} + \int_{z_{\rm ref}}^{z} d\zeta\,\{\mathbf{E}^{-1}(\zeta)\mathbf{j}(\zeta)\mathbf{E}(\zeta)\}\mathbf{L}_{r-1}(\zeta), \tag{3.95}$$

with $\mathbf{L}_0 = \mathbf{I}$. At each stage we introduce a further coupling into the parameter gradients present in $\mathbf{j}$ through $\mathbf{C}^{-1}\partial_z\mathbf{C}$ and $\partial_z\mathbf{\Phi}\mathbf{\Phi}^{-1}$. All the elements of $\mathbf{E}^{-1}\mathbf{j}\mathbf{E}$ are bounded and so the recursion (3.95) will converge. The interaction series is therefore of the form

$$\begin{aligned}\mathbf{L}(z;z_{\rm ref}) = \mathbf{I} + &\int_{z_{\rm ref}}^{z} d\zeta\,\{\mathbf{E}^{-1}\mathbf{j}\mathbf{E}\}(\zeta) + \\ &\int_{z_{\rm ref}}^{z} d\zeta\,\{\mathbf{E}^{-1}\mathbf{j}\mathbf{E}\}(\zeta)\int_{z_{\rm ref}}^{\zeta} d\eta\,\{\mathbf{E}^{-1}\mathbf{j}\mathbf{E}\}(\eta) + \ldots.\end{aligned} \tag{3.96}$$

and the terms may be identified with successive interactions of the seismic waves with the parameter gradients. For slowly varying media the elements of $\mathbf{j}$ will be small, and the series (3.96) will converge rapidly. In this case it may be sufficient to retain only the first integral term.

For *SH* waves the total gradient effects are controlled by the diagonal matrix

$$\mathbf{j}_H = (\tfrac{1}{2}\partial_z\mu/\mu + \tfrac{1}{2}\partial_{zz}\phi_\beta/\partial_z\phi_\beta)\,\mathrm{diag}[1,-1], \tag{3.97}$$

and

$$\begin{aligned}\mathsf{E}_\beta^{-1}\mathbf{j}_H\mathsf{E}_\beta = &-\frac{i}{4}(\partial_z\mu/\mu + \partial_{zz}\phi_\beta/\partial_z\phi_\beta) \\ &\times\begin{bmatrix}\mathrm{Ej}_\beta\mathrm{Fk}_\beta + \mathrm{Ek}_\beta\mathrm{Fj}_\beta & 2\mathrm{Fj}_\beta\mathrm{Fk}_\beta \\ -2\mathrm{Ej}_\beta\mathrm{Ek}_\beta & -(\mathrm{Ej}_\beta\mathrm{Fk}_\beta + \mathrm{Ek}_\beta\mathrm{Fj}_\beta)\end{bmatrix},\end{aligned} \tag{3.98}$$

where we have written Ej_β for $\mathrm{Ej}(\omega\tau_\beta)$. Asymptotically, well away from any turning level, in the propagating and evanescent regimes

$$\partial_{zz}\phi_\beta/\partial_z\phi_\beta \approx \partial_z q_\beta/q_\beta, \tag{3.99}$$

and then the gradient term behaves like γ_H (3.46). In the propagating regime, for example,

$$\mathsf{E}_\beta^{-1}\mathbf{j}_H\mathsf{E}_\beta \approx \gamma_H\begin{bmatrix}0 & \exp[-2i\omega\tau_\beta] \\ \exp[+2i\omega\tau_\beta] & 0\end{bmatrix}, \tag{3.100}$$

and the interaction series parallels the treatment for the eigenvector decomposition. Now, however, (3.98) is well behaved at a turning level.

For the *P-SV* system, the structure of the block diagonal entries of $\mathbf{j}_P$ for *P* and *S* parallel our discussion for the *SH* wave case and asymptotically the coefficients

γ_P, γ_S control the interaction with the gradients. The off-diagonal matrices lead to interconversion of *P* and *SV* waves and here asymptotically we recover γ_T and γ_R determining transmission and reflection effects.

The structure of the interaction terms **L** for *P-SV* waves is

$$\mathbf{L}_P = \begin{bmatrix} \mathbf{I} + \mathbf{L}_{\alpha\alpha} & \mathbf{L}_{\alpha\beta} \\ \mathbf{L}_{\beta\alpha} & \mathbf{I} + \mathbf{L}_{\beta\beta} \end{bmatrix}. \tag{3.101}$$

$\mathbf{L}_{\alpha\alpha}$, $\mathbf{L}_{\beta\beta}$ represent multiple interactions with the parameter gradients without change of wave type. $\mathbf{L}_{\alpha\beta}$, $\mathbf{L}_{\beta\alpha}$ allow for the coupling between *P* and *SV* waves that is not present in our choice of phase matrix and which only appears with the first integral contribution in (3.96).

From the series (3.96) we may find $\mathbf{L}(\omega, p, z; z_R)$ to any desired level of interaction with the medium and then construct an approximate **B** matrix as

$$\mathbf{B}_I(\omega, p, z) \approx \mathbf{C}(p, z)\mathbf{E}(\omega, p, z)\mathbf{L}(\omega, p, z; z_{\text{ref}}). \tag{3.102}$$

For a region (z_A, z_B) the propagator may be approximated as

$$\mathbf{P}(z_A, z_B) \approx \mathbf{C}(z_A)\mathbf{E}(z_A)\mathbf{L}(z_A; z_{\text{ref}})\mathbf{L}^{-1}(z_B; z_{\text{ref}})\mathbf{E}^{-1}(z_B)\mathbf{C}^{-1}(z_B), \tag{3.103}$$

and the kernel $\mathbf{L}(z_A; z_{\text{ref}})\mathbf{L}^{-1}(z_B; z_{\text{ref}})$ may now be identified with internal reflections in (z_A, z_B). When a turning point occurs within the region we can split the calculation above and below the turning level and then use the chain rule for the propagator.

The asymptotic series approach in the previous section leads to a solution which is most effective at high frequencies when only a few terms need be included. The interaction series approach is not restricted in its frequency coverage. Although our starting point is a high frequency approximation to the solution, this is compensated for by the presence of the same term in the kernel of the interaction series development. The number of terms required to get an adequate approximation of the wavefield depends on $\{\mathbf{E}^{-1}\mathbf{j}\mathbf{E}\}$ and thus on the size of the parameter gradients and the frequency. At low frequencies we need more terms in the interaction series to counteract the high frequency character of **E**.

3.3.3 Relation to eigenvector decomposition

If we adopt the interaction series approach, all our fundamental matrix representations include the *leading order* approximation $\mathbf{C}(p, z)\mathbf{E}(\omega, p, z)$. This form gives no coupling between *P* and *S* waves but such coupling will be introduced by the interaction term $\mathbf{L}(\omega, p, z)$.

At high frequencies, **CE** asymptotically comes to resemble the WKBJ solution (3.52), since the depth behaviour is the same although there may be constant amplitude and phase factors between the two forms. To exploit this relation we extend an idea of Richards (1976) and rearrange **CE** to a form with only diagonal phase terms.

(a) Propagating forms

We introduce the generalized vertical slownesses

$$\begin{aligned} i\eta_{\beta u}(p,\omega,z) &= -pEk(\omega\tau_\beta)/Ej(\omega\tau_\beta), \\ i\eta_{\beta d}(p,\omega,z) &= pFk(\omega\tau_\beta)/Fj(\omega\tau_\beta). \end{aligned} \tag{3.104}$$

These quantities depend on slowness p and frequency ω through the Airy terms. Since τ_β depends on the slowness structure up to the reference level, $\eta_{\beta u}$ is not just defined by the local elastic properties. We have used the subscripts u,d as opposed to U,D to indicate that the Airy elements have up and downgoing character only in the asymptotic regime, far from a turning level. In this asymptotic region, with $q_\beta^2 > 0$

$$\eta_{\beta u}(p,\omega) \sim q_\beta(p), \quad \eta_{\beta d}(p,\omega) \sim q_\beta(p), \tag{3.105}$$

and also

$$(2\rho p)^{-1/2}Ej(\omega\tau_\beta) \sim \epsilon_\beta \exp[-i\omega \int_{z_\beta}^{z} d\zeta\, q_\beta(\zeta)]. \tag{3.106}$$

With the aid of the generalized slownesses $\eta_{u,d}$ we can recast **CE** into a form where we use only Ej, Fj to describe the phase behaviour. Thus we can write, for *SH* waves

$$\mathbf{C}_H\mathbf{E}_H = \mathbf{D}_H\mathbf{E}_H = [\mathbf{b}_u^H, \mathbf{b}_d^H]\mathbf{E}_H, \tag{3.107}$$

where $\mathbf{E}_H$ is the diagonal matrix

$$\mathbf{E}_H = (2\rho p)^{-1/2}\mathrm{diag}[Ej(\omega\tau_\beta), Fj(\omega\tau_\beta)]. \tag{3.108}$$

The column vectors of $\mathbf{D}_H$ are given by

$$\mathbf{b}_{u,d}^H = [s_\beta\beta^{-1}, \mp i\rho\beta\eta_{\beta u,d}]^T, \tag{3.109}$$

where we take the upper sign with the suffix u, and $s_\beta = 1$ when the shear velocity increases with depth. The new way of writing the leading order approximation is designed to emphasise the connection with the WKBJ solution. At high frequencies we see from (3.105) that, asymptotically, the columns of $\mathbf{D}$ reduce to those of the eigenvector matrix **D**, and the phase terms reduce to (3.52).

For *P-SV* waves we take

$$\mathbf{C}_P\mathbf{E}_P = \mathbf{D}_P\mathbf{E}_P = [\mathbf{b}_u^P, \mathbf{b}_d^P, \mathbf{b}_u^S, \mathbf{b}_d^S]\mathbf{E}_P, \tag{3.110}$$

with

$$\mathbf{E}_P = (2\rho p)^{-1/2}\mathrm{diag}[Ej(\omega\tau_\alpha), Fj(\omega\tau_\alpha), Ej(\omega\tau_\beta), Fj(\omega\tau_\beta)], \tag{3.111}$$

and column vectors

$$\begin{aligned} \mathbf{b}_{u,d}^P &= [\mp i\eta_{\alpha u,d}, s_\alpha p, s_\alpha\rho(2\beta^2p^2-1), \mp 2i\rho\beta^2\eta_{\alpha u,d}]^T, \\ \mathbf{b}_{u,d}^S &= [s_\beta p, \mp i\eta_{\beta u,d}, \mp 2i\rho\beta^2\eta_{\beta u,d}, s_\beta\rho(2\beta^2p^2-1)]^T. \end{aligned} \tag{3.112}$$

Once again these expressions reduce to the WKBJ forms in the high frequency asymptotic limit.

(b) Evanescent forms

Below a turning point we work in terms of the Ai, Bi Airy functions and now take modified forms of the generalized vertical slownesses

$$\begin{aligned} i\hat{\eta}_{\beta u} &= -p\,Bk(\omega\tau_\beta)/Bj(\omega\tau_\beta), \\ i\hat{\eta}_{\beta d} &= p\,Ak(\omega\tau_\beta)/Aj(\omega\tau_\beta). \end{aligned} \tag{3.113}$$

At high frequency, in the far evanescent regime

$$\hat{\eta}_{\beta u} \sim i|q_\beta|, \quad \hat{\eta}_{\beta d} \sim i|q_\beta|, \tag{3.114}$$

and now

$$(\rho p)^{-1/2} Aj(\omega\tau_\beta) \sim (i/2)^{1/2} \epsilon_\beta \exp[-\omega|\tau_\beta|], \tag{3.115}$$

$$(\rho p)^{-1/2} Bj(\omega\tau_\beta) \sim (2i)^{1/2} \epsilon_\beta \exp[-\omega|\tau_\beta|]. \tag{3.116}$$

In this region we write the leading order approximation **CE** in a comparable form to the propagating case

$$\mathbf{C}_H \hat{\mathbf{E}}_H = \hat{\mathbf{D}}_H \hat{\mathbf{E}}_H = [\hat{\mathbf{b}}^H_u, \hat{\mathbf{b}}^H_d] \hat{\mathbf{E}}_H. \tag{3.117}$$

The column vectors $\hat{\mathbf{b}}^H$ differ from $\mathbf{b}^H$ by using the modified slownesses $\hat{\eta}$. The phase term $\hat{\mathbf{E}}$ has the same character as the WKBJ solution, but the overall amplitudes and phases differ.

For *P-SV* waves we make a similar development to the above when both *P* and *S* waves are evanescent. When only *P* waves are evanescent we use the $\hat{\mathbf{b}}^P$ forms and retain the propagating $\mathbf{b}^S$ vectors for *S* waves.

The organisation of the fundamental **B** matrix by wave type is very convenient for a discussion of turning point phenomena. When we come to consider reflection and transmission problems in Chapter 5, we shall see that an organisation by the asymptotic character of the column elements is preferable. The two fundamental matrices are related by a constant matrix multiplier and so a switch between the two forms is easily made.

For *SH* waves the fundamental matrix $\mathbf{B}_I$ (3.103) is already organised into columns whose characters are asymptotically that of up and downgoing waves. For coupled *P-SV* waves the fundamental matrix $\mathbf{B}_{ud}$ with this organisation is given by

$$\mathbf{B}_{ud} = \Xi \mathbf{B}_I \tag{3.118}$$

where

$$\Xi = \begin{bmatrix} 1 & 0 & 0 & 0 \\ 0 & 0 & 1 & 0 \\ 0 & 1 & 0 & 0 \\ 0 & 0 & 0 & 1 \end{bmatrix}, \quad \Xi = \Xi^{-1}. \tag{3.119}$$

The symmetric matrix Ξ achieves the desired reorganisation of the columns of $\mathbf{B}_I$.

Appendix: Transverse isotropy

In a transversely isotropic medium the wavespeed in directions perpendicular to the symmetry axis are all the same but differ from those parallel to the axis. If the symmetry axis is vertical, the elastic properties do not vary in a horizontal plane. We may once again use an expansion in terms of vector harmonics as in (2.55) and the corresponding stress-displacement vectors satisfy first order differential equations (2.26) with coefficient matrices $\mathbf{A}(p, z)$ which depend only on the slowness p (Takeuchi & Saito, 1972).

With this form of symmetry there are five independent elastic moduli: A, C, F, L, N. In the more restricted case of isotropy there are only two independent moduli λ, μ and then

$$A = C = \lambda + 2\mu, F = \lambda, L = N = \mu. \tag{3a.1}$$

Thus A, C and F are related to dilatation waves and L, N to shear waves. In a homogeneous medium three types of plane waves exist. For horizontal transmission, the wavespeeds of *P*, *SV* and *SH* waves are

$$\alpha_h = (A/\rho)^{1/2}, \beta_v = (L/\rho)^{1/2}, \beta_h = (N/\rho)^{1/2}. \tag{3a.2}$$

For vertical transmission the wavespeeds are

$$\alpha_v = (C/\rho)^{1/2}, \beta_v = (L/\rho)^{1/2}, \tag{3a.3}$$

and there is no distinction between the *SH* and *SV* waves. In directions inclined to the *z*-axis, there are still three possible plane waves but their velocities depend on the inclination.

In a transversly isotropic medium where the properties depend on depth, there is a separation of the stress-displacement vector equations into *P-SV* and *SH* wave sets.

The coefficient matrix for *P-SV* waves is now

$$\mathbf{A}_P = \begin{bmatrix} 0 & pF/C & 1/C & 0 \\ -p & 0 & 0 & 1/L \\ -\rho & 0 & 0 & p \\ 0 & -\rho + p^2(A - F^2/C) & -pF/C & 0 \end{bmatrix}, \tag{3a.4}$$

and its eigenvalues are $\pm iq_1, \pm iq_2$ where q is a root of

$$q^4 + q^2 G + H = 0, \tag{3a.5}$$

with

$$\begin{aligned} G &= [\rho L + \rho C - p^2(AC - F^2 - 2FL)]/LC, \\ H &= (\rho - p^2 L)(\rho - p^2 A)/LC, \end{aligned} \tag{3a.6}$$

and so

$$q^2 = \tfrac{1}{2}[G \mp (G^2 - 4H)^{1/2}]. \tag{3a.7}$$

In an isotropic medium the upper sign in the expression for q yields q_α, the lower sign q_β. The eigenvector matrix $\mathbf{D}$ may be constructed from column vectors corresponding to up and downgoing quasi-*P* and *S* waves, and analysis parallels the isotropic case.

For *SH* waves the relations are rather simpler

$$\mathbf{A}_H = \begin{bmatrix} 0 & 1/L \\ -\rho + p^2 & 0 \end{bmatrix} = \begin{bmatrix} 0 & (\rho\beta_v^2)^{-1} \\ \rho(\beta_h^2 p^2 - 1) & 0 \end{bmatrix}, \tag{3a.8}$$

and its eigenvalues are $\pm iq_h$, with

$$q_h^2 = (\rho - p^2 N)/L = (1 - p^2\beta_h^2)/\beta_v^2 \tag{3a.9}$$

which, except at vertical incidence ($p = 0$), differs from the *SV* case. The main modification to our discussions of an isotropic medium is that the vertical slownesses for *SV* and *SH* are no longer equal.

Chapter 4

Seismic Sources

A significant seismic event arises from the sudden release of some form of potential energy within the Earth or at its surface. For earthquakes the stored energy is usually associated with the strain built up across a fault zone by continuing deformation. In some deep events it is possible that volume collapse occurs with the release of configurational energy in upper mantle minerals in a phase transition. For explosions either chemical or nuclear energy is released. Surface impact sources dissipate mechanical energy.

In all these cases only a fraction of the original energy is removed from the 'hypocentre' in the form of seismic waves. Frictional resistance to faulting in an earthquake may be overcome with the melting of material on the fault surface (McKenzie & Brune, 1972). A chemical explosion compacts the material about the original charge and a cavity is normally produced. For nuclear explosions much energy is dissipated in the vaporisation of rock. At first a shock wave spreads out into the medium and the stress is non-linearly related to the considerable displacements. Only at some distance from the point of initiation do the displacements in the disturbance become small enough to be described by the linearised elastic equations we have discussed in chapter 1.

When we try to reconstruct the source characteristics from observations of linear seismic waves we are not able to extrapolate the field into the near-source region where non-linear effects are important, and so we have to be content with an equivalent description of the source in terms of our linearised wave equations.

4.1 Equivalent forces and moment tensor densities

In our treatment of seismic waves we have assumed that the behaviour of the material within the Earth is governed by linearised elastic (or nearly elastic) equations of motion, and sources will be introduced wherever the actual stress distribution departs from that predicted from our linearised model.

The true, local, equation of motion in the continuum is

$$\rho \partial_{tt} u_j = \partial_i [\sigma_{ij} + \sigma^0_{ij}] + \rho \partial_j \psi, \tag{4.1}$$

in an Eulerian viewpoint, even in the non-linear regime. As in Section 1.3, σ_{ij} represents the deviation of the local stress state from the initial stress state specified by σ^0_{ij}, and ψ is the gravitational potential. The stress will be related to the displacement **u** via some constitutive equation, which in the non-linear zone may have a complex dependence on the history of the motion.

As a model of the incremental stress we construct a field τ_{ij} based on simple assumptions about the nature of the constitutive relation between stress and strain and the variation of the elastic constants with position. In order to get our linearised equation of motion based on this model stress τ_{ij} to give the same displacement field as in (4.1), we have to introduce an additional force distribution **e** so that

$$\rho \partial_{tt} u_j = \partial_i \tau_{ij} + f_j + e_j, \tag{4.2}$$

here **f** includes any self-gravitation effects. The presence of the force **e** means that the seismic displacements predicted by (4.2) are the same as in the true physical situation represented by (4.1). We shall refer to **e** as the 'equivalent force distribution' to the original source.

From our linearised viewpoint the *source region* will be that portion of the medium in which **e** is non-zero. Where the displacements are small $\mathbf{e}(\mathbf{x}, t)$ is just the gradient of the difference between the physical stress field $\boldsymbol{\sigma}$ and our postulated field $\boldsymbol{\tau}$

$$e_j = \partial_i \sigma_{ij} - \partial_i \tau_{ij}, \tag{4.3}$$

If the source region extends to the earth's surface we have to allow for the presence of additional surface tractions $\mathbf{e}_S$ to compensate for the difference between the tractions induced by $\boldsymbol{\tau}$ and $\boldsymbol{\sigma}$. Thus we take

$$e^S_j = n_i \tau_{ij} - n_i \sigma_{ij}, \tag{4.4}$$

at the surface, where **n** is the local outward normal.

For an explosive source, the principal source region will be the sphere within the 'elastic radius', which is somewhat arbitrarily taken as the range from the point of detonation at which the radial strain in the material drops to 10^{-6}. For this volume our linearised equations will be inadequate and the equivalent forces **e** will be large. Outside the elastic radius we may expect to do a reasonable job of approximating the stress and so **e** will be small and tend to zero with increasing distance from the origin if our structural model is adequate. For small explosions, the equivalent forces will be confined to a very limited region which, compared with most seismic wavelengths, will approximate a point; we shall see later how we can represent such a source via a singular force distribution.

For an earthquake the situation is somewhat different. The two faces of the fault will be relatively undeformed during the motion and the major deformation will occur in any fault gouge present. When we use the stress estimated from the rock properties ($\boldsymbol{\tau}$) in place of that for the gouge material ($\boldsymbol{\sigma}$) we get a serious discrepancy and so a large equivalent force. This force **e** will be concentrated on the line of the

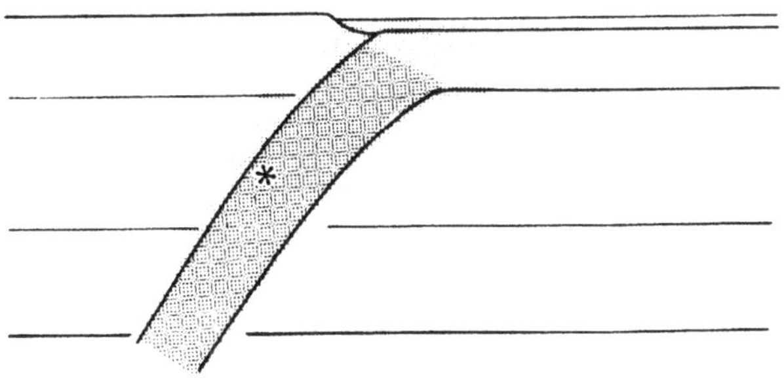

Figure 4.1. Intermediate depth event in a subduction zone; the region in which sources are introduced due to departures from the horizontally stratified reference model are indicated by shading.

fault. For long wavelengths it will appear as if we have a discontinuity across the fault with a singular force distribution concentrated on the fault plane.

In these circumstances we have a direct association of our equivalent forces **e** with the idea of localised sources. However, from (4.3) we see that we have to introduce some 'source' terms **e** whenever our model of the stress field $\boldsymbol{\tau}$ departs from the actual field. Such departures will occur whenever we have made the wrong assumptions about the character of the constitutive relation between stress and strain or about the spatial variation of the elastic parameters.

Suppose, for example, we have an *anisotropic* region and we use an *isotropic* stress-strain relation to find $\boldsymbol{\tau}$. In order to have the correct displacements predicted by (4.2) we need a force system **e**, (4.3), distributed throughout the region. This force depends on the actual displacement **u** and is therefore difficult to evaluate, but without it we get incorrect predictions of the displacement.

Another case of particular importance is provided by lateral heterogeneity in the Earth's structure when we are using a stratified model, not least because of the location of many earthquakes in zones of localised heterogeneity (subduction zones). Consider an intermediate depth event in such a downgoing slab (figure 4.1). If we try to represent the radiation from such a source using the stratified model appropriate to the continental side, our model stress $\boldsymbol{\tau}$ will not match the actual physical stress in the neighbourhood of the source itself, and will also be in error in the downgoing slab and the upper part of the oceanic structure. The affected areas are indicated by shading in figure 4.1, with a greater density in those regions where the mismatch is likely to be significant. The equivalent sources will depend on the actual displacement field and at low frequencies we would expect the principal contribution to be near the earthquake's hypocentre. The importance of the heterogeneity will increase with increasing frequency; at 0.05 Hz the shear wavelength will be comparable to the thickness of the downgoing slab and so the influence of the structure can no longer be neglected.

When an attempt is made to use observed teleseismic seismograms to invert for the source characteristics using a horizontally stratified reference model, the effective source will appear throughout the shaded region in figure 4.1 and so any localised source estimate is likely to be contaminated by the inadequacy of the reference model. At present, in the absence of effective means of solving the propagation problem for the heterogeneous situation it is difficult to assess how large a systematic error will be introduced.

For each equivalent force distribution $\mathbf{e}(\mathbf{x}, t)$ we follow Backus & Mulcahy (1976a,b) and introduce a moment tensor density $m_{ij}(\mathbf{x}, t)$ such that, in the Earth's interior

$$\partial_i m_{ij} = -e_j, \tag{4.5}$$

and if there are any surface traction effects,

$$n_i m_{ij} = e_j^s, \tag{4.6}$$

at the Earth's surface. For any *indigenous* source, i.e. any situation in which forces are not imposed from outside the Earth, there will be no net force or torque on the Earth. The total force and torque exerted by equivalent forces $\mathbf{e}$ for such a source must therefore vanish. In consequence the moment tensor density for an indigenous source is symmetric:

$$m_{ij} = m_{ji}, \tag{4.7}$$

(cf. the symmetry of the stress tensor in the absence of external couples).

The moment tensor density is not unique, but all forms share the same equivalent forces and thus the same radiation. We see from (4.3) that a suitable choice for the moment tensor density is

$$m_{ij} = \Gamma_{ij} = \tau_{ij} - \sigma_{ij}, \tag{4.8}$$

the difference between the model stress and the actual physical stress. Backus & Mulcahy refer to Γ as the 'stress glut'. This choice of m_{ij} has the convenient property that it will vanish outside our source region.

4.2 The representation theorem

We would now like to relate our equivalent source descriptions to the seismic radiation which they produce, and a convenient treatment is provided by the use of the elastodynamic representation theorem (Burridge & Knopoff 1964).

Since we wish to consider sources in dissipative media we will work in the frequency domain, and use complex moduli (1.8) in the constitutive relation connecting our model stress $\tau_{ij}(\mathbf{x}, \omega)$ to the displacement $\mathbf{u}(\mathbf{x}, \omega)$.

In the presence of a body force $\mathbf{f}(\mathbf{x}, \omega)$ we then have the equation of motion

$$\partial_i \tau_{ij} + \rho\omega^2 u_j = -f_j. \tag{4.9}$$

We now introduce the Green's tensor $G_{jp}(\mathbf{x}, \xi, \omega)$ for our reference medium.

For a unit force in the pth direction at ξ, with a Dirac delta function time dependence, the time transform of the displacement in the jth direction at $\mathbf{x}$ (subject to some boundary conditions) is then $G_{jp}(\mathbf{x}, \xi, \omega)$. This Green's tensor satisfies the equation of motion

$$\partial_i H_{ijp} + \rho\omega^2 G_{jp} = -\delta_{jp}\delta(\mathbf{x} - \xi), \tag{4.10}$$

where H_{ijp} is the stress tensor at $\mathbf{x}$ due to the force in the pth direction at ξ.

We take the scalar product of (4.9) with $G_{ji}(\mathbf{x}, \xi, \omega)$ and subtract the scalar product of (4.10) with u_j, and then integrate over a volume V enclosing the point ξ. We obtain

$$\begin{aligned} u_p(\xi, \omega) = &\int_V d^3\mathbf{x}\, G_{jp}(\mathbf{x}, \xi, \omega) f_j(\mathbf{x}, \omega) \\ &+ \int_V d^3\mathbf{x}\, [G_{jp}(\mathbf{x}, \xi, \omega)\partial_i \tau_{ij}(\mathbf{x}, \omega) - u_j(\mathbf{x}, \omega)\partial_i H_{ijp}(\mathbf{x}, \xi, \omega)]. \end{aligned} \tag{4.11}$$

If ξ is excluded from the integration volume, the right hand side of (4.12) must vanish. The second integral in (4.12) can be converted into an integral over the surface ∂V of V, by use of the tensor divergence theorem. For the anisotropic 'elastic' constitutive relation (1.2) we have

$$\begin{aligned} u_k(\xi, \omega) = &\int_V d^3\xi\, G_{qk}(\xi, \mathbf{x}, \omega) f_q(\xi, \omega) \\ &+ \int_{\partial V} d^2\xi\, n_p [G_{qk}(\xi, \mathbf{x}, \omega)\tau_{pq}(\xi, \omega) - u_q(\xi, \omega) H_{pkq}(\xi, \mathbf{x}, \omega)], \end{aligned} \tag{4.12}$$

where $\mathbf{n}$ is the outward normal to ∂V, and for subsequent convenience we have interchanged the roles of $\mathbf{x}$ and ξ.

This representation theorem applies to an arbitrary volume V and we may, for example, take V to be the whole Earth. With a homogeneous boundary condition for the Green's tensor on ∂V, such as vanishing traction for a free surface, we may derive the reciprocity relation

$$G_{jp}(\mathbf{x}, \xi, \omega) = G_{pj}(\xi, \mathbf{x}, \omega), \tag{4.13}$$

from (4.13), since the surface integral vanishes. This enables us to recast the representation theorem so that the Green's tensor elements correspond to a receiver at $\mathbf{x}$ and a source at ξ, thus

$$\begin{aligned} u_k(\mathbf{x}, \omega) = &\int_V d^3\xi\, G_{kq}(\mathbf{x}, \xi, \omega) f_q(\xi, \omega) \\ &+ \int_{\partial V} d^2\xi\, [G_{kq}(\mathbf{x}, \xi, \omega) t_q(\xi, \omega) - u_q(\xi, \omega) h_{kq}(\mathbf{x}, \xi, \omega)], \end{aligned} \tag{4.14}$$

in terms of the traction components t_q and the Green's tensor traction elements h_{kq} on ∂V.

In this form the representation theorem (4.14) also applies to the case of a prestressed medium with constitutive relation (1.4) if the quantities t_q, h_{kq} are calculated using the Piola-Kirchhoff stress tensor (Dahlen 1972) in the undeformed frame, as

$$t_j = n_i[\tau_{ij} + \sigma^0_{jk}(\partial_i u_k - \partial_k u_i)]. \tag{4.15}$$

4.2.1 Source representation

Now we have established our representation theorem we will use it to investigate source effects. We have seen that equivalent forces **e** are introduced whenever the model stress tensor $\boldsymbol{\tau}$ departs from the actual physical stress. In the equations of motion **e** plays the same role as any external forces **f**, and so, since we consider linearised seismic wave propagation, the additional displacement consequent on the presence of the source is

$$u_k(\mathbf{x}, \omega) = \int_V d^3\boldsymbol{\eta}\, G_{kq}(\mathbf{x}, \boldsymbol{\eta}, \omega) e_q(\boldsymbol{\eta}, \omega). \tag{4.16}$$

This radiation field may be alternatively expressed in terms of a moment tensor density by using the definitions (4.5), (4.6) and then integrating (4.16) by parts to give

$$u_k(\mathbf{x}, \omega) = \int_V d^3\boldsymbol{\eta}\, \partial_p G_{kq}(\mathbf{x}, \boldsymbol{\eta}, \omega) m_{pq}(\boldsymbol{\eta}, \omega). \tag{4.17}$$

In (4.16) the integration will in fact be restricted to the source region Y and this same region applies in (4.17) if we choose the 'stress-glut' Γ as the moment tensor density.

As a model of an explosive type of source we consider a region v_s in V bounded by a surface S on which the displacement $\mathbf{u}^s$ and traction $\mathbf{t}^s$ are specified, the associated seismic radiation will then be

$$u_k(\mathbf{x}, \omega) = \int_S d^2\boldsymbol{\xi} \{G_{kq}(\mathbf{x}, \boldsymbol{\xi}, \omega) t^s_q(\boldsymbol{\xi}, \omega) - u^s_q(\boldsymbol{\xi}, \omega) h_{kq}(\mathbf{x}, \boldsymbol{\xi}, \omega)\}. \tag{4.18}$$

If, then, we have some mathematical or physical model which allows us to predict the displacement and traction field for an explosion as a function of position we may carry the calculation out into the linear regime, e.g., to the 'elastic radius'. These calculated values for $\mathbf{u}^s$ and $\mathbf{t}^s$ may then be used to determine the seismic radiation. For nuclear explosions such models based on finite element or finite difference calculations have reached a high level of sophistication. With a Green's tensor that satisfies the free surface condition, (4.18) will give a good representation of the radiation up to the time of the return of surface reflections to the source region, if direct calculations for $\mathbf{u}^s$, $\mathbf{t}^s$ are used. Ideally, for shallow explosions the values of $\mathbf{u}^s$ and $\mathbf{t}^s$ should include the effect of surface interactions, e.g., spalling.

The Green's tensor to be used in (4.18) should correspond to propagation in the geological situation prevailing before the explosion, including any pre-existing stress fields; otherwise additional equivalent forces need to be introduced.

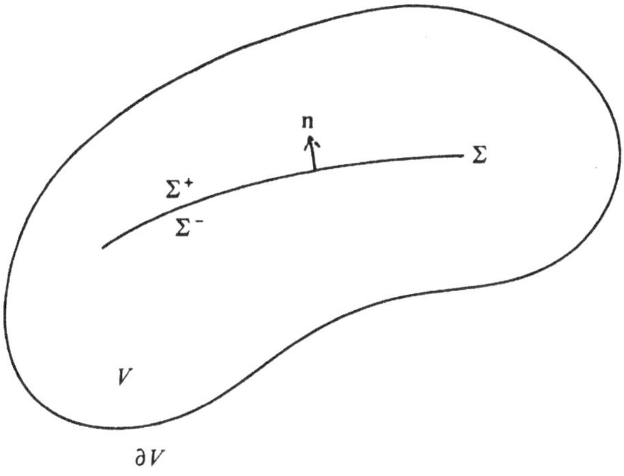

Figure 4.2. Surface of discontinuity in displacement and tractions Σ.

Unfortunately, our ability to solve propagation problems in complex media is limited and so quite often although prestress will be included in the calculation of $\mathbf{t}^s$, $\mathbf{u}^s$, it is neglected away from the surface S and a relatively simple Green's function is used to estimate the radiation via (4.18).

A source representation in terms of displacement and traction behaviour on a surface S is not restricted to explosions and could be used in association with some numerical modelling of the process of faulting in an earthquake. Commonly, however, we regard the earthquake as represented by some dynamic discontinuity in displacement across a fault surface Σ. We can derive this singular case from (4.18) by taking the surface S to consist of the two surfaces Σ^+, Σ^- lying on either side of the fault surface Σ and joined at the termination of any dislocation (figure 4.2). If we take the Green's tensor to correspond to the original configuration in V, $\mathbf{G}_k(\mathbf{x}, \xi, \omega)$ and its associated traction $\mathbf{h}_k(\mathbf{x}, \xi, \omega)$ will be continuous across Σ. The two surface integrals over Σ^+, Σ^- can therefore be combined into a single integral

$$u_k(\mathbf{x}, \omega) = -\int_\Sigma d^2\xi \{G_{kq}(\mathbf{x}, \xi, \omega)[t_q(\xi, \omega)]_-^+ - [u_q(\xi, \omega)]_-^+ h_{kq}(\mathbf{x}, \xi, \omega)\}, \tag{4.19}$$

where $[t_q(\xi, \omega)]_-^+$, $[u_q(\xi, \omega)]_-^+$ represent the jump in traction and displacement in going from Σ^- to Σ^+. The normal $\mathbf{n}$ is taken to be directed from Σ^- to Σ^+. In order to allow a flexible parameterisation of sources we do not restrict attention to tangential displacement discontinuities. As it stands (4.19) is not readily interpreted in terms of equivalent forces, as in (4.17), but it is apparent that all such forces will lie in the fault surface Σ.

We will consider the anisotropic 'elastic' case and then

$$h_{kq}(\mathbf{x}, \xi, \omega) = n_p(\xi) c_{pqrs}(\xi) \partial_r G_{ks}(\mathbf{x}, \xi, \omega), \tag{4.20}$$

where for a dissipative medium c_{pqrs} will be complex. Since we are only concerned with the values of G_{kq} and h_{kq} on Σ, we introduce the Dirac delta function $\delta_\Sigma(\xi, \eta)$ localised on the surface Σ. The integral in (4.19) can then be cast into the form

$$u_k(\mathbf{x}, \omega) = -\int_V d^3\eta\, G_{kq}(\mathbf{x}, \eta, \omega)\{[t_q(\xi, \omega)]_-^+ \delta_\Sigma(\xi, \eta) \\ + [u_s(\xi, \omega)]_-^+ n_r(\xi) c_{pqrs}(\xi) \partial_p \delta_\Sigma(\xi, \eta)\}, \tag{4.21}$$

where $-\partial_p\delta$ extracts the derivative of the function it acts upon and we have used the symmetry of $c_{rspq} = c_{pqrs}$. Equation (4.21) is now just in the form (4.17) and so we may recognise the equivalent forces $\mathbf{e}$. For the traction jump we have force elements distributed along Σ weighted by the size of the discontinuity. The displacement jump leads to force doublets along Σ which are best represented in terms of a moment tensor density m_{pq}.

We emphasise the difference in character between the two classes of discontinuity by writing

$$u_k(\mathbf{x}, \omega) = \int_V d^3\eta\, \{G_{kq}(\mathbf{x}, \eta, \omega) e_q(\eta, \omega) + \partial_p G_{kq}(\mathbf{x}, \eta, \omega) m_{pq}(\eta, \omega)\}. \tag{4.22}$$

The forces $\mathbf{e}$ are then determined by the traction jump

$$e_q(\eta, \omega) = -n_r(\xi)[\tau_{qr}(\xi, \omega)]_-^+ \delta_\Sigma(\xi, \eta). \tag{4.23}$$

The moment tensor density is specified by the displacement jump as

$$m_{pq}(\eta, \omega) = n_r(\xi) c_{pqrs}(\xi)[u_s(\xi, \omega)]_-^+ \delta_\Sigma(\xi, \eta), \tag{4.24}$$

which has the appropriate symmetry for an indigenous source contribution (4.7).

The analysis leading to equivalent forces for a dislocation in a prestressed medium is more complex but leads to comparable results (Dahlen, 1972; Walton, 1973). When we allow for the possibility of traction discontinuities, the force $\mathbf{e}$ is once again given by (4.23). The displacement jumps may be represented by the moment tensor density

$$m_{pq}(\eta, \omega) = n_r(\xi) d_{pqrs}(\xi)[u_s(\xi, \omega)]_-^+ \delta_\Sigma(\xi, \eta) \\ + n_r(\xi)[u_r(\xi, \omega)]_-^+ \sigma^0_{pq} \delta_\Sigma(\xi, \eta), \tag{4.25}$$

in terms of the moduli d_{pqrs} (1.4) which themselves depend on the prestress σ^0. The contribution to (4.25) which involves σ^0 explicitly, only appears if there is an opening crack situation when $\mathbf{e}$ is also likely to be non-zero. Only the deviatoric portion of σ^0 affects the moment tensor elements for tangential slip. In this case (4.25) will only depart significantly from (4.24) if the shear stresses are large.

The seismic radiation predicted by (4.22) is determined purely by the displacement and traction jumps across the surface Σ and the properties of the material surrounding the fault appear only indirectly through the Green's tensor G_{kq}. In many cases some assumed model of the slip behaviour on the fault is used to specify $[\mathbf{u}]_-^+$. However, a full solution for $[\mathbf{u}]_-^+$ requires the interaction of the propagating fault with its surroundings to be accounted for and this has only been achieved for a few idealised cases.

Generally earthquake models prescribe only tangential displacement jumps and then $n_r[\mathbf{u}]_-^+ = 0$. However, the opening crack model is appropriate to other observable events, e.g., rock bursts in mines.

4.2.2 Relaxation sources

Although most source models have been based on the dislocation approach described above, a number of authors have worked with an alternative approach in which the properties within a portion of a prestressed region undergo a sudden change (relaxation). The simplest case is to take a spherical region with a uniform prestress and this has been used by Randall (1966) and Archambeau (1968) among others. The effect of inhomogeneous prestress has been considered by Stevens (1980).

For illustration, we suppose a cavity with a surface S is suddenly created at a time $t = 0$. The frequency domain solution for this initial value problem yields the radiated displacement as

$$u_k(\mathbf{x}, \omega) = \int_S d^2\xi\,[G_{kq}(\mathbf{x}, \xi, \omega)t_q(\xi, \omega) - u_q(\xi, \omega)h_{kq}(\mathbf{x}, \xi, \omega)] \\ +i\omega \int_V d^3\xi\,\rho(\xi)u_q^i(\xi)G_{kq}(\mathbf{x}, \xi, \omega), \tag{4.26}$$

where $\mathbf{u}_i$ is the instantaneous displacement at $t = 0$. This expression is very similar to those we have already encountered except that now we have an integral equation for the displacement $\mathbf{u}$. For a stress-free cavity the traction $\mathbf{t}$ will vanish on S and so (4.26) is simplified. In order to avoid explicit calculation of the initial value term in (4.26) we may recognise that $\mathbf{u}^i$ is the difference between two static fields - the prestressed state without the cavity and the final equilibrium state. We may set up a representation for the static field $\mathbf{u}^i$ in terms of the dynamic Green's tensor $\mathbf{G}_k$ as

$$u_k^i = \omega^2 \int_V d^3\xi\,\rho n_q^i G_{kq} + \int_S d^2\xi\,[G_{kq}t_q^i - u_q^i h_{kq}]. \tag{4.27}$$

When we substitute from (4.27) for the volume integral in (4.26) we obtain

$$u_k'(\mathbf{x}, \omega) = -\int_S d^2\xi\,u_q'(\xi, \omega)h_{kq}(\mathbf{x}, \xi, \omega) \\ -(i\omega)^{-1} \int_S d^2\xi\,t_q^i(\xi)G_{kq}(\mathbf{x}, \xi, \omega), \tag{4.28}$$

where $\mathbf{u}'$ is the relative displacement field

$$\mathbf{u}'(\mathbf{x},\omega) = \mathbf{u}(\mathbf{x},\omega) - (i\omega)^{-1}\mathbf{u}^i(\mathbf{x}). \tag{4.29}$$

The quantity $\mathbf{t}^i$ is the traction drop from the initial prestressed state to the final traction-free configuration on S. The initial value description (4.26) is thus equivalent to imposing a stress pulse at $t = 0$ to negate the tractions on S (4.28).

If we take the cavity to be spherical we have a good model for the radiation produced by an explosion in a prestressed medium. For such a cavity embedded in a uniform, isotropic, material Stevens (1980) has shown that an exact solution to the integral equation (4.28) can be found in terms of vector spherical harmonics and a multipole expansion. For an adequate model of earthquake faulting the surface S should be rather flatter than a sphere, and then numerical solution of (4.28) would be required. Qualitatively, however, the spherical model is still useful.

4.3 The moment tensor and source radiation

In Section 4.2 we have adopted a representation of the seismic radiation in terms of a combination of distributed force elements ϵ and a moment tensor density m_{pq}

$$u_k(\mathbf{x},\omega) = \int_V d^3\eta\,\{G_{kq}(\mathbf{x},\boldsymbol{\eta},\omega)\epsilon_q(\boldsymbol{\eta},\omega) \\ +\partial_p G_{kq}(\mathbf{x},\boldsymbol{\eta},\omega)m_{pq}(\boldsymbol{\eta},\omega)\}. \tag{4.30}$$

We will henceforth restrict our attention to a 'stress-glut' moment tensor density, so that the integration in (4.30) can be restricted to the source region Y in which our model stress τ differs from the actual physical stress.

The response of our instruments and recording system imposes some upper limit (ω_u) on the frequencies for which we may recover useful information about the displacement. In general therefore, we will be considering a band-limited version of (4.30). At moderate frequencies the Green's tensor elements $G_{kq}(\mathbf{x},\boldsymbol{\eta},\omega)$ will vary smoothly as the point of excitation $\boldsymbol{\eta}$ varies with a fixed observation point $\mathbf{x}$, provided that $\mathbf{x}$ lies well away from the region containing $\boldsymbol{\eta}$. This suggests that in (4.30) we should be able to represent the Green's tensor variations within Y by a small number of terms in a Taylor series expansion of $G_{kq}(\mathbf{x},\boldsymbol{\eta},\omega)$ about the hypocentre $\mathbf{x}_S$

$$G_{kq}(\mathbf{x},\boldsymbol{\eta},\omega) = G_{kq}(\mathbf{x},\mathbf{x}_s,\omega) + (\eta_i - x_{Si})\partial_i G_{kq}(\mathbf{x},\mathbf{x}_s,\omega) \\ +\tfrac{1}{2}(\eta_i - x_{Si})(\eta_j - x_{Sj})\partial_{ij}G_{kq}(\mathbf{x},\boldsymbol{\eta},\omega) + \dots . \tag{4.31}$$

The number of significant terms will reduce as the frequency diminishes.

For extended faults the accuracy of this expansion may be improved by making the expansion about the centroid of the disturbance $\mathbf{x}_S(\omega)$ rather than the hypocentre, which is just the point of initiation.

With the expansion (4.31) the seismic radiation $\mathbf{u}$ can be approximated as a sequence of terms which represent increasingly detailed aspects of the source behaviour.

We consider first the force contribution to (4.30)

$$\begin{aligned} u_k^{\epsilon}(\mathbf{x}, \omega) &= \int_Y d^3\eta \, G_{kq}(\mathbf{x}, \boldsymbol{\eta}, \omega) \epsilon_q(\boldsymbol{\eta}, \omega), \\ &= G_{kq}(\mathbf{x}, \mathbf{x}_s, \omega) \int_Y d^3\eta \, \epsilon_q(\boldsymbol{\eta}, \omega) \\ &\quad + \partial_i G_{kq}(\mathbf{x}, \mathbf{x}_s, \omega) \int_Y d^3\eta \, (\eta_i - x_{Si}) \epsilon_q(\boldsymbol{\eta}, \omega) + \dots \end{aligned} \quad (4.32)$$

The displacement $\mathbf{u}^{\epsilon}$ can therefore be represented in terms of the polynomial moments of the distributed forces, so we may write

$$u_k^{\epsilon}(\mathbf{x}, \omega) = G_{kq}(\mathbf{x}, \mathbf{x}_s, \omega) \mathcal{E}_q(\omega) + \partial_i G_{kq}(\mathbf{x}, \mathbf{x}_s, \omega) \mathcal{E}_{q,i}^{(1)} + \dots . \quad (4.33)$$

$\mathcal{E}$ will be the total force exerted on the source region Y and $\mathcal{E}_{q,i}^{(1)}$ the tensor of force moments about the point of expansion $\mathbf{x}_S$. All of the elements in the series (4.33) will appear to be situated at $\mathbf{x}_S$. We represent the distributed force system in Y by a compound point source composed of a delta function and its derivatives at $\mathbf{x}_S$. For an *indigeneous* source the total force $\mathcal{E}$ and the moments $\mathcal{E}_{q,i}^{(1)}$ will vanish. We will, however, retain these terms since a number of practical sources, e.g., those depending on surface impact, are not indigenous.

The radiation associated with the moment tensor density may also be expanded as in (4.32),

$$\begin{aligned} u_k^{m}(\mathbf{x}, \omega) &= \int_Y d^3\eta \, \partial_p G_{kq}(\mathbf{x}, \boldsymbol{\eta}, \omega) m_{pq}(\boldsymbol{\eta}, \omega), \\ &= \partial_p G_{kq}(\mathbf{x}, \mathbf{x}_s, \omega) \int_Y d^3\eta \, m_{pq}(\boldsymbol{\eta}, \omega) \\ &\quad + \partial_{ip} G_{kq}(\mathbf{x}, \mathbf{x}_s, \omega) \int_Y d^3\eta \, (\eta_i - x_{Si}) m_{pq}(\boldsymbol{\eta}, \omega) + \dots . \end{aligned} \quad (4.34)$$

This representation of $\mathbf{u}^m$ in terms of the polynomial moments of m_{pq} means that we may write

$$\begin{aligned} u_k^{m}(\mathbf{x}, \omega) &= \partial_p G_{kq}(\mathbf{x}, \mathbf{x}_s, \omega) M_{pq}(\omega) \\ &\quad + \partial_{ip} G_{kq}(\mathbf{x}, \mathbf{x}_s, \omega) M_{pq,i}^{(1)}(\omega) + \dots . \end{aligned} \quad (4.35)$$

The integral of the moment tensor density across the source region (M_{pq}) is the quantity which is frequently referred to as *the* moment tensor

$$M_{pq}(\omega) = \int_Y d^3\eta \, m_{pq}(\boldsymbol{\eta}, \omega). \quad (4.36)$$

We see that $m_{pq}(\boldsymbol{\eta}, \omega) d^3\eta$ can be thought of as the moment tensor for the element of volume $d^3\eta$. The third order tensor $M_{pq,i}^{(1)}$ preserves more information about the

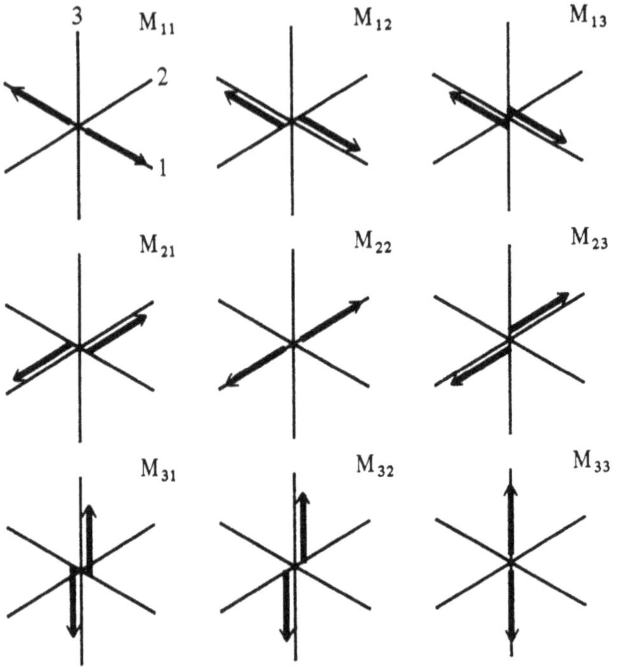

Figure 4.3. The representation of the elements of the moment tensor as weights for a set of dipoles and couples.

spatial distribution of the moment tensor density $m_{pq}(\eta, \omega)$, and the inclusion of further terms in the series (4.35) gives a higher resolution of the source behaviour.

The representation (4.35) is again an equivalent point source description of the radiation from the original source volume Y. The leading order source here may be thought of as a superposition of first derivatives of a delta function, describing force doublets. The moment tensor M_{pq} may therefore be regarded as the weighting factor to be applied to the nine elements of the array of dipoles and couples illustrated in figure 4.3. The diagonal elements of **M** correspond to dipoles and the off-diagonal elements to pure couples. The higher order tensors $\mathbf{M}^{(j)}$ are the weighting factors in a multipole expansion.

4.3.1 A small fault

We now specialise our results for a dislocation source to the case of a small fault embedded in a uniform medium.

From (4.23) and (4.30) any traction jump will, for low frequencies, be equivalent to the point force components

$$\mathcal{E}_j(\omega) = \int_\Sigma d^2\xi \, n_i(\xi)[\tau_{ij}(\xi, \omega)]_-^+ \tag{4.37}$$

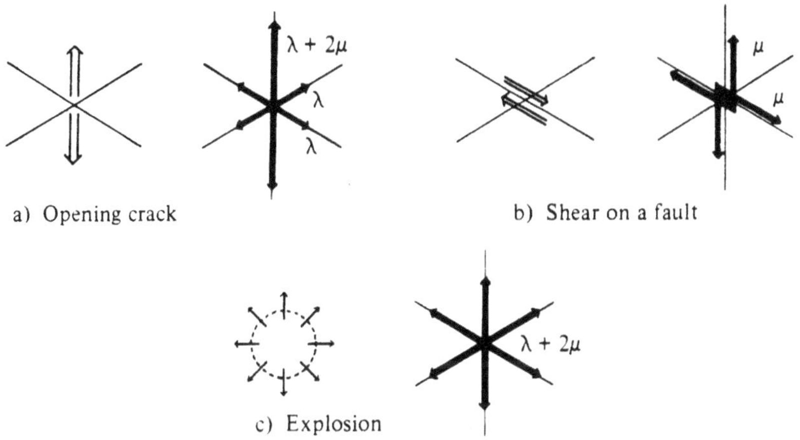

Figure 4.4. Force equivalents for simple source mechanisms: (a) an opening crack; (b) tangential slip on a fault; (c) an explosion.

and for a small fault we may represent this as an averaged traction jump over the surface area A of the fault

$$\mathcal{E}_j(\omega) = A n_i \overline{[\tau_{ij}(\omega)]}_-^+. \tag{4.38}$$

For a displacement discontinuity the equivalent leading order moment tensor is given by

$$M_{ij} = -\int_\Sigma d^2\xi\, n_k(\xi) c_{ijkl}(\xi) [u_l(\xi, \omega)]_-^+. \tag{4.39}$$

We consider a displacement discontinuity in the direction of the unit vector $\boldsymbol{\nu}$ with jump $[u]$, and then

$$M_{ij} = A\overline{[u(\omega)]}\{\lambda n_k \nu_k + \mu(n_i \nu_j + n_j \nu_i)\}, \tag{4.40}$$

for an isotropic medium (1.6), where $\overline{[u(\omega)]}$ is the averaged slip spectrum on the fault surface.

When the direction $\boldsymbol{\nu}$ lies along the normal $\mathbf{n}$, as in an opening crack - figure 4.4a,

$$M_{ij} = A\overline{[u(\omega)]}\{\lambda \delta_{ij} + 2\mu n_i n_j\}, \tag{4.41}$$

and if we rotate the coordinate system so that the 3 axis lies along the normal, M_{ij} is diagonal with components

$$M_{ij} = A\overline{[u(\omega)]}\text{diag}\{\lambda, \lambda, \lambda + 2\mu\}. \tag{4.42}$$

The equivalent force system is thus a centre of dilatation of strength λ with a dipole of strength 2μ along the normal to the fault.

For a purely tangential slip along the fault plane (figure 4.4b), $\boldsymbol{\nu}.\mathbf{n} = 0$, and so the moment tensor components depend only on the shear modulus μ and

$$M_{ij} = A\mu[\overline{u(\omega)}]\{n_i\nu_j + n_j\nu_i\}. \quad (4.43)$$

In (4.43) $\mathbf{n}$ and $\boldsymbol{\nu}$ appear in a completely symmetric role and so the zeroth order moment tensor M_{ij} does not allow one to distinguish between the fault plane with normal $\mathbf{n}$ and the perpendicular auxiliary plane with normal $\boldsymbol{\nu}$. Since $\mathbf{n}.\boldsymbol{\nu}$ vanishes, the trace of the moment tensor is zero

$$\mathrm{tr}\,\mathbf{M} = \sum_i M_{ii} = 0. \quad (4.44)$$

If we consider a coordinate frame with the 3 axis along $\mathbf{n}$ and the slip $\boldsymbol{\nu}$ in the direction of the 1-axis, the moment tensor has only two non-zero components:

$$M_{13} = M_{31} = M_0(\omega), \qquad \text{all other } M_{ij} = 0. \quad (4.45)$$

Here we have introduced the moment spectrum

$$M_0(\omega) = A\mu[\overline{u(\omega)}], \quad (4.46)$$

which defines the source characteristics as a function of time. The equivalent force distribution specified by (4.45) will be two couples of equal and opposite moment - the familiar 'double-couple' model of fault radiation.

The effect of a point explosion or implosion can be simulated by taking an isotropic moment tensor

$$M_{ij} = M_0(\omega)\delta_{ij}, \quad (4.47)$$

where once again it is convenient to work in terms of the moment spectrum, which is now equal to $(\lambda + 2\mu)A_e\overline{u_r(\omega)}$ where A_e is the surface area of a sphere with the 'elastic radius' r_e and $\overline{u_r(\omega)}$ is the average radial displacement spectrum at this radius (Müller, 1973). The equivalent force system consists of three perpendicular dipoles of equal strength (figure 4.4c).

The equivalent force systems we have described apply to faults which are small compared with the wavelengths of the recorded seismic waves. For larger faults, the zeroth order point source contributions we have just discussed do not, by themselves, provide an adequate representation of the seismic radiation. In addition we need the first order contributions which give quadrupole sources and perhaps even higher order terms.

4.3.2 Radiation into an unbounded medium

In the small source approximation the seismic displacement generated by the presence of a dislocation is given by

$$u_i(\mathbf{x}, \omega) = G_{ij}(\mathbf{x}, \mathbf{x}_s, \omega)\mathcal{E}_j(\omega) + \partial_k G_{ij}(\mathbf{x}, \mathbf{x}_s, \omega)M_{jk}(\omega). \quad (4.48)$$

We specialise to a hypocentre at the origin and consider an observation point at **R** in a direction specified by direction cosines γ_i. In the time domain we have a convolution of the temporal Green's tensor and the force or moment tensor time functions. For an unbounded isotropic elastic medium the displacement produced by a point force was first given by Stokes (1849) and the results were extended to couple sources by Love (1903) and we shall use these results as the basis of our discussion.

The force contribution to the displacement is

$$\begin{aligned}4\pi\rho u_i^{e}(\mathbf{R},t) &= (3\gamma_i\gamma_j-\delta_{ij})R^{-3}\int_{R/\alpha}^{R/\beta} ds\, s\mathcal{E}_j(t-s)\\ &+\gamma_i\gamma_j(\alpha^2R)^{-1}\mathcal{E}_j(t-R/\alpha)\\ &-(\gamma_i\gamma_j-\delta_{ij})(\beta^2R)^{-1}\mathcal{E}_j(t-R/\beta)\end{aligned} \tag{4.49}$$

The 'far-field' contribution decaying as R^{-1} follows the same time dependence as the force $\mathcal{E}$. Between the *P* and *S* wave arrivals is a disturbance which decays more rapidly, as R^{-3}, as we move away from the source.

The displacement associated with the moment tensor M_{ij} depends on the spatial derivative of the Green's tensor and is

$$\begin{aligned}4\pi\rho u_i^{m}(\mathbf{R},t) &= 3(5\gamma_i\gamma_j\gamma_k-l_{ijk})R^{-4}\int_{R/\alpha}^{R/\beta} ds\, sM_{jk}(t-s)\\ &-(6\gamma_i\gamma_j\gamma_k-l_{ijk})(\alpha^2R^2)^{-1}M_{jk}(t-R/\alpha)\\ &+(6\gamma_i\gamma_j\gamma_k-l_{ijk}-\delta_{ij}\gamma_k)(\beta^2R^2)^{-1}M_{jk}(t-R/\beta)\\ &+\gamma_i\gamma_j\gamma_k(\alpha^3R)^{-1}\partial_t M_{jk}(t-R/\alpha)\\ &-(\gamma_i\gamma_j-\delta_{ij})\gamma_k(\beta^3R)^{-1}\partial_t M_{jk}(t-R/\beta),\end{aligned} \tag{4.50}$$

with

$$l_{ijk} = \gamma_i\delta_{jk}+\gamma_j\delta_{ik}+\gamma_k\delta_{ij} \tag{4.51}$$

The 'far-field' terms which decay least rapidly with distance now behave like the time derivative of the moment tensor components.

The radiation predicted in the far-field from (4.49) and (4.50) is rather different for *P* and *S* waves. In figure 4.5 we illustrate the radiation patterns in three dimensions for some simple sources. For a single force in the 3 direction the *P* wave radiation is two-lobed with a $\cos\theta$ dependence in local spherical polar coordinates. The *S* radiation resembles a doughnut with a $\sin\theta$ dependence. The patterns for a 33 dipole are modulated with a further $\cos\theta$ factor arising from differentiating in the 3 direction; and so we have a two-lobed $\cos^2\theta$ behaviour for *P*. The corresponding *S* wave radiation pattern, depending on $\sin\theta\cos\theta$, has an attractive waisted shape. In a 31 couple a horizontal derivative is applied to the force behaviour which leads to rather different radiation. The *P* wave pattern is here

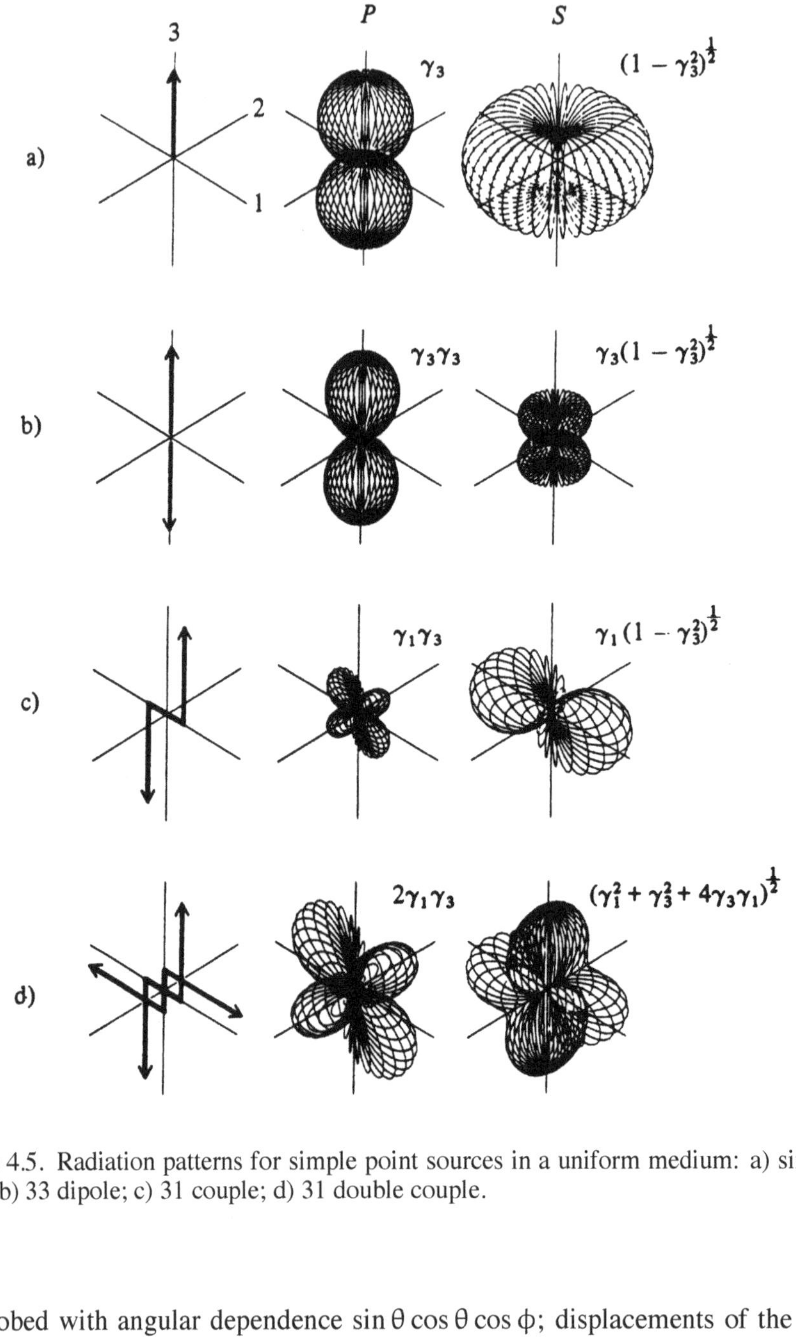

Figure 4.5. Radiation patterns for simple point sources in a uniform medium: a) single 3 force; b) 33 dipole; c) 31 couple; d) 31 double couple.

four-lobed with angular dependence $\sin\theta\cos\theta\cos\phi$; displacements of the same sign lie in opposing quadrants and the radiation maxima lie in the 13 plane. There is no *P* wave radiation in the 12 or 23 planes. The *S* wave radiation is proportional to $\sin^2\theta\cos\phi$ and this gives a two-lobed pattern with maxima along the 1 axis and a null on the 23 plane; the two lobes give an opposite sense of displacement to the medium and so impart a couple.

For an explosion, with an isotropic moment tensor $M_{jk} = M_0\delta_{jk}$, (4.50) reduces to

$$4\pi\rho u_i(\mathbf{R}, t) = \gamma_i\{(\alpha^2R^2)^{-1}M_0(t - R/\alpha) + (\alpha^3R)^{-1}\partial_t M_0(t - R/\alpha)\}. \quad (4.52)$$

This is a spherically symmetric, purely radial, *P* disturbance with no *S* wave part.

For a double couple without moment, simulating slip on a fault plane, $M_{jk} = M_0[n_i\nu_j + n_j\nu_i]$, the displacement at the receiver point **R** is given by

$$\begin{aligned} 4\pi\rho u_i(\mathbf{R}, t) = {} & (9Q_i - 6T_i)R^{-4}\int_{R/\alpha}^{R/\beta} ds\, sM_0(t - s) \\ & + (4Q_i - 2T_i)(\alpha^2R^2)^{-1}M_0(t - R/\alpha) \\ & - (3Q_i - 3T_i)(\beta^2R^2)^{-1}M_0(t - R/\beta) \\ & + Q_i(\alpha^3R)^{-1}\partial_t M_0(t - R/\alpha) \\ & + T_i(\beta^3R)^{-1}\partial_t M_0(t - R/\beta), \end{aligned} \quad (4.53)$$

where

$$Q_i = 2\gamma_i(\gamma_j\nu_j)(\gamma_k n_k), \qquad T_i = n_i(\gamma_j\nu_j) + \nu_i(\gamma_k n_k) - Q_i. \quad (4.54)$$

We note that **Q** which specifies the far-field *P* wave radiation is purely radial, lying in the direction of **R**. Further $\mathbf{Q}.\mathbf{T} = 0$, so that the far-field *S* wave radiation which depends on **T** is purely transverse.

For both the explosion and the double couple, the radiation for the whole field may be expressed in terms of the far-field factors.

It is interesting to compare the radiation pattern for a double couple with that for a single couple with moment. If we take **n** in the 3 direction and **ν** in the 1 direction then

$$Q_i = 2\gamma_i\gamma_1\gamma_3, \quad T_i = \gamma_1\delta_{i3} + \gamma_3\delta_{i1} - 2\gamma_i\gamma_1\gamma_3. \quad (4.55)$$

The far-field *P* wave radiation pattern is given by

$$|\mathbf{Q}|_{31} = 2\gamma_1\gamma_3 = \sin 2\theta \cos\phi, \quad (4.56)$$

and the far-field *S* radiation depends on

$$|\mathbf{T}|_{31} = (\gamma_1^2 + \gamma_3^2 - 4\gamma_1^2\gamma_3^2)^{1/2} = (\cos^2 2\theta\cos^2\phi + \cos^2\theta\sin^2\phi)^{1/2}. \quad (4.57)$$

The *P* wave radiation pattern for the double couple is just twice that for either of the constituent couples, so that there are still nodes in the pattern on the fault plane (12) and the auxiliary plane (23). The *S* wave radiation is now four lobed rather than two lobed and the net torque vanishes.

4.3.3 Influences on radiation patterns

The classical method of source characterisation, the fault-plane solution, relies on the assumption that the radiation leaving the source is indeed that for the same source in a uniform medium. Observations of P-wave polarity at distant stations are projected back onto the surface of a notional focal sphere surrounding the source, with allowance for the main effects of the Earth's structure using ray theory. The method requires that we can recognise the nodal planes separating dilatations and compressions, which should correspond to the fault plane and auxiliary plane. For a small source for which the localised moment tensor model is appropriate, the radiation pattern itself is not sufficient to distinguish the fault plane and other, usually geological, criteria have to be used.

A number of factors can conspire to modify the radiation pattern in the real Earth. The standard correction tables for the angle of emergence of rays from the focal sphere are based on an assumed mantle P wave velocity at the source. For crustal events, at short periods, the standard approach leads to apparent nodal planes which are not orthogonal. This however can be normally rectified by correcting the velocity at the source to a crustal P velocity.

When a source occurs in a region of velocity gradient, the radiation patterns will be modified by the presence of the structure. This effect will be frequency dependent - at low frequencies the radiation pattern will be as in a locally uniform medium, as the frequency increases the departures from the simple theory given above can become significant.

In addition many small earthquakes occur within regions which are under strain before a major earthquake. Such a prestrain has a number of effects (Walton 1974). The presence of the prestrain will lead to anisotropy due to modification of existing crack systems in the rock, in addition to weak anisotropy associated with the strain itself. These combined effects will modify the radiation pattern, in particular in the neighbourhood of the crossing of the nodal planes. The anisotropy will lead to splitting of the S wave degeneracy, two quasi-S waves will exist with different velocity. Observations of such S wave splitting in an active earthquake zone have been reported by Crampin et al (1980).

All these influences will make it difficult to get reliable fault-plane solutions for frequencies such that the P wavelengths are of the same size as the source dimensions. In addition it proves to be easier to read the first motions on teleseismic long-period records and so good results for fault-plane solutions can be obtained (Sykes 1967). For studies of small local events good results can be obtained at quite high frequencies.

The radiation patterns we have calculated depend on the assumption of an unbounded medium. As pointed out by Burridge, Lapwood & Knopoff (1964), the presence of a free surface close to the source significantly modifies the radiation pattern. If we insist on using an unbounded medium Green's tensor we cannot use the usual moment tensor representation (4.31). For sources which are shallow compared with the recorded wavelength, interference of *P* with the surface reflections *pP*, *sP* will modify the apparent first motion and give records which resemble those seen near nodal planes. Such effects can be assessed quantitatively by modelling the waveform generated by the source in a stratified half space (see Chapter 9).

4.4 The source as a stress-displacement vector discontinuity

When we wish to consider a source in a stratified medium we need to bring together our representations of the source in terms of equivalent forces or moment tensor and the wave propagation techniques based on the use of the stress-displacement vector we have introduced in chapter 2.

We will work with the cartesian components of the force system relative to a coordinate system with the origin at the epicentre. We take the $\hat{\mathbf{x}}$ axis to the North, $\hat{\mathbf{y}}$ axis to the East, and $\hat{\mathbf{z}}$ axis vertically downwards. The azimuthal angle ϕ to the x axis in our cylindrical coordinate system then follows the geographical convention.

For small sources we will use the point source representation in terms of a force $\mathcal{E}$ and moment tensor M_{jk}, for which the equivalent force system is

$$f_j = \mathcal{E}_j \delta(\mathbf{x} - \mathbf{x}_S) - \partial_k \{M_{jk} \delta(\mathbf{x} - \mathbf{x}_S)\}. \tag{4.58}$$

It is this system of forces which will now appear in the equations of motion and ultimately determine the forcing terms in the differential equations for the stress-displacement vector **b**, (2.24)-(2.25). For larger source regions we may simulate the radiation characteristics by the superposition of a number of point source contributions separated in space and time to handle propagation effects. Alternatively we can perform a volume integral over the source region, in which case each volume element $d^3\eta$ has an associated force $\epsilon d^3\eta$ and moment tensor $m_{ij} d^3\eta$.

In each case we have the problem of finding the coefficients F_z, F_V, F_H in a vector harmonic expansion (cf. 2.55)

$$\mathbf{f} = \frac{1}{2\pi} \int_{-\infty}^{\infty} d\omega\, e^{-i\omega t} \int_0^{\infty} dk\, k \sum_m [F_z \mathbf{R}_k^m + F_V \mathbf{S}_k^m + F_H \mathbf{T}_k^m]. \tag{4.59}$$

If we choose the z axis of our cylindrical coordinate system to pass through the source point $\mathbf{x}_S$, the cartesian components of the point source f_x, f_y, f_z will all be singular at the origin in the horizontal plane $z = z_S$. The coefficients F_z, F_V, F_H will only appear at the source depth z_S and may be evaluated by making use of the orthonormality of the vector harmonics, so that, e.g.,

$$F_V = \frac{1}{2\pi} \int_{-\infty}^{\infty} dt\, e^{i\omega t} \int_0^{\infty} dr\, r \int_0^{2\pi} d\phi\, [\mathbf{S}_k^m]^* . \mathbf{f}\, . \tag{4.60}$$

To evaluate the integrals over the horizontal plane we have to make use of the expansion of $J_m(kr) e^{im\phi}$ near the origin, and it is often more convenient to work in cartesian coordinates. The integrations leading to F_z, F_V, F_H will leave the z dependence of the source terms unaffected. Thus we anticipate that the force components $\mathcal{E}_x, \mathcal{E}_y, \mathcal{E}_z$ and the moment tensor elements describing doublets in the horizontal plane ($M_{xx}, M_{xy}, M_{yx}, M_{yy}$) will have a $\delta(z - z_S)$ dependence. The remaining moment tensor elements ($M_{xz}, M_{zx}, M_{yz}, M_{zy}, M_{zz}$) will appear with a $\delta'(z - z_S)$ term.

The total forcing term **F** in (2.26), for each angular order m, will therefore have a z dependence,

$$\mathbf{F}(k, m, z, \omega) = \mathbf{F}_1(k, m, \omega)\delta(z - z_S) + \mathbf{F}_2(k, m, \omega)\delta'(z - z_S), \tag{4.61}$$

which is of just the form (2.98) we have discussed in Section 2.2.2. Thus when we solve for the stress-displacement vector **b** in the presence of the general point source excitation, there will be a discontinuity in **b** across the source plane $z = z_S$

$$\begin{aligned}\mathbf{b}(k, m, z_S+, \omega) - \mathbf{b}(k, m, z_S-, \omega) &= \mathbf{S}(k, m, z_S, \omega) \\ &= \mathbf{F}_1 + \omega\mathbf{A}(p, z_S)\mathbf{F}_2.\end{aligned} \tag{4.62}$$

The presence of the $\delta'(z - z_S)$ term means that although the forcing terms **F** appear only in the equations for the stress elements in (2.24)-(2.25), the discontinuity in the **b** vector extends also to the displacement terms because of coupling via the **A** matrix.

The jump in the components of the stress-displacement vector **b** across z_S depends strongly on angular order:

$$\begin{aligned}[U]_-^+ &= M_{zz}(\rho\alpha^2)^{-1}, && m = 0,\\ [V]_-^+ &= \tfrac{1}{2}[\pm M_{xz} - iM_{yz}](\rho\beta^2)^{-1}, && m = \pm 1,\\ [W]_-^+ &= \tfrac{1}{2}[\pm M_{yz} - iM_{xz}](\rho\beta^2)^{-1}, && m = \pm 1,\end{aligned} \tag{4.63}$$

and

$$\begin{aligned}[P]_-^+ &= -\omega^{-1}\mathcal{E}_z, && m = 0\\ &= \tfrac{1}{2}p[i(M_{zy} - M_{yz}) \pm (M_{xz} - M_{zx})], && m = \pm 1,\\ [S]_-^+ &= \tfrac{1}{2}p(M_{xx} + M_{yy}) - pM_{zz}(1 - 2\beta^2/\alpha^2) && m = 0,\\ &= \tfrac{1}{2}\omega^{-1}(\mp\mathcal{E}_x + i\mathcal{E}_y), && m = \pm 1,\\ &= \tfrac{1}{4}p[M_{yy} - M_{xx}) \pm i(M_{xy} + M_{yx})], && m = \pm 2,\\ [T]_-^+ &= \tfrac{1}{2}p(M_{xy} - M_{yx}), && m = 0,\\ &= \tfrac{1}{2}\omega^{-1}(i\mathcal{E}_x \pm \mathcal{E}_y), && m = \pm 1,\\ &= \tfrac{1}{4}p[\pm i(M_{xx} - M_{yy}) + (M_{xy} + M_{yx})], && m = \pm 2,\end{aligned} \tag{4.64}$$

For our point equivalent source (4.58), the vector harmonic expansion (2.55) for the displacement will be restricted to azimuthal orders $|m| < 2$. These results for a general moment tensor generalise Hudson's (1969a) analysis of an arbitrarily oriented dislocation.

We recall that for an *indigeneous* source the moment tensor is symmetric and thus has only six independent components, and further $\mathcal{E}$ will then vanish. This leads to a significant simplification of these results. In particular, for a point source the stress variable P will always be continuous, excitation for $m = \pm 1$ is confined to the horizontal displacement terms and T will only have a jump for $m = \pm 2$.

We have already noted that the only azimuthal dependence in the displacement and traction quantities U, V, W, P, S, T arise from the azimuthal behaviour of the source, in an isotropic medium. For an indigenous source we can therefore associate the azimuthal behaviour of the displacement field in the stratification with certain combinations of the moment tensor elements.

(a) No variation with azimuth:
For the *P-SV* wavefield this is controlled by the diagonal elements $(M_{xx} + M_{yy})$, M_{zz} and is completely absent for *SH* waves.

(b) $\cos\phi, \sin\phi$ *dependence:*
This angular behaviour arises from the presence of the vertical couples $M_x z, M_{yz}$. The term M_{xz} leads to $\cos\phi$ dependence for *P-SV* and $\sin\phi$ for *SH*, whilst M_{yz} gives $\sin\phi$ behaviour for *P-SV* and $\cos\phi$ for *SH*.

(c) $\cos 2\phi, \sin 2\phi$ *dependence:*
This behaviour is controlled by the horizontal dipoles and couples M_{xx}, M_{yy}, M_{xy}. The difference $(M_{xx} - M_{yy})$ leads to $\cos 2\phi$ behaviour for *P-SV* and $\sin 2\phi$ for *SH*. The couple M_{xy} gives $\sin 2\phi$ dependence for *P-SV* and $\cos 2\phi$ for *SH*.

These azimuthal dependences do not rest on any assumptions about the nature of the propagation path through the medium and so hold for both body waves and surface waves.

4.5 Wavevector representation of a source

We have so far represented the action of sources within the stratification in terms of a discontinuity ß in the stress-displacement vector **b** at the level of the source. An alternative approach is to regard the source as giving rise to a discontinuity in the wavevector **v**. Such an approach has been used by Haskell (1964) and Harkrider (1964) to specify their sources.

Consider a source in a locally uniform region about the source plane z_S, we may then convert the stress-displacement vectors $\mathbf{b}(z_S-)$, $\mathbf{b}(z_S+)$ immediately above and below the source plane into their up and downgoing wave parts by the operation of the inverse eigenvector matrix $\mathbf{D}^{-1}(z_S)$. The corresponding wavevectors **v** will suffer a discontinuity $\mathbf{\Sigma}$ across the source plane z_S, consequent upon the jump in **b**. Thus

$$\begin{aligned}\mathbf{v}(k, m, z_S+, \omega) - \mathbf{v}(k, m, z_S-, \omega) &= \mathbf{\Sigma}(k, m, z_S, \omega) \\ &= \mathbf{D}^{-1}(p, z_S)\mathbf{S}(k, m, z_S, \omega).\end{aligned} \tag{4.65}$$

We may represent this jump vector $\boldsymbol{\Sigma}$ in terms of upgoing and downgoing parts as in (3.17)

$$\boldsymbol{\Sigma} = [-\boldsymbol{\Sigma}_U, \boldsymbol{\Sigma}_D]^T, \tag{4.66}$$

The choice of sign is taken to facilitate physical interpretation. The structure of the relation (4.66) may be seen by partitioning $\mathbf{S}$ into its displacement and traction jumps and then making use of the partitioned forms of $\mathbf{D}^{-1}(z_S)$ (3.40),

$$\begin{bmatrix} -\boldsymbol{\Sigma}_U \\ \boldsymbol{\Sigma}_D \end{bmatrix} = i \begin{bmatrix} \mathbf{n}_{DS}^T & -\mathbf{m}_{DS}^T \\ -\mathbf{n}_{US}^T & -\mathbf{m}_{US}^T \end{bmatrix} \begin{bmatrix} \mathbf{S}_W \\ \mathbf{S}_T \end{bmatrix}. \tag{4.67}$$

The elements $\boldsymbol{\Sigma}_U, \boldsymbol{\Sigma}_D$ have rather similar forms

$$\boldsymbol{\Sigma}_U = i[\mathbf{m}_{DS}^T\mathbf{S}_T - \mathbf{n}_{DS}^T\mathbf{S}_W], \qquad \boldsymbol{\Sigma}_D = i[\mathbf{m}_{US}^T\mathbf{S}_T - \mathbf{n}_{US}^T\mathbf{S}_W]. \tag{4.68}$$

The significance of these terms is most readily seen if we consider a source embedded in an unbounded medium. Above such a source we would expect only upgoing waves and below only downgoing waves, so that the wavevector will be of the form

$$\begin{aligned} \mathbf{v}(z) &= [\mathbf{v}_U(z),\ 0]^T, \qquad z < z_S, \\ &= [0,\ \mathbf{v}_D(z)]^T, \qquad z > z_S, \end{aligned} \tag{4.69}$$

with a jump $\boldsymbol{\Sigma}$ across z_S of

$$\boldsymbol{\Sigma} = [-\mathbf{v}_U(z_S), \mathbf{v}_D(z_S)]^T. \tag{4.70}$$

Comparison of the two expressions (4.66) and (4.70) for $\mathbf{S}$ shows that a source will radiate $\boldsymbol{\Sigma}_U$ upwards and $\boldsymbol{\Sigma}_D$ downwards into an unbounded medium.

For a source in a vertically inhomogeneous region we may still use the jump vector $\mathbf{S}$ if we split the medium at the source level z_S and consider each of the two halves of the stratification to be extended by uniform half spaces with the properties at z_S. This procedure will correspond to our treatment of reflection and transmission problems and the radiation components $\boldsymbol{\Sigma}_U$, $\boldsymbol{\Sigma}_D$ will enter with reflection matrices into a compact physical description of the seismic wavefield. We lose no generality by our assumption of an infinitesimal uniform region at z_S since we will construct the correct Green's tensors for the stratified regions above and below the source.

Chapter 5

Reflection and Transmission I

Nearly all recording of seismic waves is performed at the Earth's surface and most seismic sources are fairly shallow. We are therefore, of necessity, interested in waves reflected back by the Earth's internal structure. In seismic prospecting we are particularly interested in *P* waves reflected at near-vertical incidence. For longer range explosion seismology wide-angle reflections from the crust-mantle boundary are often some of the most significant features on the records. At teleseismic distances the main *P* and *S* arrivals have been reflected by the continuously varying wavespeed profile in the Earth's mantle. For deep sources we are also interested in the transmission of seismic waves to the surface.

In this chapter we consider the reflection of cylindrical waves by portions of a stratified medium and show how reflection coefficients for the different wave types can be constructed using the stress-displacement fields introduced in chapter 3. We define such reflection and transmission coefficients by the relation between up and downgoing wave amplitudes at the limits of the region of interest.

An exact decomposition of the seismic wavefield into up and downgoing parts can only be made in a uniform medium. Thus, in order to give a unique and unambiguous definition of the reflection and transmission coefficients for the region $z_A \leq z \leq z_B$, we adopt the following stratagem. We isolate this region from the rest of the stratification by introducing hypothetical half spaces in $z < z_A, z > z_B$ with properties equal to those at $z = z_A, z = z_B$ respectively. We can now visualise up and downgoing waves in these half spaces and then define reflection coefficients by relating the wavevectors in the upper and lower half spaces. The continuity of the seismic properties at z_A and z_B ensures that the reflection properties are controlled entirely by the structure within (z_A, z_B).

5.1 Reflection and transmission at an interface

We may isolate an interface across which there is a change in the elastic properties of the material by embedding it between two half spaces with the properties just at the two sides of the interface. With the assumption of welded contact at the interface, $z = z_I$ say, we will have continuity of the stress-displacement vector $\mathbf{b}$

across z_I. Any such vector will, however, have a different representation in terms of up and downgoing wave components on the two sides of the interface.

In medium '−', $z < z_I$

$$\mathbf{b}(z_I-) = \mathbf{D}_-(z_I-)\mathbf{v}_-(z_I-), \tag{5.1}$$

and in medium '+', $z > z_I$

$$\mathbf{b}(z_I+) = \mathbf{D}_+(z_I+)\mathbf{v}_+(z_I+). \tag{5.2}$$

These two expressions represent the same stresses and displacements and so we can connect the wavevectors in the upper and lower half spaces,

$$\begin{aligned}\mathbf{v}_-(z_I-) &= \mathbf{D}_-^{-1}(z_I-)\mathbf{D}_+(z_I+)\mathbf{v}_+(z_I+),\\ &= \mathbf{Q}(z_I-, z_I+)\mathbf{v}_+(z_I+).\end{aligned} \tag{5.3}$$

This relation enables us to extract reflection and transmission coefficients for the interface. We illustrate the procedure for scalar *SH* waves and then extend the treatment to coupled *P-SV* waves with the introduction of reflection and transmission matrices.

5.1.1 SH waves

We prescribe the slowness p to be the same in both half spaces and so automatically satisfy Snell's law at the interface. The inclination of the waves to the vertical depends on the vertical slownesses

$$q_{\beta-} = (\beta_-^{-2} - p^2)^{1/2}, \quad q_{\beta+} = (\beta_+^{-2} - p^2)^{1/2}, \tag{5.4}$$

with, e.g., an angle of inclination

$$j_- = \cos^{-1}(\beta_- q_{\beta-}). \tag{5.5}$$

On substituting the explicit forms for $\mathbf{v}$ (3.16) and $\mathbf{D}$ (3.25) we may write (5.3) as

$$\begin{bmatrix} H_{U-} \\ H_{D-} \end{bmatrix} = \frac{\epsilon_{\beta-}\epsilon_{\beta+}}{\beta_-\beta_+} \begin{bmatrix} \mu_- q_{\beta-} & i \\ \mu_- q_{\beta-} & -i \end{bmatrix} \begin{bmatrix} 1 & 1 \\ -i\mu_+ q_{\beta+} & i\mu_+ q_{\beta+} \end{bmatrix} \begin{bmatrix} H_{U+} \\ H_{D+} \end{bmatrix}, \tag{5.6}$$

in terms of the shear moduli μ_-, μ_+. The normalisation $\epsilon_{\beta-}/\beta_- = (2\mu_- q_{\beta-})^{1/2}$ and thus all the entries in the matrix $\mathbf{Q}$ depend on the products μq_β;

$$\begin{bmatrix} H_{U-} \\ H_{D-} \end{bmatrix} = \frac{1}{2(\mu_-\mu_+ q_{\beta-} q_{\beta+})^{1/2}} \begin{bmatrix} \mu_- q_{\beta-} + \mu_+ q_{\beta+} & \mu_- q_{\beta-} - \mu_+ q_{\beta+} \\ \mu_- q_{\beta-} - \mu_+ q_{\beta+} & \mu_- q_{\beta-} + \mu_+ q_{\beta+} \end{bmatrix} \begin{bmatrix} H_{U+} \\ H_{D+} \end{bmatrix}. \tag{5.7}$$

The combination μq_β plays the role of an impedance for the obliquely travelling *SH* waves.

Consider an *incident downgoing wave* from medium '−'. This will give a reflected upgoing wave in $z < z_I$ and a transmitted downgoing wave in $z > z_I$. There will be no upcoming wave in medium '+' and so we require H_{U+} to vanish.

We define the reflection coefficient for downward propagation R_D^I to connect the wave elements in medium '−',

$$H_{U-} = R_D^I H_{D-}, \tag{5.8}$$

and a transmission coefficient T_D^I connecting wave components across the interface,

$$H_{D+} = T_D^I H_{D-}. \tag{5.9}$$

For the incident downgoing wave (5.5) has the form

$$\begin{bmatrix} H_{U-} \\ H_{D-} \end{bmatrix} = \begin{bmatrix} Q_{UU} & Q_{UD} \\ Q_{DU} & Q_{DD} \end{bmatrix} \begin{bmatrix} 0 \\ H_{D+} \end{bmatrix}, \tag{5.10}$$

and so

$$R_D^I = Q_{UD}(Q_{DD})^{-1}, \quad T_D^I = (Q_{DD})^{-1}. \tag{5.11}$$

In terms of the *SH* wave impedances the reflection and transmission coefficients are

$$\begin{aligned} R_D^I &= (\mu_- q_{\beta-} - \mu_+ q_{\beta+})/(\mu_- q_{\beta-} + \mu_+ q_{\beta+}), \\ T_D^I &= 2(\mu_- \mu_+ q_{\beta-} q_{\beta+})^{1/2}/(\mu_- q_{\beta-} + \mu_+ q_{\beta+}). \end{aligned} \tag{5.12}$$

For propagating waves in a perfectly elastic medium the columns of **D** are normalised to unit energy flux in the z direction. The coefficients R_D^I, T_D^I are therefore measures of the reflected and transmitted energy in such propagating waves.

The structure of (5.12) is not affected if the half spaces are weakly dissipative so that we use complex *S* wavespeeds, or if the slowness p is such that we are considering evanescent waves, provided we take a consistent choice of branch cut (3.8) for the radicals $q_{\beta-}$, $q_{\beta+}$. We will therefore refer to (5.12) as the reflection and transmission coefficients for any slowness p.

As the contrast in properties across the interface becomes very small

$$R_D^I \to \tfrac{1}{2}\Delta(\mu q_\beta)/(\mu q_\beta), \quad T_D^I \to 1, \tag{5.13}$$

where $\Delta(\mu q_\beta)$ is the contrast in the impedance across the interface.

For this scalar case the reflection and transmission coefficients for upward incidence from medium '+' are most easily obtained by exchanging the suffices − and +, so that

$$R_U^I = -R_D^I, \quad T_U^I = T_D^I, \quad 1 - (R_D^I)^2 = (T_D^I)^2. \tag{5.14}$$

5.1.2 Coupled P and SV waves

Although *P* and *SV* waves propagate independently in a uniform medium, once they impinge on a horizontal interface there will be conversion to the other wave type in both reflection and transmission. By working with fixed slowness p we require the same horizontal phase behaviour for all the waves and so satisfy Snell's law both above and below the interface (figure 5.1).

The coupling between *P* and *SV* waves may be treated conveniently by using matrix methods. We split the wavevectors, on each side of the interface, into their up and downgoing parts (3.17) and partition the coupling matrix

$$\mathbf{Q}(z_I-, z_I+) = \mathbf{D}_-^{-1}(z_I-)\mathbf{D}_+(z_I+) \tag{5.15}$$

into 2×2 submatrices $\mathbf{Q}_{ij}$; so that (5.2) becomes

$$\begin{bmatrix} \mathbf{v}_{U-} \\ \mathbf{v}_{D-} \end{bmatrix} = \begin{bmatrix} \mathbf{Q}_{UU} & \mathbf{Q}_{UD} \\ \mathbf{Q}_{DU} & \mathbf{Q}_{DD} \end{bmatrix} \begin{bmatrix} \mathbf{v}_{U+} \\ \mathbf{v}_{D+} \end{bmatrix}. \tag{5.16}$$

We have already established partitioned forms of the eigenvector matrix $\mathbf{D}$ and its inverse (3.36), (3.40) and now evaluate the partitions of $\mathbf{Q}$ in terms of the displacement and stress transformation matrices $\mathbf{m}_{U-}, \mathbf{n}_{U-}$ etc:

$$\begin{aligned} \mathbf{Q}(z_I-, z_I+) &= i \begin{bmatrix} -\mathbf{n}_{D-}^T & \mathbf{m}_{D-}^T \\ \mathbf{n}_{U-}^T & -\mathbf{m}_{U-}^T \end{bmatrix} \begin{bmatrix} \mathbf{m}_{U+} & \mathbf{m}_{D+} \\ \mathbf{n}_{U+} & \mathbf{n}_{D+} \end{bmatrix}, \\ &= i \begin{bmatrix} \mathbf{m}_{D-}^T\mathbf{n}_{U+} - \mathbf{n}_{D-}^T\mathbf{m}_{U+} & \mathbf{m}_{D-}^T\mathbf{n}_{D+} - \mathbf{n}_{D-}^T\mathbf{m}_{D+} \\ \mathbf{n}_{U-}^T\mathbf{m}_{U+} - \mathbf{m}_{U-}^T\mathbf{n}_{U+} & \mathbf{n}_{U-}^T\mathbf{m}_{D+} - \mathbf{m}_{U-}^T\mathbf{n}_{D+} \end{bmatrix}. \end{aligned} \tag{5.17}$$

The partitions of (5.17) may now be recognised as having the form of the matrix propagation invariants in (2.68); since, for example, $\mathbf{n}_U$ gives the stress elements corresponding to the displacements $\mathbf{m}_U$.

Our relation (5.16) connecting the up and downgoing wave components on the two sides of the interface can therefore be written as

$$\begin{bmatrix} \mathbf{v}_{U-} \\ \mathbf{v}_{D-} \end{bmatrix} = i \begin{bmatrix} \langle \mathbf{m}_{D-}, \mathbf{m}_{U+} \rangle & -\langle \mathbf{m}_{U-}, \mathbf{m}_{U+} \rangle \\ \langle \mathbf{m}_{D-}, \mathbf{m}_{D+} \rangle & -\langle \mathbf{m}_{U-}, \mathbf{m}_{D+} \rangle \end{bmatrix} \begin{bmatrix} \mathbf{v}_{U+} \\ \mathbf{v}_{D+} \end{bmatrix}. \tag{5.18}$$

Although we have introduced (5.18) in the context of *P-SV* wave propagation, this form is quite general and if we use the *SH* wave forms for $\mathbf{m}_U$, $\mathbf{n}_U$ etc., (3.100), we will recover (5.7). For full anisotropic propagation we would use 3×3 matrices $\mathbf{m}_U, \mathbf{m}_D$.

Consider a downgoing wave system, comprising both *P* and *SV* waves, in medium '$-$'. When this interacts with the interface we get reflected *P* and *SV* waves in medium '$-$' and transmitted downgoing waves in medium '+' (figure 5.1). No upward travelling waves will be generated in medium '+' and so $\mathbf{v}_{U+} = 0$. We now define a reflection matrix $\mathbf{R}_D^I$ for downward incidence, whose entries are reflection coefficients, by

$$\mathbf{v}_{U-} = \mathbf{R}_D^I \mathbf{v}_{D-} \quad \text{i.e.} \quad \begin{bmatrix} P_{U-} \\ S_{U-} \end{bmatrix} = \begin{bmatrix} R_D^{PP} & R_D^{PS} \\ R_D^{SP} & R_D^{SS} \end{bmatrix} \begin{bmatrix} P_{D-} \\ S_{D-} \end{bmatrix}, \tag{5.19}$$

relating the up and downgoing wave elements in medium '$-$'. Similarly connecting

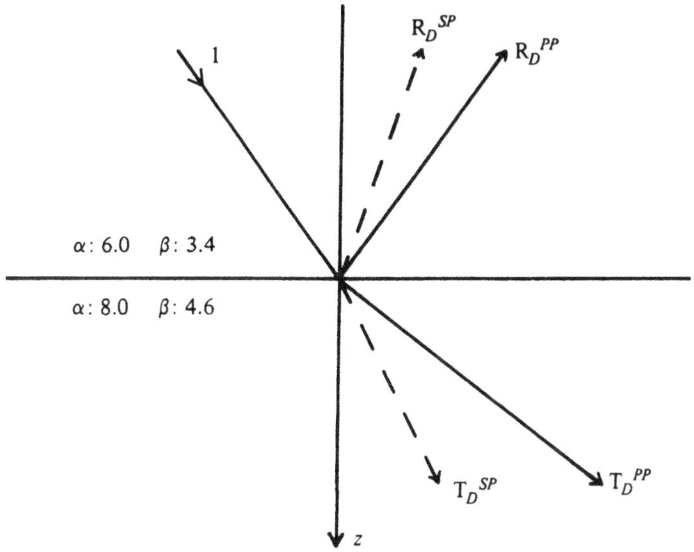

Figure 5.1. The configuration of reflected and transmitted waves at an elastic interface for slowness p = 0.1.

the downgoing wave components across the interface we introduce a transmission matrix $\mathbf{T}_D^I$ by

$$\mathbf{v}_{D+} = \mathbf{T}_D^I \mathbf{v}_{D-} \quad \text{i.e.} \quad \begin{bmatrix} P_{D+} \\ S_{D+} \end{bmatrix} = \begin{bmatrix} T_D^{PP} & T_D^{PS} \\ T_D^{SP} & T_D^{SS} \end{bmatrix} \begin{bmatrix} P_{D-} \\ S_{D-} \end{bmatrix}. \tag{5.20}$$

We have chosen the convention for the conversion coefficients R_D^{PS} etc. so that the indexing of the reflection and transmission matrices follows the standard matrix pattern, which is very useful for manipulation.

For these incident downgoing waves the wave elements are related by

$$\begin{bmatrix} \mathbf{v}_{U-} \\ \mathbf{v}_{D-} \end{bmatrix} = \begin{bmatrix} \mathbf{Q}_{UU} & \mathbf{Q}_{UD} \\ \mathbf{Q}_{DU} & \mathbf{Q}_{DD} \end{bmatrix} \begin{bmatrix} 0 \\ \mathbf{v}_{D+} \end{bmatrix}, \tag{5.21}$$

and so, the reflection and transmission matrices can be found in terms of the partitions of $\mathbf{Q}$ as

$$\mathbf{T}_D^I = (\mathbf{Q}_{DD})^{-1}, \quad \mathbf{R}_D^I = \mathbf{Q}_{UD}(\mathbf{Q}_{DD})^{-1}, \tag{5.22}$$

cf. (5.11) for *SH* waves. With the explicit forms for the partitions of $\mathbf{Q}$ (5.17), (5.18) we have

$$\begin{aligned} \mathbf{T}_D^I &= i\langle \mathbf{m}_{U-}, \mathbf{m}_{D+}\rangle^{-1}, \\ \mathbf{R}_D^I &= -\langle \mathbf{m}_{D-}, \mathbf{m}_{D+}\rangle\langle \mathbf{m}_{U-}, \mathbf{m}_{D+}\rangle^{-1}. \end{aligned} \tag{5.23}$$

The angle bracket symbol acts as a dissimilarity operator, and it is the mismatch between $\mathbf{m}_{D-}$ and $\mathbf{m}_{D+}$ which determines the reflection matrix $\mathbf{R}_D^I$.

An *incident upcoming* wave system in medium '+' will give reflected waves in medium '+' and transmitted waves in medium '−'. No downgoing waves in medium '−' will be generated, so that $\mathbf{v}_{D-} = 0$, and now we have

$$\begin{bmatrix} \mathbf{v}_{U-} \\ 0 \end{bmatrix} = \begin{bmatrix} \mathbf{Q}_{UU} & \mathbf{Q}_{UD} \\ \mathbf{Q}_{DU} & \mathbf{Q}_{DD} \end{bmatrix} \begin{bmatrix} \mathbf{v}_{U+} \\ \mathbf{v}_{D+} \end{bmatrix}. \tag{5.24}$$

We define the reflection and transmission matrices for these upward incident waves as

$$\mathbf{v}_{D+} = \mathbf{R}^I_U \mathbf{v}_{U+}, \quad \mathbf{v}_{U-} = \mathbf{T}^I_U \mathbf{v}_{U+}, \tag{5.25}$$

and from (5.24) we may construct $\mathbf{R}^I_U, \mathbf{T}^I_U$ from the partitions of $\mathbf{Q}$ as

$$\begin{aligned} \mathbf{R}^I_U &= -(\mathbf{Q}_{DD})^{-1}\mathbf{Q}_{DU}, \\ \mathbf{T}^I_U &= \mathbf{Q}_{UU} - \mathbf{Q}_{UD}(\mathbf{Q}_{DD})^{-1}\mathbf{Q}_{DU}. \end{aligned} \tag{5.26}$$

For this single interface we would, of course, obtain the same results by interchanging the suffices + and − in the expressions for $\mathbf{R}^I_D, \mathbf{T}^I_D$; but as we shall see the present method may be easily extended to more complex cases.

From the expressions for the reflection and transmission matrices in terms of the partitions of $\mathbf{Q}$ (5.21), (5.26) we can reconstruct the interface matrix itself as

$$\mathbf{Q}(z_I-, z_I+) = \mathbf{D}_-^{-1}(z_I-)\mathbf{D}_+(z_I+), \tag{5.27}$$

$$= \begin{bmatrix} \mathbf{T}^I_U - \mathbf{R}^I_D(\mathbf{T}^I_D)^{-1}\mathbf{R}^I_U & \mathbf{R}^I_D(\mathbf{T}^I_D)^{-1} \\ -(\mathbf{T}^I_D)^{-1}\mathbf{R}^I_U & (\mathbf{T}^I_D)^{-1} \end{bmatrix}. \tag{5.28}$$

The eigenvector matrices depend only on the slowness p and so $\mathbf{Q}$ is frequency independent. All the interface coefficients share this property.

The upward reflection and transmission matrices can be expressed in terms of the propagation invariants at the interface as

$$\begin{aligned} \mathbf{R}^I_U &= -\langle \mathbf{m}_{U-}, \mathbf{m}_{D+} \rangle^{-1} \langle \mathbf{m}_{U-}, \mathbf{m}_{U+} \rangle, \\ \mathbf{T}^I_U &= -i \langle \mathbf{m}_{D+}, \mathbf{m}_{U-} \rangle^{-1} = (\mathbf{T}^I_D)^T. \end{aligned} \tag{5.29}$$

We may now construct the reflection and transmission matrices for the *P-SV* wave case by using the expressions (3.37) for the displacement and stress transformation matrices $\mathbf{m}_U, \mathbf{n}_U$ etc. All the interface coefficient matrices depend on $\langle \mathbf{m}_{U-}, \mathbf{m}_{D+} \rangle^{-1}$ and so the factor $\det\langle \mathbf{m}_{U-}, \mathbf{m}_{D+} \rangle$ will appear in the denominator of every reflection and transmission coefficient. The transmission coefficients are individual elements of $\mathbf{Q}$ divided by this determinant, but the reflection coefficients take the form of ratios of second order minors of $\mathbf{Q}$.

The denominator

$$\det\langle m_{U-}, m_{D+}\rangle = \epsilon_{\alpha-}\epsilon_{\alpha+}\epsilon_{\beta-}\epsilon_{\beta+} \{[2p^2\Delta\mu(q_{\alpha-} - q_{\alpha+}) + (\rho_- q_{\alpha+} + \rho_+ q_{\alpha-})] \times [2p^2\Delta\mu(q_{\beta-} - q_{\beta+}) + (\rho_- q_{\beta+} + \rho_+ q_{\beta-})] + p^2[2\Delta\mu(q_{\alpha-}q_{\beta+} + p^2) - \Delta\rho][2\Delta\mu(q_{\beta-}q_{\alpha+} + p^2) - \Delta\rho]\}, \tag{5.30}$$

where we have introduced the contrasts in shear modulus and density across the interface $\Delta\mu = \mu_- - \mu_+$, $\Delta\rho = \rho_- - \rho_+$. It was pointed out by Stoneley (1924) that if this determinant vanishes we have the possibility of free interface waves with evanescent decay away from the interface into the media on either side. These Stoneley waves have a rather restricted range of existence, for most reasonable density contrasts the shear velocities β_- and β_+ must be nearly equal for (5.30) to be zero. The slowness of the Stoneley wave is always greater than $[\min(\beta_-, \beta_+)]^{-1}$.

The expressions we have just derived for the reflection and transmission matrices may alternatively be derived directly by making use of the propagation invariants. Consider, for example, a system of incident downgoing waves with displacements given by the matrix m_{D-}. When we equate the incident and reflected displacements in medium '−' to the transmitted displacement in medium '+' we have

$$m_{D-} + m_{U-}\mathbf{R}_D^I = m_{D+}\mathbf{T}_D^I, \tag{5.31}$$

and there is a corresponding equation for the tractions

$$n_{D-} + n_{U-}\mathbf{R}_D^I = n_{D+}\mathbf{T}_D^I \tag{5.32}$$

We may now eliminate $\mathbf{R}_D^I$ by premultiplying (5.31) by n_{U-}^T and (5.32) by m_{U-}^T and then subtracting to give

$$(m_{U-}^T n_{D-} - n_{U-}^T m_{D-}) = (m_{U-}^T n_{D+} - n_{U-}^T m_{D+})\mathbf{T}_D^I. \tag{5.33}$$

The invariant on the left hand side of (5.33) is just i times the unit matrix and so we recover our previous result (5.23) for the transmission matrix

$$\mathbf{T}_D^I = i\langle m_{U-}, m_{D+}\rangle^{-1}. \tag{5.34}$$

The reflection matrix $\mathbf{R}_D^I$ may be similarly found by elimination from (5.31), (5.32). The expressions (5.29) for $\mathbf{R}_U^I$, $\mathbf{T}_U^I$ may be found by constructing the equivalent equations to (5.31)-(5.32) for an incident upcoming wave.

5.1.3 The variation of reflection coefficients with slowness

As an illustration of the interface coefficients we have been discussing, we display in figure 5.2 the amplitude and phase behaviour of the downward reflection coefficients for a plane wave at the interface appearing in figure 5.1. We have

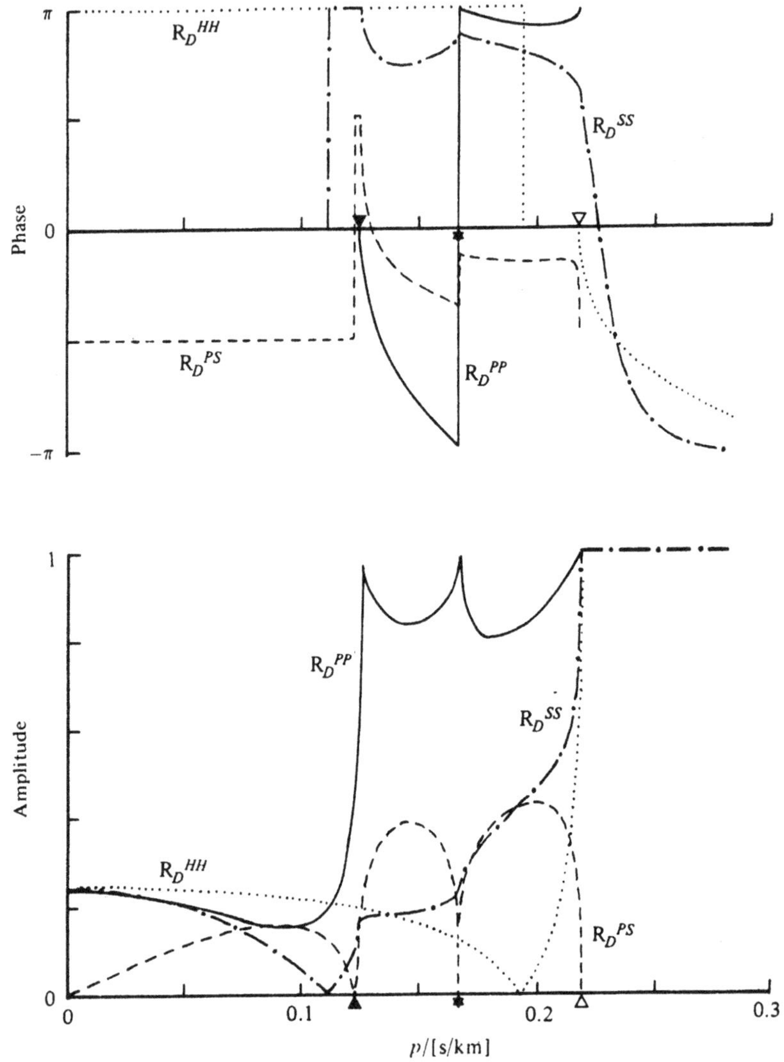

Figure 5.2. The amplitude and phase behaviour of the interfacial reflection coefficient for the model of figure 5.1 as a function of slowness: ▲ , △ critical slownesses for *P* waves (α_+^{-1}), *S* waves (β_+^{-1}); * onset of evanescence for *P*.

chosen to represent the coefficients as a function of slowness, rather than the conventional angle of incidence, because we can use a common reference for *P* and *S* waves and so display many of the characteristics more clearly.

The behaviour of the reflection coefficients is governed by the relative sizes of the waveslownesses for *P* and *S* in the media on the two sides of the interface. When $p > \alpha_+^{-1}$ (here 0.125 s/km) all the reflection coefficients are real. At vertical incidence the amplitudes of the *S* wave coefficients $|R_D^{SS}|$, $|R_D^{HH}|$ are equal and there

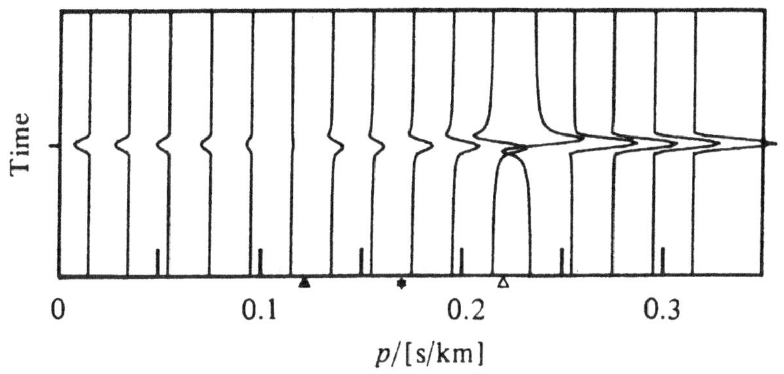

Figure 5.3. The behaviour of a reflected *SV* pulse at an interface as a function of slowness.

is no conversion from *P* to *S* waves. The behaviour of R_D^{SS} and R_D^{HH} as slowness increases is very different: the *SH* wave coefficient is fairly simple, but the *SV* wave coefficient is profoundly influenced by the *P* wave behaviour.

At $p = \alpha_+^{-1}$, *P* waves are travelling horizontally in the lower medium and we have reached the critical slowness for *P* waves. For $\alpha_+^{-1} < p < \alpha_-^{-1}$, *P* waves are reflected at the interface and give rise to only evanescent waves in the lower half space. Once $p > \alpha_+^{-1}$ all the reflection coefficients for the *P-SV* system become complex. The phase of R_D^{PP} and R_D^{PS} change fairly rapidly with slowness, but R_D^{SS} has slower change.

For $p > \alpha_-^{-1}$ (here 0.166 s/km) *P* waves become evanescent in the upper medium, but we can still define reflection coefficients for these evanescent incident waves. The amplitude of R_D^{PS} drops to zero at $p = \alpha_-^{-1}$ and then recovers before falling to zero again at $p = \beta_+^{-1}$. The amplitude of R_D^{SS} has an inflexion at $p = \alpha_-^{-1}$ and the character of the phase variation changes at this slowness.

The critical slowness for *S* waves is β_+^{-1} and for p greater than this value both *SV* and *SH* waves are totally reflected. The phase for the *SV* wave coefficient for this interface varies more rapidly with slowness than that for *SH* waves.

When the reflection coefficients are real, an incident plane wave pulse with slowness p is merely scaled in amplitude on reflection. Once the coefficients become complex, the shape of the reflected pulse is modified (see, e.g., Hudson, 1962). The real part of the coefficient gives a scaled version of the original pulse and the imaginary part introduces a scaling of the Hilbert transform of the pulse, which for an impulse has precursory effects. The consequent pulse distortion is illustrated in figure 5.3 for the reflected *SV* wave pulse from an incident *SV* wave impulse, as a function of slowness. For $p < \alpha_+^{-1}$ the pulse shape is unchanged, but once $p > \alpha_+^{-1}$ and R_D^{SS} becomes complex the pulse shape is modified. The large amplitude of the reflection at and beyond the critical slowness for *S* waves is clearly seen and the steady phase change for $p > \beta_+^{-1}$ continues to modify the shape of the reflected pulse. Once the incident S wave becomes evanescent ($p > \beta_-^{-1}$), the reflection coefficient is real and the original pulse shape is restored.

5.2 A stratified region

We now look at the reflection and transmission response of a portion (z_A, z_C) of a stratified medium by embedding this region between uniform half spaces in $z < z_A$, $z > z_C$ with continuity of elastic properties at z_A and z_C. In the uniform half spaces we can represent a stress-displacement field in terms of up and downgoing waves by means of the eigenvector matrix $\mathbf{D}$ introduced in Section 3.1.

5.2.1 The wave-propagator

The stress-displacement vectors at z_A and z_C are connected by the propagator for the intervening region

$$\mathbf{b}(z_A) = \mathbf{P}(z_A, z_C)\mathbf{b}(z_C). \tag{5.35}$$

In the two uniform half spaces we make a decomposition of the stress-displacement field into up and downgoing P and S waves and using the continuity of the $\mathbf{b}$ vector we write,

$$\begin{aligned} \mathbf{b}(z_A) &= \mathbf{D}(z_A)\mathbf{v}(z_A-), \\ \mathbf{b}(z_C) &= \mathbf{D}(z_C)\mathbf{v}(z_C+), \end{aligned} \tag{5.36}$$

since we have continuity of elastic properties at z_A,z_C. The wavevectors $\mathbf{v}$ in the upper and lower uniform half spaces are therefore related by

$$\mathbf{v}(z_A-) = \mathbf{D}^{-1}(z_A)\mathbf{P}(z_A, z_C)\mathbf{D}(z_C)\mathbf{v}(z_C+), \tag{5.37}$$

when we combine (5.35) and (5.36). In terms of a single matrix $\mathbf{Q}(z_A, z_C)$

$$\mathbf{v}(z_A-) = \mathbf{Q}(z_A, z_C)\mathbf{v}(z_C+), \tag{5.38}$$

and by analogy with (5.35) we call $\mathbf{Q}$ the wave-propagator.

The wave-propagator has similar properties to the stress-displacement propagator $\mathbf{P}$. From the chain rule (2.89)

$$\begin{aligned} \mathbf{Q}(z_A, z_C) &= \mathbf{D}^{-1}(z_A)\mathbf{P}(z_A, z_B)\mathbf{P}(z_B, z_C)\mathbf{D}(z_C), \\ &= \mathbf{D}^{-1}(z_A)\mathbf{P}(z_A, z_B)\mathbf{D}(z_B)\mathbf{D}^{-1}(z_B)\mathbf{P}(z_B, z_C)\mathbf{D}(z_C). \end{aligned} \tag{5.39}$$

By using the decomposition of the unit matrix into the product of $\mathbf{D}(z_B)$ and its inverse, we have in effect introduced an infinitesimal uniform region at z_B. This is sufficient for us to recognise the wave-propagators for the regions (z_A, z_B), (z_B, z_C) and so give a chain rule

$$\mathbf{Q}(z_A, z_C) = \mathbf{Q}(z_A, z_B)\mathbf{Q}(z_B, z_C). \tag{5.40}$$

A consequence of (5.40) is that we have a simple expression for the inverse of the wave-propagator

$$\mathbf{Q}(z_A, z_C) = \mathbf{Q}^{-1}(z_C, z_A). \tag{5.41}$$

Although $\mathbf{P}(z_A, \xi)$ will be continuous across a plane $z = \xi$, the wave-propagator will not be unless the elastic parameters are continuous across ξ. Thus we must choose which side of an interface we wish to be on when we split the stratification for the chain rule (5.40).

Our expression (5.38) relating the wavevector in the bounding half spaces via the wave-propagator $\mathbf{Q}(z_A, z_C)$ has the same form as (5.3) for the interface problem. Indeed we may identify $\mathbf{Q}(z_I-, z_I+)$ as the wave-propagator for the interface. We have therefore already established the formal basis for the construction of the reflection and transmission coefficients for our stratified region in the previous section.

We split the wavevectors in the uniform half spaces into their up and downgoing wave parts and partition $\mathbf{Q}(z_A, z_C)$ so that (5.38) becomes

$$\begin{bmatrix} \mathbf{v}_U(z_A-) \\ \mathbf{v}_D(z_A-) \end{bmatrix} = \begin{bmatrix} \mathbf{Q}_{UU} & \mathbf{Q}_{UD} \\ \mathbf{Q}_{DU} & \mathbf{Q}_{DD} \end{bmatrix} \begin{bmatrix} \mathbf{v}_U(z_C+) \\ \mathbf{v}_D(z_C+) \end{bmatrix}, \tag{5.42}$$

which has just the same structure as (5.16). Thus for incident *downgoing* waves from the half space $z < z_A$ the reflected and transmission matrices have the representation (5.22) in terms of the partitions of the wave-propagator $\mathbf{Q}(z_A, z_C)$:

$$\begin{aligned} \mathbf{T}_D^{AC} &= \mathbf{T}_D(z_A, z_C) = (\mathbf{Q}_{DD})^{-1} \\ \mathbf{R}_D^{AC} &= \mathbf{R}_D(z_A, z_C) = \mathbf{Q}_{UD}(\mathbf{Q}_{DD})^{-1}. \end{aligned} \tag{5.43}$$

With incident *upgoing* waves in $z > z_C$ the transmission and reflection matrices are given by

$$\begin{aligned} \mathbf{T}_U^{AC} &= \mathbf{Q}_{UU} - \mathbf{Q}_{UD}(\mathbf{Q}_{DD})^{-1}\mathbf{Q}_{DU}, \\ \mathbf{R}_U^{AC} &= -(\mathbf{Q}_{DD})^{-1}\mathbf{Q}_{DU}. \end{aligned} \tag{5.44}$$

For a single interface, at fixed slowness p, $\mathbf{Q}(z_I-, z_I+)$ is independent of frequency, but for a stratified region $\mathbf{Q}(z_A, z_C)$ includes the frequency dependent propagator term $\mathbf{P}(z_A, z_C)$ and so $\mathbf{T}_D^{AC}, \mathbf{R}_D^{AC}$ etc. depend on the frequency ω.

The expressions (5.43), (5.44) can be recast to express the partitions of $\mathbf{Q}(z_A, z_C)$ in terms of the reflection and transmission matrices for both upward and downward incident waves. The wave-propagator takes the form

$$\mathbf{Q}(z_A, z_C) = \begin{bmatrix} \mathbf{T}_U^{AC} - \mathbf{R}_D^{AC}(\mathbf{T}_D^{AC})^{-1}\mathbf{R}_U^{AC} & \mathbf{R}_D^{AC}(\mathbf{T}_D^{AC})^{-1} \\ -(\mathbf{T}_D^{AC})^{-1}\mathbf{R}_U^{AC} & (\mathbf{T}_D^{AC})^{-1} \end{bmatrix}. \tag{5.45}$$

When we choose to look at the stratified region from below, we prefer to work with $\mathbf{Q}(z_C, z_A)$ and from (5.41) this can be found by constructing the inverse of the partitioned matrix (5.45), so that

$$\mathbf{Q}(z_C, z_A) = \begin{bmatrix} (\mathbf{T}_U^{AC})^{-1} & -(\mathbf{T}_U^{AC})^{-1}\mathbf{R}_D^{AC} \\ \mathbf{R}_U^{AC}(\mathbf{T}_U^{AC})^{-1} & \mathbf{T}_D^{AC} - \mathbf{R}_U^{AC}(\mathbf{T}_U^{AC})^{-1}\mathbf{R}_D^{AC} \end{bmatrix}. \tag{5.46}$$

If we exchange the subscripts U and D and reflect the matrix (5.38) blockwise about the diagonal, we recover the matrix (5.37). This structure arises because upward reflection and transmission matrices are in fact the downward matrices when the stratified region is inverted.

Our definitions of the reflection and transmission matrices presuppose that $z_C \geq z_A$. When we wish to represent a wave-propagator $\mathbf{Q}(z_E, z_F)$ in terms of the reflection and transmission properties of (z_E, z_F) we use the form (5.45) if $z_E \geq z_F$, but if z_E is less than z_F we will employ the representation (5.46).

The stress-displacement propagator $\mathbf{P}(z_A, z_C)$ can be recovered from the wave-propagator $\mathbf{Q}(z_A, z_C)$ as

$$\mathbf{P}(z_A, z_C) = \mathbf{D}(z_A)\mathbf{Q}(z_A, z_C)\mathbf{D}^{-1}(z_C). \tag{5.47}$$

This relation gives some insight into the physical nature of the propagator. The action of $\mathbf{D}^{-1}(z_C)$ is to break the stress and displacement field at z_C into its up and downgoing parts. The corresponding up and downgoing waves at z_A are generated by the action of the wave-propagator $\mathbf{Q}$ which requires a knowledge of the propagation characteristics in both directions through the stratification. The eigenvector matrix $\mathbf{D}(z_A)$ then reconstructs the displacements and tractions at z_A from the wave components.

A particularly simple case of the wave-propagator is provided by a *uniform* medium for which

$$\begin{aligned} \mathbf{Q}_{un}(z_A, z_C) &= \exp\{i\omega(z_A - z_C)\mathbf{\Lambda}\} \\ &= \begin{bmatrix} \mathsf{E}_D^{AC} & 0 \\ 0 & (\mathsf{E}_D^{AC})^{-1} \end{bmatrix}, \end{aligned} \tag{5.48}$$

where E_D^{AC} is the phase income matrix for downward propagation from z_A to z_C introduced in (3.45). Since the off-diagonal partitions of $\mathbf{Q}_{un}$ are null, both $\mathbf{R}_D^{AC}$ and $\mathbf{R}_U^{AC}$ vanish, as would be expected. We may also identify the transmission matrices

$$\mathbf{T}_D^{AC} = \mathsf{E}_D^{AC}, \quad \mathbf{T}_U^{AC} = \mathsf{E}_D^{AC}, \tag{5.49}$$

for such a uniform zone.

5.2.2 Displacement matrix representations

We now introduce a fundamental stress-displacement matrix $\mathbf{B}_V$ whose columns are the $\mathbf{b}$ vectors corresponding to up and downgoing waves at some level in the stratification. We will be particularly interested in the displacement and traction matrix partitions of $\mathbf{B}_V$ and will use these to generalise the concept of reflection and transmission matrices.

At a plane $z = z_G$ within the stratification we construct the displacements which would be produced by unit amplitude *P* and *S* waves in a uniform medium with the elastic properties at z_G:

$$\mathcal{W}_{UG} = \mathfrak{m}_{UG}, \tag{5.50}$$

where $\mathfrak{m}_{UG}$ is a displacement partition of the eigenvector matrix $\mathbf{D}(z_G)$. The corresponding traction components are given by the matrix

$$\mathcal{T}_{UG} = \mathfrak{n}_{UG}. \tag{5.51}$$

In a similar way we can construct displacement and traction matrices for unit amplitude downgoing *P* and *S* waves

$$\mathcal{W}_{DG} = \mathfrak{m}_{DG}, \quad \mathcal{T}_{DG} = \mathfrak{n}_{DG}. \tag{5.52}$$

We have here, in effect, introduced an infinitesimal uniform region at z_G in which we can define up and downgoing waves in just the same way as in the derivation of the chain rule for the wave-propagator.

From the displacement and traction matrices introduced in (5.50)-(5.52) we construct a fundamental stress-displacement matrix

$$\mathbf{B}_{VG} = \begin{bmatrix} \mathcal{W}_{UG} & \mathcal{W}_{DG} \\ \mathcal{T}_{UG} & \mathcal{T}_{DG} \end{bmatrix}, \tag{5.53}$$

and at z_G this matrix reduces to $\mathbf{D}(z_G)$. Away from the level z_G we may construct $\mathbf{B}_V$G by using the propagator matrix for the stratified medium operating on $\mathbf{D}(z_G)$,

$$\mathbf{B}_{VG}(z_J) = \mathbf{P}(z_J, z_G)\mathbf{D}(z_G). \tag{5.54}$$

The propagator $\mathbf{P}(z_J, z_G)$ may be represented in terms of the wave-propagator $\mathbf{Q}(z_J, z_G)$ by (5.47) and so

$$\mathbf{B}_{VG}(z_J) = \mathbf{D}(z_J)\mathbf{Q}(z_J, z_G). \tag{5.55}$$

The explicit forms of the displacement matrices $\mathcal{W}_{DG}(z_J)$, $\mathcal{W}_{UG}(z_J)$ depend on the relative location of z_J and z_G.

If z_J lies above z_G we construct $\mathbf{B}_{VG}(z_J)$ from the partitioned representations, (5.45) for the wave-propagator and (3.36) for the eigenvector matrix $\mathbf{D}(z_J)$ with entries $\mathsf{m}_{DJ}, \mathsf{m}_{UJ}$ etc. The displacement matrices $\mathsf{W}_{UG}, \mathsf{W}_{DG}$ are then given in terms of the reflection and transmission matrices for (z_J, z_G) by

$$\begin{aligned} \mathsf{W}_{UG}(z_J) &= \mathsf{m}_{UJ}\mathbf{T}_U^{JG} - (\mathsf{m}_{DJ} + \mathsf{m}_{UJ}\mathbf{R}_D^{JG})(\mathbf{T}_D^{JG})^{-1}\mathbf{R}_U^{JG}, \\ \mathsf{W}_{DG}(z_J) &= (\mathsf{m}_{DJ} + \mathsf{m}_{UJ}\mathbf{R}_D^{JG})(\mathbf{T}_D^{JG})^{-1}. \end{aligned} \tag{5.56}$$

The traction matrices $\mathsf{T}_{UG}, \mathsf{T}_{DG}$ have a comparable form with $\mathsf{n}_{UJ}, \mathsf{n}_{DJ}$ replacing $\mathsf{m}_{UJ}, \mathsf{m}_{DJ}$. From (5.56) we see that

$$\mathsf{W}_{UG}(z_J) + \mathsf{W}_{DG}(z_J)\mathbf{R}_U^{JG} = \mathsf{m}_{UJ}\mathbf{T}_U^{JG}, \tag{5.57}$$

the field on the left hand side of the equation can be recognised as that produced by an incident upcoming wave system on the the region (z_G, z_J) which at z_J will consist of just transmitted waves.

When z_J lies below z_G we use (5.38) for the wave-propagator and now the displacement matrices $\mathsf{W}_{UG}, \mathsf{W}_{DG}$ are given by

$$\begin{aligned} \mathsf{W}_{UG}(z_J) &= (\mathsf{m}_{UJ} + \mathsf{m}_{DJ}\mathbf{R}_U^{GJ})(\mathbf{T}_U^{GJ})^{-1}, \\ \mathsf{W}_{DG}(z_J) &= \mathsf{m}_{DJ}\mathbf{T}_D^{GJ} - (\mathsf{m}_{UJ} + \mathsf{m}_{DJ}\mathbf{R}_U^{GJ})(\mathbf{T}_U^{GJ})^{-1}\mathbf{R}_D^{GJ}, \end{aligned} \tag{5.58}$$

where the reflection and transmission matrices are for a region below the level z_G. In this case

$$\mathsf{W}_{DG}(z_J) + \mathsf{W}_{UG}(z_J)\mathbf{R}_D^{GJ} = \mathsf{m}_{DJ}\mathbf{T}_D^{GJ}, \tag{5.59}$$

corresponding to an incident downward wave system on the region (z_G, z_J).

For the displacement fields W_{UG}, W_{DG} the matrix $<\mathsf{W}_{UG}, \mathsf{W}_{DG}>$ will be independent of depth and may be conveniently evaluated at z_G itself

$$<\mathsf{W}_{UG}, \mathsf{W}_{DG}> = <\mathsf{m}_{UG}, \mathsf{m}_{DG}> = i\mathbf{I}. \tag{5.60}$$

We have just seen that the displacement matrices $\mathsf{W}_{UG}, \mathsf{W}_{DG}$ are closely related to the reflection and transmission properties of the stratification. In fact if we can construct the fundamental matrices $\mathbf{B}_{VA}$ and $\mathbf{B}_{VC}$ for two different starting levels z_A and z_C, we can find the reflection and transmission matrices for the zone (z_A, z_C).

Consider an incident *downgoing* field at z_A, on the region $z_A \le z \le z_C$, this will give rise to a reflected contribution specified by $\mathbf{R}_D^{AC}$. The resultant displacement field may be represented as

$$\mathsf{W}_R(z) = \mathsf{W}_{DA}(z) + \mathsf{W}_{UA}(z)\mathbf{R}_D^{AC}, \tag{5.61}$$

and we see from (5.59) that at z_C this displacement field takes the form $\mathsf{m}_{DC}\mathbf{T}_D^{AC}$, with a comparable form for traction components. We recall that $\mathsf{W}_{DC}(z)$ is the displacement field arising from displacement m_{DC} and traction n_{DC} at z_C and so we have an alternative representation for $\mathsf{W}_R(z)$:

$$\mathsf{W}_R(z) = \mathsf{W}_{DC}\mathbf{T}_D^{AC}. \tag{5.62}$$

At any level in (z_A, z_C) we must be able to equate these displacement and traction representations based on viewpoints at z_A and z_C so that

$$\begin{aligned} \mathsf{W}_{DA}(z) + \mathsf{W}_{UA}(z)\mathbf{R}_D^{AC} &= \mathsf{W}_{DC}(z)\mathbf{T}_D^{AC}, \\ \mathsf{T}_{DA}(z) + \mathsf{T}_{UA}(z)\mathbf{R}_D^{AC} &= \mathsf{T}_{DC}(z)\mathbf{T}_D^{AC}. \end{aligned} \tag{5.63}$$

A special case of these equations has appeared in the interface problem (5.31)-(5.32) and our method of solution parallels that case. We make use of the properties of the matrix invariants in Section 2.2 to eliminate variables between the displacement and traction equations (5.63), and solve for $\mathbf{R}_D^{AC}$, $\mathbf{T}_D^{AC}$. If we eliminate $\mathbf{R}_D^{AC}$ we have

$$<\mathsf{W}_{UA}, \mathsf{W}_{DA}> = <\mathsf{W}_{UA}, \mathsf{W}_{DC}>\mathbf{T}_D^{AC}, \tag{5.64}$$

and from (5.60) we may simplify the solution for $\mathbf{T}_D^{AC}$ to give

$$\mathbf{T}_D^{AC} = \mathrm{i}<\mathsf{W}_{UA}, \mathsf{W}_{DC}>^{-1}. \tag{5.65}$$

There are two equivalent forms for $\mathbf{R}_D^{AC}$. Firstly in terms of $\mathbf{T}_D^{AC}$ we have

$$<\mathsf{W}_{DA}, \mathsf{W}_{UA}>\mathbf{R}_D^{AC} = -\mathrm{i}\mathbf{R}_D^{AC} = <\mathsf{W}_{DA}, \mathsf{W}_{DC}>\mathbf{T}_D^{AC} \tag{5.66}$$

so that

$$\mathbf{R}_D^{AC} = -<\mathsf{W}_{DA}, \mathsf{W}_{DC}><\mathsf{W}_{UA}, \mathsf{W}_{DC}>^{-1}. \tag{5.67}$$

Alternatively we may eliminate $\mathbf{T}_D^{AC}$ between the equations (5.63) to give

$$<\mathsf{W}_{DC}, \mathsf{W}_{DA}> + <\mathsf{W}_{DC}, \mathsf{W}_{UA}>\mathbf{R}_D^{AC} = 0, \tag{5.68}$$

and so

$$\mathbf{R}_D^{AC} = -<\mathsf{W}_{DC}, \mathsf{W}_{UA}>^{-1}<\mathsf{W}_{DC}, \mathsf{W}_{DA}>. \tag{5.69}$$

Under transposition,

$$<\mathsf{W}_{DC}, \mathsf{W}_{UA}>^{\mathsf{T}} = -<\mathsf{W}_{UA}, \mathsf{W}_{DC}>, \tag{5.70}$$

with the result that the second form for $\mathbf{R}_D^{AC}$ may be recognised as the transpose of the first. The downward reflection matrix is therefore symmetric

$$\mathbf{R}_D^{AC} = (\mathbf{R}_D^{AC})^{\mathsf{T}}. \tag{5.71}$$

For an incident *upcoming* field at z_C, we have displacement and traction equations involving upward reflection and transmission matrices

$$\begin{aligned} \mathsf{W}_{UC}(z) + \mathsf{W}_{DC}\mathbf{R}_U^{AC} &= \mathsf{W}_{UA}(z)\mathbf{T}_U^{AC}, \\ \mathsf{T}_{UC}(z) + \mathsf{T}_{DC}\mathbf{R}_U^{AC} &= \mathsf{T}_{UA}(z)\mathbf{T}_U^{AC}. \end{aligned} \tag{5.72}$$

The upward transmission matrix is given by

$$\mathbf{T}_U^{AC} = -\mathrm{i}<\mathsf{W}_{DC}, \mathsf{W}_{UA}>^{-1} = (\mathbf{T}_D^{AC})^{\mathsf{T}}. \tag{5.73}$$

The upward reflection matrix

$$\mathbf{R}_U^{AC} = -\langle \mathbf{W}_{UA}, \mathbf{W}_{DC} \rangle^{-1} \langle \mathbf{W}_{UA}, \mathbf{W}_{UC} \rangle, \\ = -\langle \mathbf{W}_{UC}, \mathbf{W}_{UA} \rangle \langle \mathbf{W}_{DC}, \mathbf{W}_{UA} \rangle^{-1}, \tag{5.74}$$

and like the downward matrix, $\mathbf{R}_U^{AC}$, is symmetric.

If we can construct the displacements and tractions within the stratified region (z_A, z_C) corresponding to the displacements and tractions generated by upward and downward travelling waves in uniform half spaces with the properties at z_A and z_C, we are able to form all the reflection and transmission matrices for the region. The expressions we have derived for the reflection and transmission properties do not depend on any assumptions about the nature of the parameter distribution within (z_A, z_C). For an arbitrary attenuative region we have therefore established the symmetry relations

$$\mathbf{R}_D^{AC} = (\mathbf{R}_D^{AC})^T, \quad \mathbf{R}_U^{AC} = (\mathbf{R}_U^{AC})^T, \quad \mathbf{T}_U^{AC} = (\mathbf{T}_D^{AC})^T. \tag{5.75}$$

These relations have previously been demonstrated by Kennett, Kerry & Woodhouse (1978) with a rather different approach. When the medium is perfectly elastic we can make further use of invariants to derive unitarity relations for the reflection and transmission matrices and this is considered in the appendix to this chapter.

5.2.3 Generalisation of reflection matrices

With the aid of the displacement fields $\mathbf{W}_{UG}$, $\mathbf{W}_{DG}$ we are able to extend the concept of reflection matrices to accommodate general linear boundary conditions on the seismic wavefield.

As an example we consider the free-surface condition of vanishing traction at $z = 0$. We construct a linear superposition of $\mathbf{W}_{UG}, \mathbf{W}_{DG}$:

$$\mathbf{W}_{1G}(z) = \mathbf{W}_{UG}(z) + \mathbf{W}_{DG}(z)\mathbf{R}_U^{fG}, \tag{5.76}$$

and choose the free-surface reflection matrix $\mathbf{R}_U^{fG}$ so that the associated traction vanishes at $z = 0$, i.e.

$$\mathsf{T}_{1G}(0) = \mathsf{T}_{UG}(0) + \mathsf{T}_{DG}(0)\mathbf{R}_U^{fG} = 0. \tag{5.77}$$

Thus we have a symmetric reflection matrix

$$\mathbf{R}_U^{fG} = -[\mathsf{T}_{DG}(0)]^{-1}\mathsf{T}_{UG}(0), \tag{5.78}$$

and the traction matrices can alternatively be represented as partitions of the product of the propagator from the surface to z_G and the eigenvector matrix at z_G: $[\mathbf{P}(0, z_G)\mathbf{D}(z_G)]$. We may regard $\mathbf{W}_{1G}(z)$ as the resultant displacement field due to an incident upcoming wave system at z_G from a uniform half space with the properties at z_G. A useful associated quantity is the surface displacement matrix $\mathbf{W}_{1G}(0)$ due to the incident upward wave which we shall denote as $\mathbf{W}_U^{fG}$.

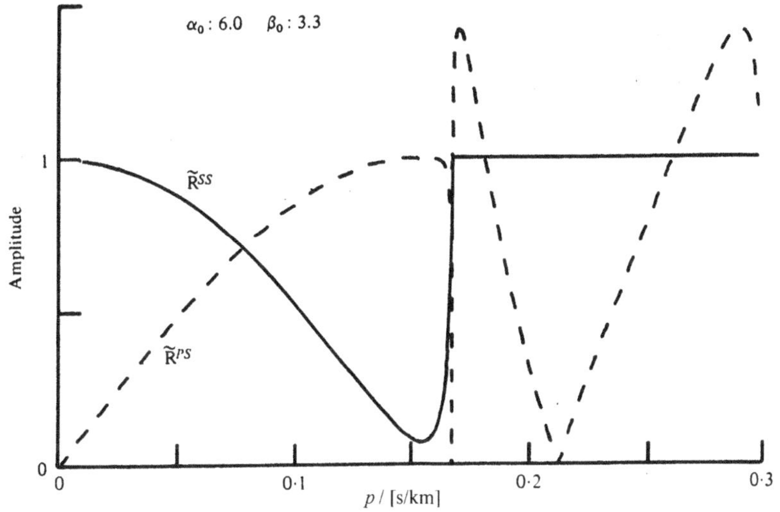

Figure 5.4. The amplitude of the free-surface reflection coefficients R_F^{SS}, R_F^{PS} as a function of slowness.

This matrix shares some of the attributes of a transmission matrix and has the representation

$$\mathbf{W}_U^{fG} = -i[\mathbf{T}_{DG}(0)]^{-T}, \tag{5.79}$$

which may be found by eliminating $\mathbf{R}_U^{fG}$ between the surface displacement and traction equations.

If we move the level z_G up to just below the surface i.e. $z_G = 0+$, the resulting free-surface reflection matrix

$$\mathbf{R}_F = \mathbf{R}_U^{f0} = -\mathbf{n}_{D0}^{-1}\mathbf{n}_{U0}, \tag{5.80}$$

is frequency independent at fixed slowness p. For *SH* waves, from (3.38),

$$R_F^{HH} = 1, \tag{5.81}$$

and for *P-SV* waves, from (3.37),

$$\begin{bmatrix} R_F^{PP} & R_F^{SP} \\ R_F^{SP} & R_F^{SS} \end{bmatrix} = \frac{1}{4p^2 q_{\alpha 0} q_{\beta 0} + \upsilon^2} \begin{bmatrix} 4p^2 q_{\alpha 0} q_{\beta 0} - \upsilon^2 & 4ip\upsilon(q_{\alpha 0} q_{\beta 0})^{1/2} \\ 4ip\upsilon(q_{\alpha 0} q_{\beta 0})^{1/2} & 4p^2 q_{\alpha 0} q_{\beta 0} - \upsilon^2 \end{bmatrix}, \tag{5.82}$$

where

$$\upsilon = (2p^2 - \beta_0^{-2}). \tag{5.83}$$

With our choice of normalisation $\mathbf{R}_F$ is a symmetric matrix, and we note

$$R_F^{PP} = R_F^{SS} \tag{5.84}$$

These surface coefficients become singular at a slowness p_R such that the denominator vanishes i.e.

$$(2p_R^2 - \beta_0^{-2})^2 + 4p_R^2 q_{\alpha 0} q_{\beta 0} = 0, \tag{5.85}$$

and this is just the condition for the existence of free Rayleigh surface waves on a uniform half space with the surface properties (cf., Section 11.2). In a Rayleigh wave both *P* and *S* waves are evanescent throughout the half space and so $p_R > \beta_0^{-1}$. For a Poisson solid $(\alpha = \sqrt{3}\beta)$ $p_R = 1.0876\beta_0^{-1}$.

The free-surface reflection elements R_F^{SS}, R_F^{PS} are shown in figure 5.4 as a function of slowness p, for $\alpha_0 = 6.0$ km/s, $\beta_0 = 3.33$ km/s.

At vertical incidence $(p = 0)$ R_F^{SS} is unity and falls to a minimum just before the *P* waves go evanescent at $p = 0.167$ s/km. For larger slowness R_F^{SS} has modulus unity until *S* waves becomes evanescent at $p = 0.3$ s/km. The R_F^{PS} coefficient is zero at vertical incidence, but the efficiency of conversion increases with slowness until $p = \alpha_0^{-1}$ at which there is a null. For larger slowness there is a rapid increase to a value greater than unity when propagating *S* waves and evanescent *P* waves are coupled at the surface. R_F^{PS} drops once again to zero when $p = \beta_0^{-1}/\sqrt{2}$.

At the free surface, the displacement matrix due to an incident upgoing wave is

$$\mathbf{W}_F = (\mathbf{m}_{U0} + \mathbf{m}_{D0}\mathbf{R}_F). \tag{5.86}$$

For *SH* waves we have a simple scalar multiplication

$$W_F^{HH} = 2, \tag{5.87}$$

but for *P-SV* waves the matrix $\mathbf{W}_F$ has a more complex form

$$\mathbf{W}_F = \begin{bmatrix} -iq_{\alpha 0}\epsilon_{\alpha 0}C_1 & p\epsilon_{\beta 0}C_2 \\ p\epsilon_{\alpha 0}C_2 & -iq_{\beta 0}\epsilon_{\beta 0}C_1 \end{bmatrix}, \tag{5.88}$$

in which the elements of $\mathbf{m}_{U0}$ are modified by the presence of the conversion factors from infinite medium to free-surface displacements

$$\begin{aligned} C_1 &= 2\beta_0^{-2}(2p^2 - \beta_0^{-2})/\{4p^2 q_{\alpha 0} q_{\beta 0} + \upsilon^2\}, \\ C_2 &= 4\beta_0^{-2} q_{\alpha 0} q_{\beta 0}/\{4p^2 q_{\alpha 0} q_{\beta 0} + \upsilon^2\}. \end{aligned} \tag{5.89}$$

These conversion factors are appropriate to both propagating and evanescent waves.

To satisfy the boundary condition at the base of the stratification, for example, we would construct a displacement field

$$\mathbf{W}_{2G}(z) = \mathbf{W}_{DG}(z) + \mathbf{W}_{UG}(z)\mathbf{R}_D^{GL}, \tag{5.90}$$

where $\mathbf{R}_D^{GL}$ is chosen so that a specified linear combination of displacement and traction vanishes at some depth. If the stratification is underlain by a uniform half space for $z > z_L$ we would choose $\mathbf{R}_D^{GL}$ in the usual way so that no upgoing waves are present in this region; the boundary condition here would be

$$\mathbf{m}_{DL}^T \mathbf{T}_{2G}(z_L) - \mathbf{n}_{DL}^T \mathbf{W}_{2G}(z_L) = <\mathbf{W}_{DL}, \mathbf{W}_{2G}> = 0. \tag{5.91}$$

For wavespeed distributions which increase steadily with depth below z_L we would now choose $\mathbf{R}_D^{GL}$ so that W_2 tends to zero as $z \to \infty$.

5.2.4 Reflection matrices for spherical stratification

For spherically stratified media we may once again specify the reflection and transmission properties of a spherical shell by surrounding this region by uniform media with continuity of properties at the internal and external radii.

In a uniform medium we can separate *P* and *SV* wave contributions for which the displacement solutions in the (l, m, ω) transform domain have a radial dependence in terms of spherical Bessel functions. There is now the added complication that the character of the solution for each wave type switches from oscillatory to exponential across a 'turning radius' which for *P* waves is

$$R_\alpha = \bar{p}\alpha = (l + \tfrac{1}{2})\alpha/\omega. \tag{5.92}$$

Above this level R_α we take solutions depending on $h_l^{(1)}(\omega R/\alpha)$ for upgoing *P* waves and $h_l^{(2)}(\omega R/\alpha)$ for downgoing *P* waves. Below R_α we switch to solutions which give a better representation of the evanescent character: $j_l(\omega R/\alpha)$ which decays away from R_α and $y_l(\omega R/\alpha)$ which grows exponentially. The nature of the turning level R_α is best seen from a physical ray picture. In a uniform medium a *P* ray path is a straight line and for angular slowness $\bar{p}$ the closest approach to the origin is at a radius R_α.

For a particular wave type we will designate as 'downgoing', the actual downgoing waves above the turning level and the evanescently decaying solution below this level. Similarly, we will use 'upgoing' to mean upward travelling waves above the turning level and the exponentially growing solution below. In this way we achieve the same specification as was possible in horizontal stratification by our choice of physical Riemann sheet (3.8).

We may now set up a fundamental **B** matrix at a radius *R* in a uniform medium with a character determined by the relative location of *R* and the turning levels R_α, R_β. The reflection and transmission matrices for a spherical shell or interface may then be found by following the development of Section 5.1 with $\mathbf{D}(z_A)$ etc. replaced by the appropriate fundamental matrices.

For models composed of uniform shells it is worthwhile to follow Chapman & Phinney (1972) and extract from the fundamental matrices a diagonal term which represents the main dependence on *R*. This procedure allows the propagator for a uniform layer to be written as

$$\mathbf{P}(R_1, R_2) = \mathbf{F}(R_1)\mathbf{E}(R_1, R_2)\mathbf{F}^{-1}(R_2), \tag{5.93}$$

where the radial phase behaviour is concentrated in a ratio of spherical Bessel functions in $\mathbf{E}(R_1, R_2)$.

The propagation invariants (2.36), (2.68) carry over to the spherical case and so for the region (R_C, R_A) we can find, for example, the reflection and transmission matrices for downward incidence at R_A as,

$$\begin{aligned} \mathbf{T}_D^{AC} &= <W_{UA}, W_{DC}>^{-1}<W_{UA}, W_{DA}>, \\ \mathbf{R}_D^{AC} &= -<W_{DC}, W_{UA}>^{-1}<W_{DC}, W_{UA}>. \end{aligned} \tag{5.94}$$

Here W_{UA}, W_{DA} are displacement matrices with upgoing and downgoing character respectively at R_A. For a zone including the centre of the sphere we use (5.94) with W_{DC} replaced by a displacement matrix whose columns are regular at the origin.

Appendix: Unitary relations for reflection and transmission

In the course of this chapter we have made extensive use of the matrix invariant $<W_1, W_2>$ for two displacement fields. This form is appropriate for attenuative media, but when the material properties are *perfectly elastic* we can introduce a further invariant

$$W_1^{T*}T_2 - T_1^{T*}W_2 = \{W_1, W_2\}. \tag{5a.1}$$

We introduce the quantities j_α, j_β which specify whether *P* and *S* waves are propagating or evanescent; we take, e.g.,

$$\begin{aligned} j_\alpha &= 1 \quad P \text{ propagating}, \\ &= 0 \quad P \text{ evanescent}, \end{aligned} \tag{5a.2}$$

so that j_α acts as a projection operator onto propagating *P* waves. It is also convenient to consider $1 - j_\alpha$ which projects onto evanescent *P* waves, so we write

$$\bar{j}_\alpha = 1 - j_\alpha. \tag{5a.3}$$

The elastic invariants for the partitions of the eigenvector matrix for *P-SV* waves are

$$\{m_U, m_U\} = -\{m_D, m_D\} = i \begin{bmatrix} j_\alpha & 0 \\ 0 & j_\beta \end{bmatrix} = iJ, \tag{5a.4}$$

and

$$\{m_U, m_D\} = \{m_D, m_U\} = \begin{bmatrix} \bar{j}_\alpha & 0 \\ 0 & \bar{j}_\beta \end{bmatrix} = \bar{J}. \tag{5a.5}$$

We construct a fundamental stress-displacement matrix

$$\mathbf{B}(z) = \begin{bmatrix} W_1 & W_2 \\ T_1 & T_2 \end{bmatrix}, \tag{5a.6}$$

and choose

$$\begin{aligned} W_1(z) &= W_{DA}(z) + W_{UA}(z)\mathbf{R}_D^{AB} = W_{DB}(z)\mathbf{T}_D^{AB}, \\ W_2(z) &= W_{UA}(z)\mathbf{T}_U^{AB} = W_{UB}(z) + W_{DB}(z)\mathbf{R}_U^{AB}, \end{aligned} \tag{5a.7}$$

in terms of the displacement matrices W_{DA} etc. introduced in Section 5.2. The symmetry of the governing equations (2.40) requires that $\mathbf{B}^{T*}\mathbf{N}\mathbf{B}$ will be independent of depth; with partitioned form

$$\mathbf{B}^{T*}\mathbf{N}\mathbf{B} = \begin{bmatrix} \{W_1, W_1\} & \{W_1, W_2\} \\ \{W_2, W_1\} & \{W_2, W_2\} \end{bmatrix}. \tag{5a.8}$$

If we evaluate (5a.8) at $z = z_A$ and make use of the results (5a.4)-(5a.5) we find

$$\mathbf{B}^{T*}\mathbf{NB} = \begin{bmatrix} -iJ_A + (\mathbf{R}_D^{AB})^{T*} iJ_A \mathbf{R}_D^{AB} \\ +(\mathbf{R}_D^{AB})^{T*}\bar{J}_A - \bar{J}_A \mathbf{R}_D^{AB} & \bar{J}_A \mathbf{T}_U^{AB} + (\mathbf{R}_D^{AB})^{T*} iJ_A \mathbf{T}_U^{AB} \\ (\mathbf{T}_U^{AB})^{T*}\bar{J}_A + (\mathbf{T}_U^{AB})^{T*} iJ_A \mathbf{R}_D^{AB} & (\mathbf{T}_U^{AB})^{T*} iJ_A \mathbf{T}_U^{AB} \end{bmatrix}. \tag{5a.9}$$

We get a comparable form for $\mathbf{B}^{T*}\mathbf{NB}$ at $z = z_B$ with the roles of up and downgoing waves interchanged, and equating the two expressions for $\mathbf{B}^{T*}\mathbf{NB}$ we have

$$i\begin{bmatrix} (\mathbf{R}_D^{AB})^{T*} & (\mathbf{T}_D^{AB})^{T*} \\ (\mathbf{T}_U^{AB})^{T*} & (\mathbf{R}_U^{AB})^{T*} \end{bmatrix} \begin{bmatrix} J_A & 0 \\ 0 & J_B \end{bmatrix} \begin{bmatrix} \mathbf{R}_D^{AB} & \mathbf{T}_U^{AB} \\ \mathbf{T}_D^{AB} & \mathbf{R}_U^{AB} \end{bmatrix} - i\begin{bmatrix} J_A & 0 \\ 0 & J_B \end{bmatrix}$$
$$+\begin{bmatrix} (\mathbf{R}_D^{AB})^{T*} & (\mathbf{T}_D^{AB})^{T*} \\ (\mathbf{T}_U^{AB})^{T*} & (\mathbf{R}_U^{AB})^{T*} \end{bmatrix} \begin{bmatrix} \bar{J}_A & 0 \\ 0 & \bar{J}_B \end{bmatrix} - \begin{bmatrix} \bar{J}_A & 0 \\ 0 & \bar{J}_B \end{bmatrix} \begin{bmatrix} \mathbf{R}_D^{AB} & \mathbf{T}_U^{AB} \\ \mathbf{T}_D^{AB} & \mathbf{R}_U^{AB} \end{bmatrix} = 0. \tag{5a.10}$$

If we now introduce the matrices

$$\mathcal{R} = \begin{bmatrix} \mathbf{R}_D^{AB} & \mathbf{T}_U^{AB} \\ \mathbf{T}_D^{AB} & \mathbf{R}_U^{AB} \end{bmatrix}, \quad \mathbf{J} = \begin{bmatrix} J_A & 0 \\ 0 & J_B \end{bmatrix}, \quad \bar{\mathbf{J}} = \mathbf{I} - \mathbf{J}, \tag{5a.11}$$

we can express (5a.10) as

$$i\mathcal{R}^{T*}\mathbf{J}\mathcal{R} - i\mathbf{J} + [\mathcal{R}^{T*}\bar{\mathbf{J}} - \mathcal{R}\bar{\mathbf{J}}] = \mathbf{0}; \tag{5a.12}$$

an equation previously derived by a rather different approach by Kennett, Kerry & Woodhouse (1978).

The matrix $\mathbf{J}$ has the role of a projection operator onto any travelling waves at the top and bottom of the region (z_A, z_B), and $\bar{\mathbf{J}}$ projects onto evanescent waves. The joint operators $\mathbf{J}\bar{\mathbf{J}}$, $\bar{\mathbf{J}}\mathbf{J}$ vanish.

When we apply the projector $\mathbf{J}$ to (5a.12) we obtain

$$(\mathbf{J}\mathcal{R}\mathbf{J})^{T*}(\mathbf{J}\mathcal{R}\mathbf{J}) = \mathbf{J}, \tag{5a.13}$$

which shows that the portion of the overall reflection and transmsission matrix $\mathcal{R}$ corresponding to travelling waves is unitary and this reflects the conservation of energy amongst the travelling waves. Using the $\bar{\mathbf{J}}$ projector we may also show that

$$\begin{aligned} (\mathbf{J}\mathcal{R}\bar{\mathbf{J}})^{T*}(\mathbf{J}\mathcal{R}\bar{\mathbf{J}}) &= 2\,\mathrm{Im}\{\bar{\mathbf{J}}\mathcal{R}\bar{\mathbf{J}}\}, \\ (\mathbf{J}\mathcal{R}\mathbf{J})^{T*}(\mathbf{J}\mathcal{R}\bar{\mathbf{J}}) &= i(\mathbf{J}\mathcal{R}\bar{\mathbf{J}})^{*}. \end{aligned} \tag{5a.14}$$

The set of relations (5a.12)–(5a.14) enable us to establish a number of important interrelations between the reflection and transmission coefficients for perfectly elastic stratification, and we shall present those which will be useful in subsequent discussions.

(a) Propagating waves

If the radicals q_α, q_β are real at the top and bottom of the stratification, all *P* and *S* waves have propagating form and so $\mathbf{J} = \mathbf{I}$, $\bar{\mathbf{J}} = \mathbf{0}$ and thus

$$\mathcal{R}^{T*}\mathcal{R} = \mathrm{I}. \tag{5a.15}$$

The overall reflection and transmission matrix is therefore unitary and from (5.75) is also symmetric.

(b) Evanescent waves

When both z_A and z_B lie in the evanescent regime for both *P* and *S* waves $\mathbf{J} = \mathbf{0}$, $\bar{\mathbf{J}} = \mathbf{I}$ and

$$\mathcal{R} = \mathcal{R}^{T*}, \tag{5a.16}$$

and since $\mathcal{R}$ is also symmetric, $\mathcal{R}$ must be real i.e. all reflection and transmission coefficients will be real.

(c) Turning points for both P and S waves

When P and S waves propagate at the level $z = z_A$: $J_A = \mathbf{I}$, $\bar{J}_A = 0$. With evanescent behaviour for both P and S at z_B: $J_B = 0$, $\bar{J}_B = \mathbf{I}$. In this case we see from (5a.10) that

$$(\mathbf{R}_D^{AB})^{T*}(\mathbf{R}_D^{AB}) = \mathbf{I}, \tag{5a.17}$$

the downward reflection matrix is therefore unitary and symmetric. This unitary property requires that

$$|R_D^{PP}| = |R_D^{SS}|, \quad |R_D^{PP}|^2 + |R_D^{SP}|^2 = |R_D^{SS}|^2 + |R_D^{PS}|^2 = 1, \tag{5a.18}$$

where R_D^{PP} etc. are the components of $\mathbf{R}_D^{AB}$; also

$$\begin{aligned} \arg R_D^{PS} &= \tfrac{1}{2}\pi + \tfrac{1}{2}(\arg R_D^{PP} + \arg R_D^{SS}), \\ \det \mathbf{R}_D^{AB} &= \exp\{i(\arg R_D^{PP} + \arg R_D^{SS})\}. \end{aligned} \tag{5a.19}$$

(d) Turning point for S, Evanescent P

If only S waves are propagating at z_A and all wave types are evanescent at z_B, $\mathbf{J} = \mathrm{diag}\{0, 1, 0, 0\}$ and $\bar{\mathbf{J}} = \mathrm{diag}\{1, 0, 1, 1\}$. Now (5a.12) reduces to

$$|R_D^{SS}| = 1 \tag{5a.20}$$

and from (5a.14)

$$\begin{aligned} |R_D^{PS}|^2 &= 2\,\mathrm{Im}\, R_D^{PP}, \\ \arg R_D^{PS} &= \tfrac{1}{4}\pi + \tfrac{1}{2}R_D^{SS}, \\ \det \mathbf{R}_D^{AB} &= R_D^{SS}(R_D^{PP})^*. \end{aligned} \tag{5a.21}$$

The results we have presented in this appendix have been derived for a portion of the stratification bounded by two uniform half spaces. They may however be extended to, for example, free-surface reflections.

Chapter 6

Reflection and Transmission II

In the last chapter we showed how to define reflection and transmission matrices for portions of a stratified medium bordered by uniform half spaces or a free surface.

We now demonstrate how the reflection and transmission properties of two or more such regions can be combined to give the overall reflection and transmission matrices for a composite region. These addition rules form the basis of efficient recursive construction schemes for the reflection matrices. These recursive schemes may be developed for stacks of uniform layers or for piecewise smooth structures with gradient zones separated by discontinuities in the elastic parameters or their gradients.

6.1 Reflection and transmission matrices for composite regions

When a stratified region consists of two or more types of structural elements, we will often wish to calculate the reflection and transmission matrices for these elements separately and then to combine these results to give the reflection matrices for the entire region. We use a set of addition rules for the reflection and transmission properties which can be derived from the chain rule for the wave-propagator or from the use of displacement matrices.

6.1.1 Superimposed stratification

We consider, as before, the region (z_A, z_C) bounded by uniform half spaces with continuity of elastic parameters at z_A and z_C. This region is then divided by splitting the stratification at a level z_B between z_A and z_C. We envisage the introduction of an infinitesimal uniform region at z_B, as in the derivation of the chain rule (5.38) for the wave-propagator

$$\mathbf{Q}(z_A, z_C) = \mathbf{Q}(z_A, z_B)\mathbf{Q}(z_B, z_C). \tag{6.1}$$

Since $z_A \leq z_B \leq z_C$ we can adopt the representation (5.45) for each of the wave-propagators in (6.1). The left hand side of (6.1) consists of partitions of $\mathbf{Q}(z_A, z_C)$ represented in terms of the overall matrices $\mathbf{R}_D^{AC}$, $\mathbf{T}_D^{AC}$ etc, and these are to be equated to partitioned matrices involving products of $\mathbf{R}_U^{AB}$, $\mathbf{R}_D^{BC}$ and the other reflection matrices for the upper region 'AB' and the lower region 'BC'. Thus

$$\begin{aligned}
(\mathbf{T}_D^{AC})^{-1} &= (\mathbf{T}_D^{AB})^{-1}[\mathbf{I} - \mathbf{R}_U^{AB}\mathbf{R}_D^{BC}](\mathbf{T}_D^{BC})^{-1},\\
\mathbf{R}_D^{AC}(\mathbf{T}_D^{AC})^{-1} &= \mathbf{T}_U^{AB}\mathbf{R}_D^{BC}(\mathbf{T}_D^{BC})^{-1} + \mathbf{R}_D^{AB}(\mathbf{T}_D^{AB})^{-1}[\mathbf{I} - \mathbf{R}_U^{AB}\mathbf{R}_D^{BC}](\mathbf{T}_D^{BC})^{-1},\\
(\mathbf{T}_D^{AC})^{-1}\mathbf{R}_U^{AC} &= (\mathbf{T}_D^{AB})^{-1}\mathbf{R}_U^{AB}\mathbf{T}_U^{BC} + (\mathbf{T}_D^{AB})^{-1}[\mathbf{I} - \mathbf{R}_U^{AB}\mathbf{R}_D^{BC}](\mathbf{T}_D^{BC})^{-1}\mathbf{R}_U^{BC}.
\end{aligned} \tag{6.2}$$

There is a rather more involved expression for the remaining partition of $\mathbf{Q}(z_A, z_C)$.

With the aid of the expression for $(\mathbf{T}_D^{AC})^{-1}$ we can recover the overall reflection and transmission matrices. The downward matrices are

$$\begin{aligned}
\mathbf{T}_D^{AC} &= \mathbf{T}_D^{BC}[\mathbf{I} - \mathbf{R}_U^{AB}\mathbf{R}_D^{BC}]^{-1}\mathbf{T}_D^{AB},\\
\mathbf{R}_D^{AC} &= \mathbf{R}_D^{AB} + \mathbf{T}_U^{AB}\mathbf{R}_D^{BC}[\mathbf{I} - \mathbf{R}_U^{AB}\mathbf{R}_D^{BC}]^{-1}\mathbf{T}_D^{AB};
\end{aligned} \tag{6.3}$$

and the upward matrices have a similar structure

$$\begin{aligned}
\mathbf{T}_U^{AC} &= \mathbf{T}_U^{AB}[\mathbf{I} - \mathbf{R}_D^{BC}\mathbf{R}_U^{AB}]^{-1}\mathbf{T}_U^{BC},\\
\mathbf{R}_U^{AC} &= \mathbf{R}_U^{BC} + \mathbf{T}_D^{BC}\mathbf{R}_U^{AB}[\mathbf{I} - \mathbf{R}_D^{BC}\mathbf{R}_U^{AB}]^{-1}\mathbf{T}_U^{BC}.
\end{aligned} \tag{6.4}$$

These addition rules for reflection and transmission matrices enable us to build up the response of a stratified medium a segment at a time (see Section 6.2).

6.1.2 Generalisation of addition rules

In Section 5.2 we have shown the interrelation of reflection and transmission matrices and the fundamental stress-displacement matrix $\mathbf{B}_{VC}$ whose columns correspond to up and downgoing waves at the level z_C. We now exploit this relation to extend the addition rules to general reflection matrices.

At the level z_A we may construct $\mathbf{B}_{VC}(z_A)$ as

$$\begin{aligned}
\mathbf{B}_{VC}(z_A) &= \mathbf{P}(z_A, z_C)\mathbf{D}(z_C),\\
&= \mathbf{P}(z_A, z_B)\mathbf{P}(z_B, z_C)\mathbf{D}(z_C),
\end{aligned} \tag{6.5}$$

using the propagator chain rule (2.89). We may now rewrite (6.5) in terms of the wave-propagator between z_B and z_C, using (5.47), as

$$\begin{aligned}
\mathbf{B}_{VC}(z_A) &= \mathbf{P}(z_A, z_B)\mathbf{D}(z_B)\mathbf{Q}(z_B, z_C),\\
&= \mathbf{B}_{VB}(z_A)\mathbf{Q}(z_B, z_C),
\end{aligned} \tag{6.6}$$

where we recognise the fundamental matrix $\mathbf{B}_{VB}$ corresponding to up and downgoing waves at z_B. In partitioned form we see that the two fundamental matrices are connected by the reflection and transmission properties of the region (z_B, z_C) appearing in $\mathbf{Q}(z_B, z_C)$. Thus

$$\begin{aligned}
\boldsymbol{W}_{UC}(z_A) &= \boldsymbol{W}_{UB}(z_A)\mathbf{T}_U^{BC} - \{\boldsymbol{W}_{DB}(z_A) + \boldsymbol{W}_{UB}(z_A)\mathbf{R}_D^{BC}\}(\mathbf{T}_D^{BC})^{-1}\mathbf{R}_U^{BC},\\
\boldsymbol{W}_{DC}(z_A) &= \{\boldsymbol{W}_{DB}(z_A) + \boldsymbol{W}_{UB}(z_A)\mathbf{R}_D^{BC}\}(\mathbf{T}_D^{BC})^{-1},
\end{aligned} \tag{6.7}$$

with corresponding forms for traction components.

Consider now imposing a free-surface boundary condition at $z = z_A$, from (5.63) the upward reflection matrix between z_C at the free surface is

$$\mathbf{R}_U^{fC} = -\mathsf{T}_{DC}^{-1}(z_A)\mathsf{T}_{UC}(z_A), \tag{6.8}$$

in terms of the traction components of $\mathbf{B}_{VC}$. If we now substitute for $\mathbf{B}_{VC}$ in terms of $\mathbf{B}_{VB}$ and the 'BC' reflection and transmission matrices, we have

$$\mathbf{R}_U^{fC} = \mathbf{T}_D^{BC}\{\mathsf{T}_{DB}(z_A) + \mathsf{T}_{UB}(z_A)\mathbf{R}_D^{BC}\}^{-1}\mathsf{T}_{UB}(z_A)\mathbf{T}_U^{BC} + \mathbf{R}_U^{BC}, \tag{6.9}$$

so that from (6.8)

$$\mathbf{R}_U^{fC} = \mathbf{R}_U^{BC} + \mathbf{T}_D^{BC}\mathbf{R}_U^{fB}[\mathbf{I} - \mathbf{R}_D^{BC}\mathbf{R}_U^{fB}]^{-1}\mathbf{T}_U^{BC}. \tag{6.10}$$

This addition relation has just the same form as our previous relation for upward matrices (6.4). The structure of the addition relations for reflection will be the same for all boundary conditions on the seismic field such that a linear combination of displacement and traction vanishes at some depth. The calculation of the corresponding reflection matrices requires the evaluation of an expression like (6.8) with T replaced by some other combination of displacement and traction. Since the same linear operator will be applied to both sides of (6.7) we see that we will always extract a reverberation operator from the expression in braces (cf. 6.10).

When we require the free-surface traction to vanish we get a further generalisation of the addition rule for transmission, since the surface displacement $\mathbf{W}_U^{fC}$ may be represented as

$$\mathbf{W}_U^{fC} = \mathbf{W}_U^{fB}[\mathbf{I} - \mathbf{R}_D^{BC}\mathbf{R}_U^{fB}]^{-1}\mathbf{T}_U^{BC}, \tag{6.11}$$

when we use (2.68), and this has the same form as (6.4) for upward transmission.

A particularly useful form for this displacement operator is obtained if we bring z_B to lie just beneath the free surface $(z = 0)$

$$\mathbf{W}_U^{fC} = \mathbf{W}_F[\mathbf{I} - \mathbf{R}_D^{0C}\mathbf{R}_F]^{-1}\mathbf{T}_U^{0C}, \tag{6.12}$$

where $\mathbf{W}_F = \mathsf{m}_{U0} + \mathsf{m}_{D0}\mathbf{R}_F$, in terms of the partitions of $\mathbf{D}(0+)$. This displacement operator plays an important role in the expressions for the response of a half space to excitation by a buried source, as we shall see in Section 7.3.

6.1.3 Interpretation of addition rules

The addition rules (6.3), (6.4) enable us to construct the overall reflection and transmission matrices for a region 'AC' but we have yet to give any physical interpretation to the way in which the properties of the regions 'AB' and 'BC' are combined.

As an illustration we take the downward matrices

$$\begin{aligned}\mathbf{T}_D^{AC} &= \mathbf{T}_D^{BC}[\mathbf{I} - \mathbf{R}_U^{AB}\mathbf{R}_D^{BC}]^{-1}\mathbf{T}_D^{AB},\\ \mathbf{R}_D^{AC} &= \mathbf{R}_D^{AB} + \mathbf{T}_U^{AB}\mathbf{R}_D^{BC}[\mathbf{I} - \mathbf{R}_U^{AB}\mathbf{R}_D^{BC}]^{-1}\mathbf{T}_D^{AB};\end{aligned} \tag{6.13}$$

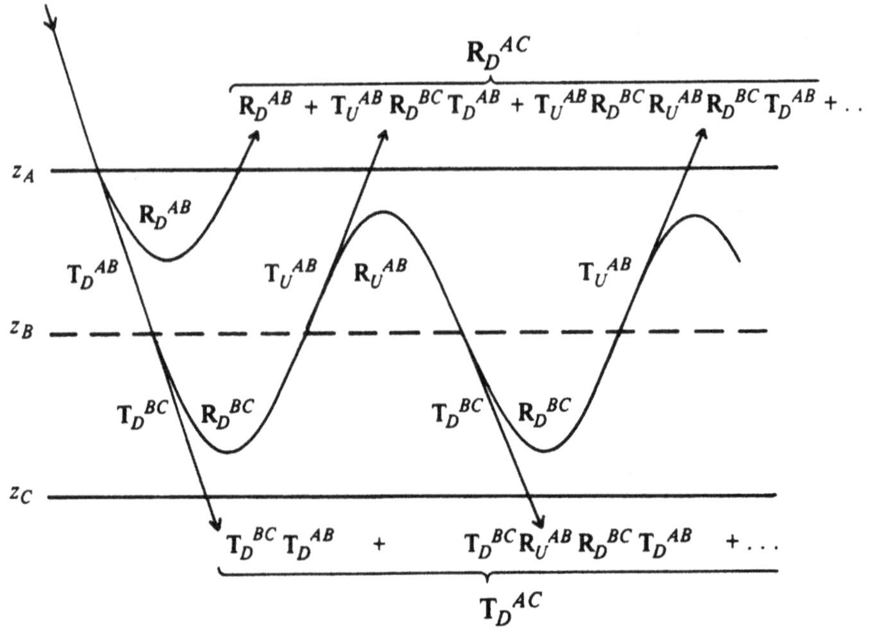

Figure 6.1. Schematic representation of the first few terms of the expansion of the addition rules for reflection and transmission matrices: showing the interactions with the regions 'AB' and 'BC'.

where the properties of the upper and lower regions are coupled through the action of the matrix inverse $[\mathbf{I} - \mathbf{R}_U^{AB}\mathbf{R}_D^{BC}]^{-1}$.

For *SH* waves $\mathbf{R}_U^{AB}\mathbf{R}_D^{BC}$ is just a product of reflection coefficients

$$\mathbf{R}_U^{AB}\mathbf{R}_D^{BC} = (R_U^{HH})(\mathbf{R}_D^{HH}), \tag{6.14}$$

where for simplicity we have dropped the superscripts AB, BC. For *P-SV* waves the matrix product generates

$$\mathbf{R}_D^{AB}\mathbf{R}_D^{BC} = \begin{pmatrix} R_U^{PP}R_D^{PP} + R_U^{PS}R_D^{SP} & R_U^{PP}R_D^{PS} + R_U^{PS}R_D^{SS} \\ R_U^{SP}R_D^{PP} + R_U^{SS}R_D^{SP} & R_U^{SP}R_D^{PS} + R_U^{SS}R_U^{SS} \end{pmatrix}, \tag{6.15}$$

and each of the components represents a physically feasible combination of reflection elements.

When we make a direct evaluation of the inverse $[\mathbf{I} - \mathbf{R}_U^{AB}\mathbf{R}_D^{BC}]^{-1}$ in terms of the determinant and the matrix adjugate we run into difficulties in all coupled wave cases. For example in the expansion of the determinant for *P-SV* waves we generate the reflection combinations

$$R_U^{SP}R_D^{PS}R_U^{PS}R_D^{SP}, \quad R_U^{PP}R_D^{PP}R_U^{SS}R_D^{SS}, \tag{6.16}$$

which because of the switch in wavetype cannot be acheived by any physical process (Cisternas, Betancourt & Leiva, 1973). The difficulty may be resolved if instead we make a series expansion of the matrix inverse as

$$[\mathbf{I} - \mathbf{R}_U^{AB}\mathbf{R}_D^{BC}]^{-1} = \mathbf{I} + \mathbf{R}_U^{AB}\mathbf{R}_D^{BC} + \mathbf{R}_U^{AB}\mathbf{R}_D^{BC}\mathbf{R}_U^{AB}\mathbf{R}_D^{BC} + \dots, \tag{6.17}$$

for which all the combinations of interactions are physically feasible. With the expansion of this inverse, the downward reflection and transmission matrices have the representation

$$\begin{aligned} \mathbf{R}_D^{AC} &= \mathbf{R}_D^{AB} + \mathbf{T}_U^{AB}\mathbf{R}_D^{BC}\mathbf{T}_D^{AB} + \mathbf{T}_U^{AB}\mathbf{R}_D^{BC}\mathbf{R}_U^{AB}\mathbf{R}_D^{BC}\mathbf{T}_D^{AB} + \dots, \\ \mathbf{T}_D^{AC} &= \mathbf{T}_D^{BC}\mathbf{T}_D^{AB} + \mathbf{T}_D^{BC}\mathbf{R}_U^{AB}\mathbf{R}_D^{BC}\mathbf{T}_D^{AB} + \dots, \end{aligned} \tag{6.18}$$

and higher terms in these series involve further powers of $\mathbf{R}_U^{AB}\mathbf{R}_D^{BC}$.

The sequences (6.18) are illustrated schematically in figure 6.1 where we have envisaged some incident downward travelling wave at z_A giving rise to both reflection and transmission terms. The physical content of each of the terms in (6.17) can be found by reading it from right to left.

For the reflection series, $\mathbf{R}_D^{AB}$ is just the reflection matrix for the region 'AB' in isolation. The second term $\mathbf{T}_U^{AB}\mathbf{R}_D^{BC}\mathbf{T}_D^{AB}$ represents waves which have been transmitted through the upper region and then been reflected by the region 'BC' and finally transmitted back through the upper zone. In $\mathbf{T}_U^{AB}\mathbf{R}_D^{BC}\mathbf{R}_U^{AB}\mathbf{R}_D^{BC}\mathbf{T}_D^{AB}$ we have the same set of interactions as in the previous term but in addition we include waves reflected down from the upper region to z_B and then reflected back from the region 'BC' before passage to z_A. The higher terms in the series include further internal interactions between the zones 'AB' and 'BC'. The total response (6.3) includes all these internal reverberations so that we refer to $[\mathbf{I} - \mathbf{R}_U^{AB}\mathbf{R}_D^{BC}]^{-1}$ as the *reverberation* operator for the region 'AC'.

A similar pattern can be seen in the transmission series (6.18). The first term $\mathbf{T}_D^{BC}\mathbf{T}_D^{AB}$ corresponds to direct transmission down through the entire region 'AC'. The second and higher terms in the series include successive internal reverberations represented by the powers of $\mathbf{R}_U^{AB}\mathbf{R}_D^{BC}$.

With this interpretation of the reverberation operator we see that if we truncate the expansion of $[\mathbf{I} - \mathbf{R}_U^{AB}\mathbf{R}_D^{BC}]^{-1}$ to $M + 1$ terms, we will include M internal reverberations in our approximations (6.18) for $\mathbf{R}_D^{AC}$, $\mathbf{T}_D^{AC}$. If these reflection and transmission quantities are themselves used in a further application of the addition rule, the possible internal reverberations are cumulative and a maximum of nM such legs can occur,where n is the number of times $\mathbf{R}_D^{AC}$ or $\mathbf{T}_D^{AC}$ appears in a full expansion of the response.

Truncated expansions have been used to study multiple reflections and to limit attention to portions of the seismic wave field by Kennett (1975,1979a) and Stephen (1977). For studies of complex multiples it is particularly convenient to suppress all internal reverberations in a region by using the approximation

$$^{0}\mathbf{R}_D^{AC} = \mathbf{R}_D^{AB} + \mathbf{T}_U^{AB}\mathbf{R}_D^{BC}\mathbf{T}_D^{AB}. \tag{6.19}$$

With only a single internal reverberation allowed we have

$$ {}^{1}\mathbf{R}_{D}^{AC} = \mathbf{R}_{D}^{AB} + \mathbf{T}_{U}^{AB}\mathbf{R}_{D}^{BC}[\mathbf{I} + \mathbf{R}_{U}^{AB}\mathbf{R}_{D}^{BC}]\mathbf{T}_{D}^{AB}, \tag{6.20} $$

and when the full reverberation sequence is included with the matrix inverse

$$ \mathbf{R}_{D}^{AC} = \mathbf{R}_{D}^{AB} + \mathbf{T}_{U}^{AB}\mathbf{R}_{D}^{BC}[\mathbf{I} - \mathbf{R}_{U}^{AB}\mathbf{R}_{D}^{BC}]^{-1}\mathbf{T}_{D}^{AB}. \tag{6.21} $$

The same combination of terms appear in (6.20),(6.21) and for *P-SV* waves we have only a $2\times\ 2$ matrix inverse to calculate. When more than a single internal interaction within 'AC' is desired, it is therefore more convenient, and efficient, to compute the full response (6.21).

In the discussion above we have allowed for the possibility of conversion between *P* and *SV* waves during reflection from the regions 'AB' and 'BC'. If such reflections are negligible, the reverberation operator essentially factors into separate operators for *P* and *SV* waves:

$$ [\mathbf{I} - \mathbf{R}_{U}^{AB}\mathbf{R}_{D}^{BC}]^{-1} \approx \begin{bmatrix} [1 - R_{U}^{AB}R_{D}^{BC}]_{PP}^{-1} & 0 \\ 0 & [1 - R_{U}^{AB}R_{D}^{BC}]_{SS}^{-1} \end{bmatrix}. \tag{6.22} $$

The situation is more complex if conversion is likely in only the region 'AB', since there will now be partial coupling between *P* and *SV* internal multiples. The reverberation sequence for each wave type will be modified by the presence of contributions arising from converted multiples of the type: $R_{U}^{PS}R_{D}^{SS}R_{U}^{PS}R_{D}^{PP}$.

6.2 Reflection from a stack of uniform layers

The absence of reflections from within any uniform layer means that the reflection matrix for a stack of uniform matrix depends heavily on the interface coefficients. Transmission through the layers gives phase terms which modulate the interface effects. For a stack of uniform layers the addition rules introduced in the previous section may be used to construct the reflection and transmission matrices in a two-stage recursive process. The phase delays through a layer and the interface terms are introduced alternately.

6.2.1 Recursive construction scheme

Consider a uniform layer in $z_1 < z < z_2$ overlying a pile of such layers in $z_2 < z < z_3$. We suppose the reflection and transmission matrices at z_2- just into the layer are known and write e.g.

$$ \mathbf{R}_{D}(z_2-) = \mathbf{R}_{D}(z_2-, z_3+) \tag{6.23} $$

We may then add in the phase terms corresponding to transmission through the uniform layer using the addition rule and (5.49), to calculate the reflection and transmission matrices just below the interface at z_1+. Thus

$$\begin{aligned}\mathbf{R}_D(z_1+) &= E_D^{12}\mathbf{R}_D(z_2-)E_D^{12}\\ \mathbf{T}_D(z_1+) &= \mathbf{T}_D(z_2-)E_D^{12}\\ \mathbf{T}_U(z_1+) &= E_D^{12}\mathbf{T}_U(z_2-)\end{aligned} \tag{6.24}$$

where E_D^{12} is the phase income for downward propagation through the layer (3.45).

A further application of the addition rules allows us to include the reflection and transmission matrices for the interface z_1, e.g.,

$$\mathbf{R}_D^1 = \mathbf{R}_D(z_1-, z_1+). \tag{6.25}$$

The downward reflection and transmission matrices just above the z_1 interface depend on the interface terms and the previously calculated downward quantities

$$\begin{aligned}\mathbf{R}_D(z_1-) &= \mathbf{R}_D^1 + \mathbf{T}_U^1\mathbf{R}_D(z_1+)[1 - \mathbf{R}_U^1\mathbf{R}_D(z_1+)]^{-1}\mathbf{T}_D^1,\\ \mathbf{T}_D(z_1-) &= \mathbf{T}_D(z_1+)[1 - \mathbf{R}_U^1\mathbf{R}_D(z_1+)]^{-1}\mathbf{T}_D^1,\end{aligned} \tag{6.26}$$

The upward matrices take a less simple form since we are in effect adding on a layer at the most complex level of the wave propagation system

$$\begin{aligned}\mathbf{R}_U(z_1-) &= \mathbf{R}_U(z_2-) + \mathbf{T}_D(z_1+)\mathbf{R}_U^1[1 - \mathbf{R}_D(z_1+)\mathbf{R}_U^1]^{-1}\mathbf{T}_U(z_1+),\\ \mathbf{T}_U(z_1-) &= \mathbf{T}_U^1[1 - \mathbf{R}_D(z_1+)\mathbf{R}_U^1]^{-1}\mathbf{T}_U(z_1+).\end{aligned} \tag{6.27}$$

If desired, truncated reverberation sequences can be substituted for the matrix inverses to give restricted, approximate results.

These two applications of the addition rule may be used recursively to calculate the overall reflection and transmission matrices. We start at the base of the layering at z_3 and calculate the interfacial matrices e.g. $\mathbf{R}_D^3$ which will also be $\mathbf{R}_D(z_3-)$. We use (6.24) to step the stack reflection and transmission matrices to the top of the lowest layer in the stack. Then we use the interfacial addition relations (6.26)-(6.27) to bring the stack matrices to the upper side of this interface. The cycle (6.24) followed by (6.26)-(6.27) allows us to work up the stack, a layer at a time, for an arbitrary number of layers.

For the downward matrices $\mathbf{R}_D, \mathbf{T}_D$ (6.24), (6.26) require only downward stack matrices to be held during the calculation. When upward matrices are needed it is often more convenient to calculate them separately starting at the top of the layering and working down a layer at a time. The resulting construction scheme has a similar structure to (6.24), (6.26) and is easily adapted to free-surface reflection matrices by starting with free-surface coefficients rather than those for an interface.

At fixed slowness p all the interfacial matrices $\mathbf{R}_D^i$, $\mathbf{T}_D^i$, etc. are frequency independent so that at each layer step frequency dependence enters via the phase term E_D^{12}. If the interfacial coefficients are stored, calculations may be performed rapidly at many frequencies for one slowness p.

When waves go evanescent in any layer our choice of branch cut for the vertical slowness q_α, q_β mean that the terms in E_D^{12} are such that

$$\exp\{i\omega q_\alpha(z_2 - z_1)\} = \exp\{-\omega|q_\alpha|(z_2 - z_1)\} \tag{6.28}$$

when $q_\alpha^2 < 0$. We always have $z_2 > z_1$ and so no exponential terms which grow with frequency will appear. This means that the recursive scheme is numerically stable even at high frequencies.

The scalar versions of the recursive forms for downward reflection and transmission coefficients have been known for a long time and are widely used in acoustics and physical optics. The extension to coupled waves seems first to have been used in plasma studies (e.g., Altman & Cory, 1969) and was independently derived for the seismic case by Kennett (1974).

6.2.2 Comparison between recursive and propagator methods

For a stack of uniform layers we may also use the analytic form of the stress-displacement propagator (3.42) in each layer and then find the reflection and transmission coefficients for the entire stack from the overall wave-propagator (5.11). The elements of the propagator depend on, for example, $C_\alpha = \cos\omega q_\alpha h$. When all wave types are propagating throughout the stack the propagator approach is very effective and for perfectly elastic media offers the advantage of working with real quantities.

Once waves become evanescent in any part of the layering the direct propagator approach is less effective. In the recursive scheme we can avoid any exponentially growing terms and so these must be absent in any representation of the reflection coefficients. However for evanescent *P* waves, for example, $C_\alpha = \cosh\omega|q_\alpha|h$ and the growing exponentials must cancel in the final calculation of the reflection coefficients. With finite accuracy computations the cancellation is not complete since the growing exponentials swamp the significant part of the calculation. Compared with this problem the complex arithmetic needed for the recursive scheme seems a small handicap.

The difficulty with the propagator method can be removed by working directly with the minors which appear in the reflection coefficients (Molotkov, 1961; Dunkin, 1965). Unfortunately for transmission coefficients some difficulties still arise since individual matrix elements are needed as well as a minor (Fuchs, 1968; Cerveny, 1974). An efficient computational procedure for the minor matrix method has been given by Kind (1976). An alternative development which aims to minimize the computational effort depending on the character of the solution has been given by Abo-Zena (1979).

Comparable numerical instability problems associated with growing solutions of the differential equations occur in the construction of propagator matrices by

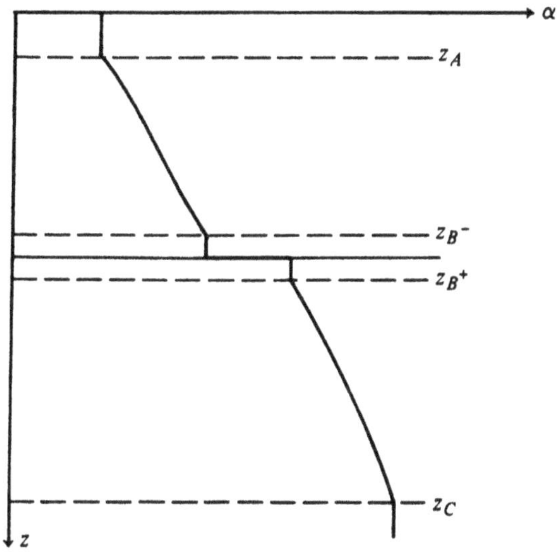

Figure 6.2. Division of a piecewise smooth medium into gradient zones $z_A < z < z_B-$, $z_B+ < z < z_C$ bordered by uniform media and an interface at z_B between uniform media.

direct numerical integration of the governing equations (2.26). Once again these difficulties can be cured by working directly in terms of minors (Gilbert & Backus, 1966; Takeuchi & Saito, 1972). However, since ratios of minors are really the quantities of interest, Abramovici (1968) has worked with the non-linear Ricatti differential equations for these ratios and found the same benefits as with the recursive approach to reflection matrix calculation.

For a fully anisotropic medium the propagator matrix itself is a 6×6 matrix and suffers from the same numerical difficulties as in the isotropic case. Once again it is possible to work with minor matrices but now we have a 27×27 matrix of 3×3 minors to be manipulated, which is rather inconvenient. The recursive method, on the other hand, involves only products and inverses of 3×3 matrices and proves to be very effective for this anisotropic case (Booth & Crampin, 1981).

6.3 Reflection matrices for piecewise smooth models.

We have just seen how we can make a recursive development for the reflection and transmission properties of a stack of uniform layers. For much of the earth a more appropriate representation of the wavespeed distribution is to take regions of smoothly varying properties interrupted by only a few major discontinuities. Such a model can be approximated by a fine cascade of uniform layers but then the process of continuous refraction by parameter gradients is represented by high order multiple reflections within the uniform layers. With a large number of such layers the computational cost can be very high.

Fortunately we can adopt a more direct approach by using a model composed of gradient zones (figure 6.2). Following Kennett & Illingworth (1981) we split the stratification into interfaces at which there is a discontinuity in elastic parameter or parameter gradients and zones of parameter gradients sandwiched between uniform media. We once again make a two-stage recursion, with alternate interface and propagation cycles, but now the transmission delays for a uniform layer are replaced by the reflection and transmission effects of a gradient zone.

6.3.1 Reflection from a gradient zone

We consider a region (z_A, z_B) in which the wavespeeds increase uniformly with depth. In order to construct reflection and transmission matrices for this zone we weld on uniform half spaces at z_A and z_B. Although we have continuity in the elastic parameters at z_A and z_B, there will normally be a discontinuity in parameter gradient at these boundaries.

We now build the wave-propagator for (z_A, z_B) from the fundamental matrix $\mathbf{B}_I$ we constructed for a smoothly varying medium in Section 3.3.2, so that

$$\mathbf{Q}(z_A, z_B) = \mathbf{D}^{-1}(z_A)\mathbf{B}_I(z_A)\mathbf{B}_I^{-1}(z_B)\mathbf{D}(z_B) \tag{6.29}$$

In order to emphasise the asymptotic relation of $\mathbf{B}_I$ to up and downgoing waves we combine the representations (3.102), (3.107) and (3.110) and then

$$\begin{aligned}\mathbf{Q}(z_A, z_B) = \mathbf{D}^{-1}(z_A)\mathsf{D}(z_A)\\ \mathsf{E}(z_A)\mathbf{L}(z_A; z_r)\mathbf{L}^{-1}(z_B; z_r)\mathsf{E}^{-1}(z_B)\\ \mathsf{D}^{-1}(z_B)\mathbf{D}(z_B).\end{aligned} \tag{6.30}$$

The wave-propagator $\mathbf{Q}$ may be factored into the terms governing the entry and exit of plane waves from the gradient zone $\mathbf{D}^{-1}(z_A)\mathsf{D}(z_A), \mathsf{D}^{-1}(z_B)\mathbf{D}(z_B)$ and the matrix

$$\mathbf{F}(z_A, z_B) = \mathsf{E}(z_A)\mathbf{L}(z_A, z_B; z_r)\mathsf{E}^{-1}(z_B), \tag{6.31}$$

which represents all the propagation characteristics within the gradient zone. The phase terms arise from the Airy function terms in $\mathsf{E}(z_A)$ and $\mathsf{E}(z_B)$ and the interaction sequence for the entire gradient zone

$$\mathbf{L}(z_A, z_B; z_r) = \mathbf{L}(z_A; z_r)\mathbf{L}^{-1}(z_B; z_r). \tag{6.32}$$

Here z_r is the reference level from which the arguments of the Airy functions are calculated. $\mathbf{L}$ can, in principle, be found from the interaction series (3.96) to any required order of interaction with (z_A, z_B). $\mathsf{E}(z_A)$ and $\mathsf{E}(z_B)$ are diagonal matrices which for *P-SV* waves are organised into blocks by wave type. The interaction series for $\mathbf{L}$ begins with the unit matrix and if all subsequent contributions are neglected, *P* and *S* waves appear to propagate independently within the gradient zone (cf. 3.101). Once the higher terms in the interaction series are included, the *P* and *SV* wave components with slowness p are coupled together. When the wavespeeds vary slowly with depth, this coupling is weak at moderate frequencies (Richards, 1974).

At the limits of the gradient zone we have introduced discontinuities in wavespeed gradient, and $\mathbf{D}^{-1}(z_A)\mathsf{D}(z_A)$ depends on the difference between the generalized slownesses for the gradient zone $\eta_{u,d}(\omega, p, z_A)$ and the corresponding radicals in the uniform region. For *SH* waves,

$$\mathbf{D}^{-1}(z_A)\mathsf{D}(z_A) = \mathbf{G}_\beta(z_A) = \rho\epsilon_\beta \begin{bmatrix} q_\beta + \eta_{\beta u} & q_\beta - \eta_{\beta d} \\ q_\beta - \eta_{\beta u} & q_\beta + \eta_{\beta d} \end{bmatrix}. \tag{6.33}$$

For *P-SV* waves we have to take account of the differing organisation of D and $\mathbf{D}$; we have arranged D by wave type and $\mathbf{D}$ by up and downgoing wave character. Thus

$$\mathbf{D}_P^{-1}(z_A)\mathsf{D}_P(z_A) = \Xi \begin{bmatrix} \mathbf{G}_\alpha & 0 \\ 0 & \mathbf{G}_\beta \end{bmatrix}, \tag{6.34}$$

where Ξ is the matrix introduced in (3.118), to connect fundamental $\mathbf{B}$ matrices organised by wave type or asymptotic character. We see from (6.34) that the interface terms do not couple *P* and *SV* waves.

The wave-propagator $\mathbf{Q}$ may be expressed in terms of the interface matrices $\mathbf{G}$ as

$$\mathbf{Q}(z_A, z_B) = \Xi\mathbf{G}(z_A)\mathbf{F}(z_A, z_B)\mathbf{G}^{-1}(z_B)\Xi. \tag{6.35}$$

for *SH* waves we take Ξ to be the unit matrix. The matrices $\mathbf{G}, \mathbf{F}$ are organised by wave type but under similarity transformation $\Xi\mathbf{F}\Xi$ takes the form

$$\Xi\mathbf{F}\Xi = \bar{\mathbf{F}} = \begin{bmatrix} \mathsf{F}_{uu} & \mathsf{F}_{ud} \\ \mathsf{F}_{ud} & \mathsf{F}_{dd} \end{bmatrix}, \tag{6.36}$$

and in the asymptotic regime far from turning points, at least, the new partitions connect up and downgoing elements rather than a single wave type. Since Ξ is its own inverse, we can recast (6.35) into the form

$$\begin{aligned} \mathbf{Q}(z_A, z_B) &= \Xi\mathbf{G}(z_A)\Xi.\Xi\mathbf{F}(z_A, z_B)\Xi.\Xi\mathbf{G}^{-1}(z_B)\Xi, \\ &= \bar{\mathbf{G}}(z_A)\bar{\mathbf{F}}(z_A, z_B)\bar{\mathbf{G}}^{-1}(z_B), \end{aligned} \tag{6.37}$$

and each of the factors in (6.37) have a strong resemblance to wave-propagators.

For each of these matrices we introduce a set of *generalized* reflection and transmission matrices $\mathbf{r}_d, \mathbf{t}_d$ etc. such that

$$\bar{\mathbf{G}}(z_A) = \begin{bmatrix} \mathbf{t}_u^G - \mathbf{r}_d^G(\mathbf{t}_d^G)^{-1}\mathbf{r}_u^G & \mathbf{r}_d^G(\mathbf{t}_d^G)^{-1} \\ -(\mathbf{t}_d^G)^{-1}\mathbf{r}_u^G & (\mathbf{t}_d^G)^{-1} \end{bmatrix}. \tag{6.38}$$

The generalized elements for a matrix product $\bar{\mathbf{G}}\bar{\mathbf{F}}$ can be determined by an extension of the addition rules (6.3), (6.4), for example

$$\mathbf{r}_d^{GF} = \mathbf{r}_d^G + \mathbf{t}_u^G\mathbf{r}_d^F[\mathbf{I} - \mathbf{r}_u^G\mathbf{r}_d^F]^{-1}\mathbf{t}_d^G. \tag{6.39}$$

The reflection and transmission matrices $\mathbf{R}_D^{AB}$, $\mathbf{T}_D^{AB}$ etc. for the whole gradient zone (z_A, z_B) can therefore be built up from the $\mathbf{r}_d$, $\mathbf{t}_d$ matrices for the factors of the wave-propagator.

For the interface matrix $\bar{\mathbf{G}}$, the generalized elements $\mathbf{r}_d^G$, $\mathbf{r}_u^G$ have a simple form: for *SH* waves

$$\begin{aligned}
\mathbf{r}_d^G|_{HH} &= (q_\beta - \eta_{\beta d})/(q_\beta + \eta_{\beta d}),\\
\mathbf{r}_u^G|_{HH} &= (\eta_{\beta u} - q_\beta)/(q_\beta + \eta_{\beta d}),\\
\mathbf{t}_u^G\mathbf{t}_d^G|_{HH} &= 2q_\beta(\eta_{\beta u} + \eta_{\beta d})/(q_\beta + \eta_{\beta d})^2.
\end{aligned} \tag{6.40}$$

The individual transmission terms are not symmetric because $\mathbf{D}$ is not normalised in the same way as $\mathbf{D}$. Since $\bar{\mathbf{G}}$ does not couple *P* and *SV* waves, the *SS* elements of $\mathbf{r}_d$ etc. are equal to the HH elements and the PP elements are obtained by exchanging α for β. The reflection terms depend on the off-diagonal parts of $\mathbf{G}_\alpha, \mathbf{G}_\beta$. We recall that $\eta_{\beta u,d}$ depend on the wavespeed distribution within the gradient zone and so the reflections should not be envisaged as just occurring at z_A.

In order to give a good approximation for the phase matrix $\mathbf{E}$ it is convenient to extrapolate the wavespeed distribution outside the gradient zone until turning points are reached and then to calculate the phase delays τ_α and τ_β from the *P* and *S* wave turning levels. When these turning points lie well outside the gradient zone (z_A, z_B), the generalized vertical slownesses $\eta_{\beta u,d}$ tend asymptotically to q_β and so $\mathbf{G}_\beta$ tends to a diagonal matrix. In this asymptotic regime there will therefore be no reflection associated with the discontinuities in wavespeed gradient. When turning levels lie close to z_A or z_B, the differences $(\eta_{\beta u,d} - q_\beta)$ become significant and noticeable reflected waves can occur as in the work of Doornbos (1981).

The nature of the generalized reflection and transmission terms $\mathbf{r}_d^F$, $\mathbf{t}_d^F$ associated with the propagation matrix within the gradient zone $\bar{\mathbf{F}}(z_A, z_B)$ depends strongly on the locations of *P* and *S* wave turning points relative to the gradient zone.

(a) Above all turning points:

For slownesses p such that both *P* and *S* wave turning points lie below z_B, we use the propagating forms Ej, Fj for the Airy function entries at both z_A and z_B. We split $\mathbf{E}(z_A)$ and $\mathbf{E}(z_B)$ into the parts which asymptotically have upgoing wave character $(\mathbf{e}_u^A, \mathbf{e}_u^B)$ and downgoing character $(\mathbf{e}_d^A, \mathbf{e}_d^B)$. The $\mathbf{e}_{u,d}$ matrices are diagonal and the total propagation term $\bar{\mathbf{F}}(z_A, z_B)$ may be expressed in partitioned form as

$$\bar{\mathbf{F}}(z_A, z_B) = \begin{bmatrix} \mathbf{e}_u^A & 0 \\ 0 & \mathbf{e}_d^A \end{bmatrix} \begin{bmatrix} \mathbf{I} + \mathbf{L}_{uu} & \mathbf{L}_{ud} \\ \mathbf{L}_{du} & \mathbf{I} + \mathbf{L}_{dd} \end{bmatrix} \begin{bmatrix} (\mathbf{e}_u^B)^{-1} & 0 \\ 0 & (\mathbf{e}_d^B)^{-1} \end{bmatrix}, \tag{6.41}$$

with a reordered interaction term $\bar{\mathbf{L}}$. The relatively simple structure of (6.32) means that we can relate the $\mathbf{r}^F_d$, $\mathbf{t}^F_d$ terms directly to those derived solely from the interaction series $\bar{\mathbf{L}}$: $(\mathbf{r}^L_d, \mathbf{t}^L_d)$. Thus

$$\begin{aligned} \mathbf{t}^F_d &= (\mathbf{e}^B_d)\mathbf{t}^L_d(\mathbf{e}^A_d)^{-1}, \quad & \mathbf{t}^F_u &= (\mathbf{e}^A_u)\mathbf{t}^L_u(\mathbf{e}^B_u)^{-1}, \\ \mathbf{r}^F_d &= (\mathbf{e}^A_u)\mathbf{r}^L_d(\mathbf{e}^A_d)^{-1}, \quad & \mathbf{r}^F_u &= (\mathbf{e}^B_d)\mathbf{r}^L_u(\mathbf{e}^B_u)^{-1}. \end{aligned} \tag{6.42}$$

When the interaction matrix $\bar{\mathbf{L}}$ departs significantly from the unit matrix, it is preferable to construct $\mathbf{r}^L_d$, $\mathbf{t}^L_d$ directly, rather than first calculate $\bar{\mathbf{L}}$. Kennett & Illingworth (1981) have shown how $\mathbf{r}^L_d$, $\mathbf{t}^L_d$ can be calculated by numerical solution of a coupled set of matrix Ricatti equations. This problem has good numerical stability and is simpler than calculating many terms in the interaction series.

If the contributions L_{uu}, L_{ud} etc. are small, the generalized reflection and transmission elements are given to first order by

$$\begin{aligned} \mathbf{t}^L_d &= \mathbf{I} - \mathsf{L}_{dd}, \quad & \mathbf{t}^L_u &= \mathbf{I} + \mathsf{L}_{uu}, \\ \mathbf{r}^L_d &= \mathsf{L}_{ud}, \quad & \mathbf{r}^L_u &= -\mathsf{L}_{du}. \end{aligned} \tag{6.43}$$

We will illustrate these relations for the *SH* wave case, by making use of our previous expressions (3.98)-(3.100) for the kernel in the interaction series (3.95). From (3.98) we see that L_{uu} has the opposite sign to L_{dd}, so that the transmission terms have the same character. In the asymptotic regime far above the *S* wave turning point

$$\begin{aligned} \mathbf{t}^L_d|_{HH} &\sim 1, \quad \mathbf{t}^L_u|_{HH} \sim 1, \\ \mathbf{r}^L_d|_{HH} &\sim \int_{z_A}^{z_B} d\zeta\, \gamma_H(\zeta) e^{-2i\omega\tau_\beta(\zeta)}, \\ \mathbf{r}^L_u|_{HH} &\sim \int_{z_A}^{z_B} d\zeta\, \gamma_H(\zeta) e^{2i\omega\tau_\beta(\zeta)}. \end{aligned} \tag{6.44}$$

When we add in the phase terms at z_A, z_B via the $\mathbf{e}^A_u$, $\mathbf{e}^B_d$ matrices to construct $\mathbf{r}^F_d$, $\mathbf{t}^F_d$ etc. we can recognise (6.44) as corresponding to first order reflections from the parameter gradients. For a thin slab of thickness Δz, the approximate *SH* reflection coefficient will have magnitude $\gamma_H \Delta z$.

For *P-SV* waves the matrices L_{uu}, L_{du}, L_{ud}, L_{dd} lead to coupling between *P* and *S* waves. The SS elements have the same structure as (6.44) with γ_H replaced by γ_S; the PP elements have γ_P in place of γ_H and also the *P* wave phase delay τ_α must be used. To this same first order asymptotic approximation

$$\begin{aligned} \mathbf{t}^L_d|_{PS} &\sim -i\int_{z_A}^{z_B} d\zeta\, \gamma_T(\zeta) e^{i\omega(\tau_\beta(\zeta)-\tau_\alpha(\zeta))}, \\ \mathbf{r}^L_d|_{PS} &\sim -\int_{z_A}^{z_B} d\zeta\, \gamma_R(\zeta) e^{i\omega(\tau_\beta(\zeta)+\tau_\alpha(\zeta))}. \end{aligned} \tag{6.45}$$

The upgoing elements have the opposite phase. These terms lead to a transfer between *P* and *S* waves, of the same slowness p, as they traverse the gradient zone.

These first order asymptotic results are equivalent to those discussed by Chapman (1974a) and Richards & Frasier (1976) based on the use of the WKBJ approximation. For moderate gradients the latter authors have shown that significant conversion of wave types can be generated by the coupling terms (6.45).

When the asymptotic limit is not appropriate, the first order approximation may still be used but the expressions for $\mathbf{r}_d^L$, $\mathbf{t}_d^L$ have a more complex form in terms of the Airy function entries Ej, Fj.

Once we have constructed $\mathbf{r}_d^F$, $\mathbf{t}_d^F$ the reflection and transmission matrices $\mathbf{R}_D^{AB}$, $\mathbf{T}_D^{AB}$ for the whole gradient zone can be found by successive applications of the addition rules (6.3), (6.4) for the $\mathbf{r}_d$, $\mathbf{t}_d$ matrices. We start by adding the matrices corresponding to $\bar{\mathbf{F}}(z_A, z_B)$ to those for $\bar{\mathbf{G}}^{-1}(z_B)$ and then the effect of $\bar{\mathbf{G}}(z_A)$ is added to these composite matrices.

(b) Below all turning points:

When the turning points for both *P* and *S* waves lie above z_A, we adopt the phase terms $\hat{\mathsf{E}}$ in terms of the Airy functions entries Bj, Aj. At the interfaces the generalized slownesses $\hat{\eta}_{u,d}$ will now appear. The analysis for the propagation effects parallels the purely propagating results, but in the asymptotic results $\exp(i\omega\tau)$ is replaced by $\exp(-\omega|\tau|)$. Direct calculation of $\mathbf{r}_d^L$ for strong gradients avoids numerical stability problems due to growing exponential terms.

(c) Turning points:

When the slowness p is such that turning points for either *P* or *S* waves occur within the zone (z_A, z_B), we have to take account of the differences in the nature of the wavefield at the top and bottom of the gradient zone. At z_A we would wish to use the propagating elements for the wave type which has the turning point. Whilst at z_B a better description is provided by using the evanescent forms.

We therefore split the gradient zone at the turning level z_T. We use the terms for propagating waves in the fundamental $\mathbf{B}$ matrix above z_T. In the region below z_T we take the fundamental matrix $\hat{\mathbf{B}}$ with the appropriate terms for evanescence. The wave-propagator can then be built up using the chain rule as

$$\begin{aligned}\mathbf{Q}(z_A, z_B) &= \mathbf{D}^{-1}(z_A)\mathbf{B}(z_A)\mathbf{B}^{-1}(z_T)\hat{\mathbf{B}}(z_T)\hat{\mathbf{B}}^{-1}(z_B)\mathbf{D}(z_B),\\ &= \mathbf{D}^{-1}(z_A)\mathbf{B}(z_A)\mathbf{H}\hat{\mathbf{B}}^{-1}(z_B)\mathbf{D}(z_B).\end{aligned} \tag{6.46}$$

The coupling matrix $\mathbf{H}$ arises from the differing functional forms of $\mathbf{B}$, $\hat{\mathbf{B}}$. The particular form of $\mathbf{H}$ depends, as we shall see, on the character of the turning point.

With the split representation within the gradient zone we will use different forms for the interface matrices $\mathbf{G}(z_A)$, $\hat{\mathbf{G}}(z_B)$ at the top and bottom of the zone. Taking z_T as the reference level for Airy function arguments, the propagation effects within the gradient zone are represented by

$$\mathbf{F}(z_A, z_B) = \mathsf{E}(z_A)\mathbf{L}(z_A; z_T)\mathbf{H}\hat{\mathbf{L}}^{-1}(z_B; z_T)\hat{\mathsf{E}}^{-1}(z), \tag{6.47}$$

and now the principal contribution to reflection from the zone will come from the presence of the coupling matrix **H**.

The simplest situation is provided by a turning point for *SH* waves. The coupling matrix has the explicit form

$$\mathsf{h} = \mathsf{E}_\beta^{-1}(z_r)\mathbf{C}_H^{-1}(z_r)\mathbf{C}_H(z_r)\hat{\mathsf{E}}_\beta(z_r), \tag{6.48}$$

in terms of the full phase matrices (3.75), (3.81). There is continuity of material properties at z_r and so

$$\mathsf{h} = \mathsf{E}_\beta^{-1}(z_r)\hat{\mathsf{E}}_\beta(z_r) = 2^{-1/2}\begin{bmatrix} e^{i\pi/4} & e^{-i\pi/4} \\ e^{-i\pi/4} & e^{i\pi/4} \end{bmatrix}. \tag{6.49}$$

If we neglect all gradient contributions by comparison with the coupling term h (i.e. take **L** and $\hat{\mathbf{L}}$ to be unit matrices) we have the approximate propagation matrix

$$\mathbf{F}_0(z_A, z_B) = \mathsf{E}_\beta(z_A)\mathbf{h}\hat{\mathsf{E}}_\beta^{-1}(z_B). \tag{6.50}$$

The matrix h accounts for total internal reflection at the turning point. The generalized reflection and transmission elements for $\mathbf{F}_0$ are then

$$\begin{aligned}
\mathbf{r}_d^{F_0}|_{HH} &= \mathrm{Ej}_\beta(z_A)e^{-i\pi/2}[\mathrm{Fj}_\beta(z_A)]^{-1} \\
\mathbf{t}_d^{F_0}|_{HH} &= \sqrt{2}\mathrm{Aj}_\beta(z_B)e^{-i\pi/4}[\mathrm{Fj}_\beta(z_A)]^{-1}, \\
\mathbf{r}_u^{F_0}|_{HH} &= \mathrm{Aj}_\beta(z_B)e^{i\pi/2}[\mathrm{Bj}_\beta(z_B)]^{-1}, \\
\mathbf{t}_u^{F_0}|_{HH} &= \sqrt{2}\mathrm{Ej}_\beta(z_A)e^{i\pi/4}[\mathrm{Bj}_\beta(z_B)]^{-1}.
\end{aligned} \tag{6.51}$$

All these elements are well behaved numerically since the only exponentially increasing term Bj_β appears as an inverse.

When the limits of the gradient zone lie well away from the turning point we may make an asymptotic approximation to the Airy functions. We obtain the 'full-wave' approximation to the reflection (see, e.g., Budden, 1961)

$$\mathbf{r}_d^{F_0}|_{HH} \sim \exp\{2i\omega\int_{z_A}^{z_r} d\zeta\, q_\beta(\zeta) - i\pi/2\}. \tag{6.52}$$

This corresponds to complete reflection with a phase shift of $\pi/2$ compared with the phase delay for propagation down to the turning level and back. This simple result forms the basis of much further work which seeks to extend ray theory (e.g., Richards, 1973; Chapman, 1978). The approximation will be most effective at high frequencies; and, for neglect of the interaction terms, requires only slight wavespeed gradients throughout (z_A, z_B). The corresponding approximation in transmission is

$$\mathbf{t}_d^{F_0}|_{HH} \sim \frac{1}{\sqrt{2}}\exp\{i\omega\int_{z_A}^{z_r} d\zeta\, q_\beta(\zeta) - i\pi/4\}\exp\{-\omega\int_{z_r}^{z_B} d\zeta\, |q_\beta(\zeta)|\}, \tag{6.53}$$

illustrating the damping of *SH* waves below the turning level.

For the *P-SV* wave system the situation is more complicated since we now have the possibility of both *P* and *S* wave turning levels.

With just a *P* wave turning point at z_T, we would choose to build the phase matrix above z_T from the propagating forms E_α and E_β. Below z_T we would take $\hat{\mathsf{E}}_\alpha$ and E_β, with the result that the coupling matrix $\mathbf{H}$ is given by

$$\mathbf{H} = \begin{bmatrix} \mathsf{E}_\alpha^{-1} & 0 \\ 0 & \mathsf{E}_\beta^{-1} \end{bmatrix} \begin{bmatrix} \hat{\mathsf{E}}_\alpha & 0 \\ 0 & \mathsf{E}_\beta \end{bmatrix} = \begin{bmatrix} \mathsf{h} & 0 \\ 0 & \mathsf{I} \end{bmatrix}, \tag{6.54}$$

where h was introduced in (6.40). The reflection and transmission elements for the propagation matrix $\mathbf{F}$ will be determined by the character of

$$\Xi \mathbf{L}(z_A; z_T)\mathbf{H}\hat{\mathbf{L}}^{-1}(z_B; z_T)\Xi, \tag{6.55}$$

which includes total reflection of *P* waves at z_T and coupling between *P* and *S* waves away from the turning level. When the wavespeed gradients are weak the coupling matrix $\mathbf{H}$ will dominate and we will have almost independent propagation of *P* and *SV* waves. The PP reflection and transmission elements will therefore have the form (6.51) with β replaced by α. The *SS* coefficient will have the form discussed above for propagating waves.

When the turning point for *P* waves lies above z_A we have evanescent character for *P* throughout (z_A, z_B). The character of the *S* wave phase terms will now need to be modified across the *S* wave turning level at z_T. The corresponding coupling matrix is

$$\mathbf{H} = \begin{bmatrix} \hat{\mathsf{E}}_\alpha^{-1} & 0 \\ 0 & \mathsf{E}_\beta^{-1} \end{bmatrix} \begin{bmatrix} \hat{\mathsf{E}}_\alpha & 0 \\ 0 & \hat{\mathsf{E}}_\beta \end{bmatrix} = \begin{bmatrix} \mathsf{I} & 0 \\ 0 & \mathsf{h} \end{bmatrix}, \tag{6.56}$$

In these circumstances there is normally only a very small reflection or transmission contribution from the evanescent *P* waves and coupling between *P* and *SV* waves is negligible. The SS coefficients will therefore match the HH coefficients in (6.51).

When both *P* and *S* wave turning points occur within the same gradient zone we split the calculation at both turning levels, but then we have to be rather careful about the nature of the coupling terms (Kennett & Illingworth, 1981). For realistic earth models the *P* and *S* wave turning levels are widely separated and so the dominant contributions will correspond to isolated turning levels.

In this discussion we have assumed that both *P* and *S* wavespeeds increase with depth, so that s_α and s_β are positive. If either wavespeed is actually smoothly decreasing with depth we have a similar development with the roles of up and downgoing waves interchanged for that wave type.

The approach we have used in this section has been based on a high frequency approximation to the seismic wavefield in terms of Airy functions with arguments determined by the delay times τ in the model. A comparable development may be made for spherical stratification (Richards, 1976; Woodhouse, 1978) and the leading order approximation will be essentially the same since the flattening transformation (1.29) preserves delay times. To this level of approximation, it therefore makes no real difference whether we work with spherical stratification or a flattened wavespeed distribution. The structure of the correction terms in $\mathbf{L}$ will differ between the two cases, since there is additional frequency dependence in the spherical case, and there appears to be no density transformation which is optimum for both *P* and *SV* waves (Chapman, 1973).

6.3.2 Recursive construction scheme

For a model composed of smooth gradient zones interrupted by discontinuities in the elastic parameters or their gradients, we can build up the overall reflection and transmission matrices by a recursive application of the addition rules.

We suppose that the model is ultimately underlain by a uniform half space in $z > z_C$ with continuity of elastic properties at z_C (figure 6.2). We start the calculation from this level and consider the overlying gradient zone in (z_B+, z_C), extended above by a uniform half space with the properties at z_B+, just below the next higher interface. For this region we construct the downward reflection and transmission matrices $\mathbf{R}_D^{B+C}, \mathbf{T}_D^{B+C}$ by successive applications of the addition rules for the $\mathbf{r}_d, \mathbf{t}_d$ matrices introduced in the previous section. $\mathbf{R}_D^{B+C}, \mathbf{T}_D^{B+C}$ can then be recognised as the reflection and transmission matrices as seen from a uniform half space at z_B+: $\mathbf{R}_D(z_B+), \mathbf{T}_D(z_B+)$.

The effect of the interface can then be introduced, as in the uniform layer scheme, by adding in the interface matrices $\mathbf{R}_D^B$, $\mathbf{T}_D^B$ etc. for an interface between two uniform media. This yields the reflection and transmission matrices as seen at z_B- so that, for example,

$$\mathbf{R}_D(z_B-) = \mathbf{R}_D^B + \mathbf{T}_U^B \mathbf{R}_D(z_B+)[\mathbf{I} - \mathbf{R}_U^B \mathbf{R}_D(z_B+)]^{-1}\mathbf{T}_D^B. \tag{6.57}$$

The calculation can then be incremented to the top of the next gradient zone z_A by introducing the reflection and transmission effects of the region (z_A, z_B-). These may be calculated, as before, by an inner recursion over the $\mathbf{r}_d, \mathbf{t}_d$ matrices for the gradient factors. Whereas in the uniform layer case only phase delays were involved, for a gradient zone all the matrices $\mathbf{R}_D^{AB-}, \mathbf{R}_U^{AB-}, \mathbf{T}_D^{AB-}, \mathbf{T}_U^{AB-}$ are needed.

The reflection and transmission matrices for the region below z_A will then be derived from the matrices $\mathbf{R}_D(z_B-), \mathbf{T}_D(z_B-)$ as

$$\begin{aligned} \mathbf{R}_D(z_A) &= \mathbf{R}_D^{AB-} + \mathbf{T}_U^{AB-} \mathbf{R}_D(z_B-)[\mathbf{I} - \mathbf{R}_U^{AB-} \mathbf{R}_D(z_B-)]^{-1}\mathbf{T}_D^{AB-}, \\ \mathbf{T}_D(z_A) &= \mathbf{T}_D(z_B-)[\mathbf{I} - \mathbf{R}_U^{AB-} \mathbf{R}_D(z_B-)]^{-1}\mathbf{T}_D^{AB-}. \end{aligned} \tag{6.58}$$

If there is a further interface at z_A we once again add in the interface matrices via the addition rule, and then continue the calculation through the next gradient zone.

Over the main structural elements we have a two-stage recursion over interfaces and gradient zones to build the overall reflection and transmission matrices. The reflection elements for the gradients are themselves found by recursive application of the same addition rules on the *generalized* coefficients.

A useful property of this approach to calculating the reflection properties is that we can separate the effect of an interface from the structure surrounding it.

At a discontinuity in parameter gradients, the elastic parameters are continuous and so the interface reflection matrices will vanish and the transmission matrices will just be the unit matrix. In the overall reflection matrix the effect of the discontinuity will appear from the contributions of $\mathbf{D}^{-1}(z_B-)\mathbf{D}(z_B)$ and $\mathbf{D}^{-1}(z_B)\mathbf{D}(z_B+)$ to the reflection properties of the regions above and below the interface. We construct the discontinuity in gradient by superimposing the effect of the transition from the gradients on either side of the interface into a uniform medium. The procedure may be visualised by shrinking the jump in properties across z_B in figure 6.2 to zero.

An alternative development to the one we have just described may be obtained by forming the wave-propagator for the entire region, for example for figure 6.2.

$$\begin{aligned}\mathbf{Q}(z_A, z_C) = \mathbf{D}^{-1}(z_A)\mathbf{D}(z_A) \\ \mathbf{F}(z_A, z_B-)\mathbf{D}^{-1}(z_B-)\mathbf{D}(z_B+)\mathbf{F}(z_B+, z_C) \\ \mathbf{D}^{-1}(z_C)\mathbf{D}(z_C).\end{aligned} \tag{6.59}$$

The overall reflection and transmission matrices can be constructed by using the addition rule to bring in alternately the effects of interfaces and propagation using the *generalized* matrices. At z_B the generalized interface coefficients are found from

$$\Xi\mathbf{D}^{-1}(z_B-)\mathbf{D}(z_+)\Xi \tag{6.60}$$

and depend on frequency ω as well as slowness p. Such generalized coefficients have been used by Richards (1976), Cormier & Richards (1977) and Choy (1977), when allowing for gradients near the core-mantle boundary.

Computationally there is little to choose between recursive schemes based on (6.57) and (6.58), and a comparable development from (6.59). The first scheme has the merit that the intermediate results at each stage are themselves reflection and transmission matrices.

6.3.3 Gradient zone calculations

The calculation scheme for the reflection and transmission matrices which we have described is based on building a fundamental matrix from Airy functions which provide a uniform asymptotic approximation across a single turning point.

When there are two close turning points we no longer have a uniform approximation to the full response. This situation will occur in the presence of a wavespeed inversion. For a slowness p lying within the range of values corresponding to the inversion there will be three turning levels at which, e.g., $p = \beta^{-1}(z)$. The shallowest will depend on the structure outside the inversion.

As the slowness increases to the maximum in the inversion, the two deeper turning levels tend towards coalescence. For our purposes the nearness of turning points depends on the size of the Airy function arguments, and so problems are more severe at low frequencies. For perfectly elastic models the unitarity relations for reflection and transmission coefficients, discussed in the appendix to Chapter 5, provide a check on the accuracy of any approximation. Numerical studies by Kennett & Illingworth (1981) have shown that for a very narrow band of slownesses, corresponding to the minimum wavespeeds in the inversion, there can be significant error with the Airy function treatment.

Within such an inversion a uniform asymptotic treatment can be achieved by building the phase matrix **E** from parabolic cylinder function entries (Woodhouse, 1978). When an inversion is the dominant feature of the model, as in the oceanic sound speed profile, the more accurate treatment is essential (Ahluwahlia & Keller, 1977).

Even when we only need to consider a single turning point at a time we can economise on computational effort by choosing the reference level for the calculation of the phase matrices **E** to match the physical character of the solution. It is for this reason that it is a good idea to extrapolate the wavespeed distribution from a region with, say, a linear gradient to create turning points if they are not present in the region. Any deficiences in our choice for **E** will mean that further terms are needed in the interaction series.

Chapman (1981) has proposed a similar development to the one we have discussed but for reflected waves he suggests that the reference level for Airy function arguments should be taken at the reflection level. Unless this reflection level is close to a turning point the phase matrix **E** will not have the character of the actual fields.

The calculation of the reflection and transmission effects of a gradient zone are greatly simplified if the interaction terms can be neglected. With a good choice of phase matrix the contributions to **L** will be small for weak gradients in the elastic parameter. Kennett & Illingworth (1981) have shown that an upper bound on the interaction series contribution to $\mathbf{r}_{\mathrm{d}}^{\mathrm{F}}$ for a region (z_A, z_B) can be found as, e.g.,

$$\mathbf{r}_{\mathrm{d}}^{\mathrm{F}}|_{PP} \leq C\epsilon_0(z_B - z_A), \tag{6.61}$$

where C is a constant of order unity and ϵ_0 is an upper bound on the elements of **j** (3.93). The combination $\epsilon_0(z_B - z_A)$ is thus a measure of the change in the parameter gradient terms across the zone.

Although this contribution may be small, in the absence of turning points it is the sole reflection return from within a gradient zone. It can only be neglected if the dominant features of the response of the *whole* model under consideration arise from turning points and discontinuities in elastic parameters with a weaker contribution from discontinuities in parameter gradient.

In order to try to meet these conditions Kennett & Illingworth restricted their gradient zones so that the relative change in elastic parameters was no more than

10 per cent. For a large portion of a model this may mean breaking the structure into portions with separate interpolation of the wavespeed distribution in each part. At each break a weak change in parameter gradients will be introduced and the contribution from these features helps to compensate for the lack of interaction terms.

When the interaction contributions are neglected, wave types propagate independently within gradients. The only coupling of *P* and *SV* waves occurs at discontinuities in elastic parameters. Good results can be achieved with this approximation, provided that care is taken to satisfy the conditions for its validity.

Combining the numerical results of Kennett & Nolet (1979) and Kennett & Illingworth (1981), we can estimate the frequencies necessary to get only slow change in a wavelength and thus have minimal reflection return from the structure. For the relatively gently varying lower mantle the frequency should be greater than 0.02 Hz, but in the upper mantle the more rapid structural variation means that away from upper mantle discontinuities we need frequencies above 0.06 Hz.

For moderate gradient zones, and at low frequencies, an improved approximation may be made by retaining the first term in the interaction series. Richards & Frasier (1976) have made a comparable development with the WKBJ solution and shown the importance of conversion when the wavespeeds change rapidly with depth. If attention is concentrated on the reflection return at small angles of incidence (as in prospecting situations) then once again at least the first order term in $\mathbf{L}$ should be retained.

When a very sharp change in elastic parameters occurs across a gradient zone, a direct numerical solution for the elements $\mathbf{r}_{\mathrm{d}}^{\mathrm{L}}$, $\mathbf{t}_{\mathrm{d}}^{\mathrm{L}}$ would be needed. Commonly such zones are approximated by a cascade of fine uniform layers or a combination of interfaces and gradients. Calculations by Kennett & Illingworth (1981) show that for the upper mantle discontinuities several per cent error in the reflection coefficients can be produced by neglect of conversions in a model with only closely spaced changes in wavespeed gradient. With a good starting approximation the variational method of Lapwood & Hudson (1975) can be used to improve the results for this strong gradient case.

One region in which there are known to be strong positive gradients in elastic parameters with depth, is in the uppermost part of the sediments on the ocean bottom. The coupling of *P* and *S* waves by the gradients in this region has been discussed by Fryer (1981), who has simulated the gradient zone by a sequence of thin uniform layers, and investigated the consequence of neglect of all PS coupling at the minor jumps in parameters introduced by the approximation scheme. For these sediments with low shear wavespeed, the coupling process was most efficient in converting *S* waves generated at the ocean bottom back to *P* waves, and was most important below 1 Hz. For higher frequencies the partial reflections from the gradients for any individual wave type can be significant up to about 7 Hz. However these effects are most important for small offsets from the source and cause very

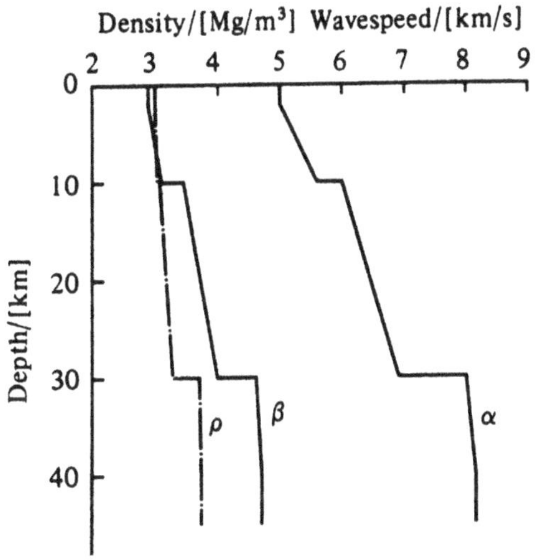

Figure 6.3. Crustal model used for reflection calculations.

little error for medium and long-range propagation (> 20 km range for a 5 km deep ocean).

6.4 The time dependence of reflections from a stratified region

As an illustration of the techniques we have discussed in this chapter we consider the reflection response of a simple stratified crustal model (figure 6.3).

For each slowness p we construct a complex reflection matrix $\mathbf{R}_D(p, \omega)$ for a suite of frequencies ω. This arrangement is convenient because we can exploit the frequency independence of the reflection and transmission coefficients at a discontinuity in the elastic parameters. We use the recursive construction schemes we have just described and build up the reflection matrix $\mathbf{R}_D$ by starting from the base of the model.

Since it is rather difficult to give an informative display of the phase behaviour of $\mathbf{R}_D(p, \omega)$, we have constructed a temporal reflection response $\check{\mathbf{R}}_D(p, t)$ by inverting the Fourier transform over frequency. The elements of $\check{\mathbf{R}}_D(p, t)$ represent the seismograms which would be obtained by illuminating the stratification from above with a plane wave of horizontal slowness p. Ideally we would have a delta function in time for the plane wave form, but our computations are, of necessity, band limited and so resolution is slightly reduced.

The model which has been used is shown in figure 6.3. It consists of a uniform layer 2 km thick underlain by a gradient zone to 10 km depth at which there is a small jump in elastic wavespeeds. A further linear gradient zone lies beneath this discontinuity and extends to the Moho at 30 km depth. There is a slight wavespeed gradient below the Moho and the model is terminated by a uniform half space. The

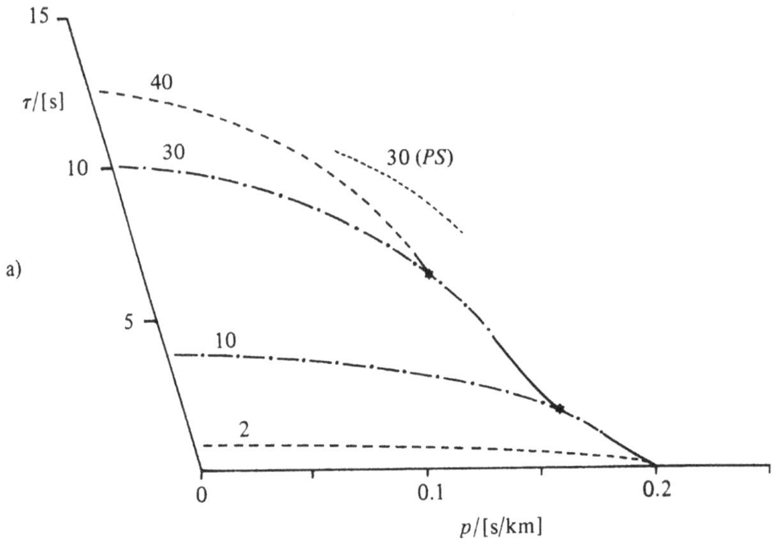

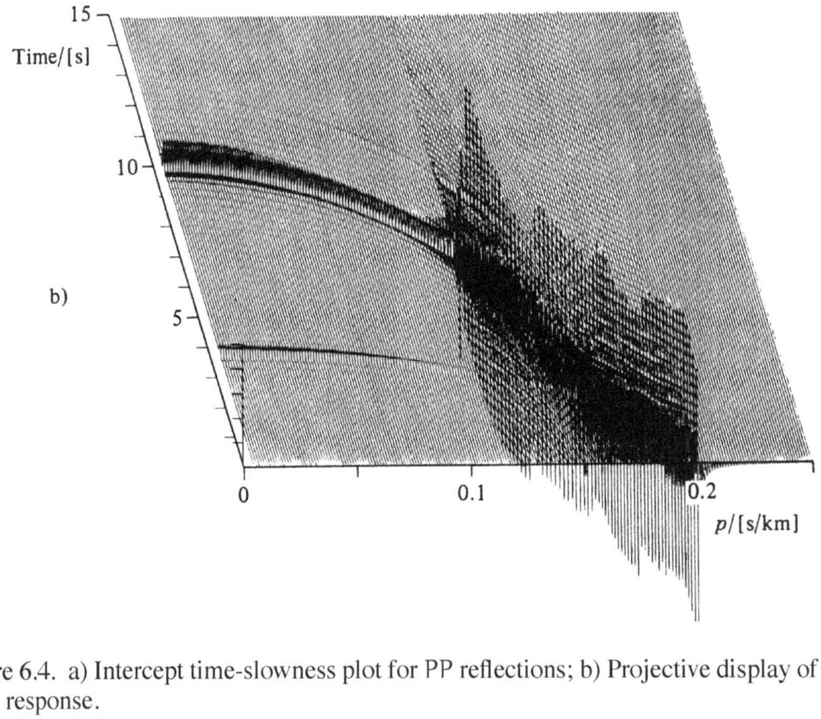

Figure 6.4. a) Intercept time-slowness plot for PP reflections; b) Projective display of temporal response.

main features of the response of the model are reflections from the discontinuities and continuous refraction through the gradient zones.

In figure 6.4a we illustrate the geometrical ray characteristics of the PP reflections for this model. We have plotted the intercept time $\tau(p)$ for various reflections and refractions against slowness. $\tau(p)$ is the integrated phase delay from the surface to the depth Z_p at which a ray is turned back, by reflection or continuous refraction. For a P ray, without any conversions,

$$\tau(p) = 2\int_0^{Z_p} dz\, q_\alpha(z). \tag{6.62}$$

The $\tau(p)$ relation in figure 6.4a is shown as a solid line for the continuous refraction in the velocity gradients. Reflections from the discontinuities at 10 km and 30 km are indicated by chain dotted lines and from the velocity gradient jumps at 2 km and 40 km by dashed lines. The depth of reflection is indicated in each case. The critical points of 0.1786 s/km for the 10 km discontinuity and 0.125 s/km for the Moho are indicated by stars.

In figure 6.4b we give a projective display of the temporal response for the PP reflection coefficient from the crustal structure, for a zero phase wavelet with a pass band from 0-4 Hz.

For slowness less than 0.122 s/km the dominant features are the precritical reflections from the discontinuities. When we compare the ray times in figure 6.4a with the amplitude display in figure 6.4b we see the rapid increase in amplitude of the Moho reflection as the critical point (0.125 s/km) is approached. The large amplitude at the critical point is associated with the existence of head waves along the Moho interfae and is reinforced by the confluence of the continuous refraction in the gradient beneath the Moho. The consequent 'interference head wave' is represented by a single negative excursion at slowness 0.123 s/km in figure 6.4b. From 0.125 s/km to 0.145 s/km we have postcritical reflection from the Moho with a progressive change of phase in the waveform and a slight change in amplitude. For the precritical reflection we have a band-passed delta function time dependence, but in the post critical range we have major positive and negative excursions in the waveform.

The onset of continuous refraction at 0.145 s/km is marked by a stabilisation of the phase until the critical point for the 10 km discontinuity at 0.164 s/km. A significant precritical reflection from this interface is seen at small slowness and once again there is a rapid increase in amplitude just before the critical point (cf. figure 5.2). Post critical reflections occupy the slowness range from 0.167 s/km to 0.1786 s/km and then we get continuous refraction in the gradient zone from 2–10 km up to 0.2 s/km.

There are no obvious reflections from the velocity gradient change at 2 km but a small amplitude internal multiple generated by reflection, from below, at this interface can be discerned at 10s for slowness 0.125 s/km. The τ values corresponding to this multiple and that generated at the 10 km interface are indicated by a dotted line in figure 6.4a.

The PP reflection behaviour is not strongly influenced by conversion but a small reflected phase converted from P to S at the Moho and then transmitted through the

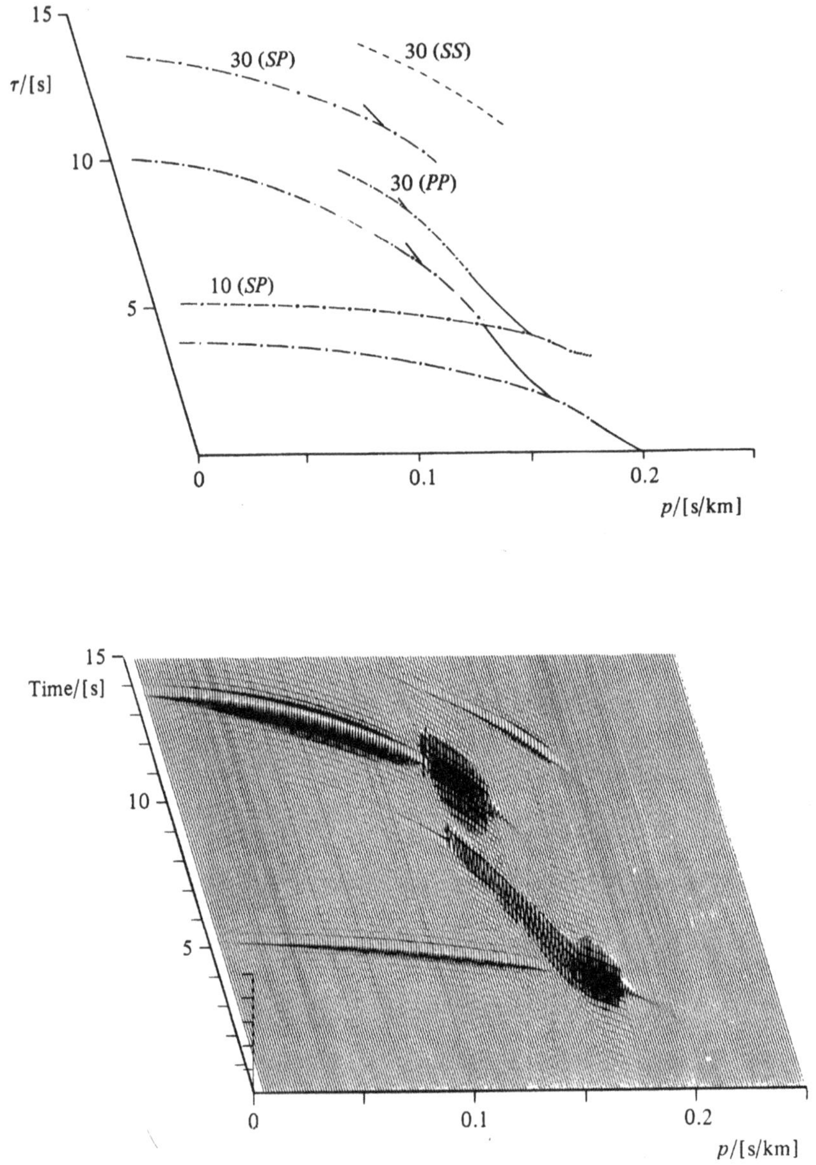

Figure 6.5. a) Intercept time-slowness plot for PS reflections; b) Projective display of temporal response.

10 km discontinuity as *P* parallel the main Moho reflection. This phase is indicated by short dashed lines in figure 6.4a.

The converted (PS) reflection for this model is illustrated in figure 6.5. The $\tau(p)$ relation for the conversions are indicated in figure 6.5a and for reference we have repeated the major PP phase behaviour. Conversion from *P* to *S* occurs at the discontinuities at 10 km and 30 km and we have indicated the depths of conversion and reflection in figure 6.5a.

A projective plot of the converted behaviour is given in figure 6.5b, and we see a number of features associated with the conversion at 10 km depth. In addition to PS reflection from the interface we see a mirror of the PP behaviour associated with waves which are transmitted downwards through the interface as *P* and then after reflection are converted to *S* on upward transmission. There is also an S wave reflection from the Moho arising from conversion at 10 km. For slowness greater than 0.179 s/km most of the P wave energy is turned back by the velocity gradient above the 10 km discontinuity, but a small amount of the *P* waves tunnel through the evanescent region to convert into propagating *S* waves at 10 km. This portion of the PS reflection is indicated by dots in figure 6.5a and since lower frequencies decay less rapidly in the evanescent region these dominant in the amplitude plot in this region.

The PS reflection at the Moho which appeared weakly in figure 6.4b is now very much stronger and we can see the change in character associated with the *P* wave critical slowness (0.125 s/km) for the sub-Moho region.

The temporal-slowness plots we have just presented enable us to relate the features of the reflection response directly to the nature of the velocity distribution. If we calculate the displacements associated with an incident *P* wave we will bring together the features we have seen on the PP and PS reflections modulated by the transformation matrix $\mathbf{m}_D$.

As we shall see in Section 7.3.2 the slowness-time plane plays an important note in the slowness method of seismogram construction. For large ranges theoretical seismograms are formed by integrating along linear trajectories through the slowness-time map.

Appendix: Mixed solid and fluid stratification

We have hitherto confined our attention to wave propagation in stratified solids but we also need to be able to allow for fluid zones within, or bounding the stratification. This is necessary for oceanic regions and for the earth's core which may be taken to behave as a fluid.

Within the fluid regions we have no shear strength and only *P* waves propagate but at solid-fluid boundaries we have the possibility of conversion to *SV* waves in the solid. No *SH* wave propagation is possible within a fluid and so solid/liquid interfaces behave like a free surface since shear stress vanishes.

For *P-SV* waves we can use reflection and transmission matrices throughout the stratification, with care as to the treatment when shear waves are absent. The simplest consistent formalism is to maintain a 2 × 2 matrix system throughout the stratification, and in fluid regions just have a single non-zero entry, e.g.,

$$\mathbf{R}_D|_{\text{fluid}} = \begin{bmatrix} R_D^{PP} & 0 \\ 0 & 0 \end{bmatrix}. \tag{6a.1}$$

If the inverse of such a reflection on transmission matrix is required it should be interpreted as a matrix with a single inverse entry, e.g.,

$$\mathbf{T}_D^{-1}|_{\text{fluid}} = \begin{bmatrix} (T_D^{PP})^{-1} & 0 \\ 0 & 0 \end{bmatrix}. \tag{6a.2}$$

Within fluid stratification these reflection and transmission matrices used in the 2×2 matrix addition rules (6.3), (6.4) will yield the correct behaviour. When a solid-fluid boundary is encountered, e.g., in the recursive schemes described in Sections 6.2 and 6.3, then the interfacial reflection and transmission coefficients used must be those appropriate to such a boundary. These forms may be obtained by a careful limiting process from the solid-solid coefficients discussed in Section 5.1, by forcing the shear wavespeed in one of the media to zero. With these coefficient the upward and downward transmission matrices for the interface will have only a single non-zero row or column and link the reflection matrices in the fluid (6.51) directly to the 2×2 forms in the solid.

Chapter 7

The Response of a Stratified Half Space

We now bring together the results we have established in previous chapters to generate the displacement field in a stratified half space ($z > 0$), due to excitation by a source at a level z_S. We suppose that we have a free surface at $z = 0$ at which the traction vanishes; and ultimately the stratification is underlain by a uniform half space in $z > z_L$ with continuity of properties at z_L. In this uniform region we impose a radiation condition that the wavefield should consist of either downward propagating waves or evanescent waves which decay with depth, the character depending on the slowness. As we shall see this lower boundary condition is not at all restrictive, and it is easy to modify the response to suit other conditions if these are more appropriate.

A formal solution for the displacement field can be found by starting with the radiation conditions and then projecting the displacement and traction to the surface using a propagator matrix. The jumps in displacement and traction across the source plane are also projected to the surface and then the displacement field is constructed so that there is no net surface traction. The physical character of this solution is seen more clearly when the response is expressed in terms of the reflection matrix for the entire stratification below the free surface. For a deep source a more convenient representation may be obtained in terms of the reflection and transmission matrices for the regions above and below the source.

7.1 The equivalent surface source representation

We consider the half space illustrated in figure 7.1 with a source at the level z_S and a uniform half space beneath z_L. At the free surface the traction must vanish and so, for any angular order m, frequency ω and slowness p, the surface stress-displacement vector must satisfy

$$\mathbf{b}(0) = [\mathbf{w}_0, 0]^T. \tag{7.1}$$

We adopt a point source representation, as discussed in Chapter 4, and so from (4.62) there will be a jump in the stress-displacement vector across the source plane

$$\mathbf{b}(z_S+) - \mathbf{b}(z_S-) = \mathbf{S}(z_S). \tag{7.2}$$

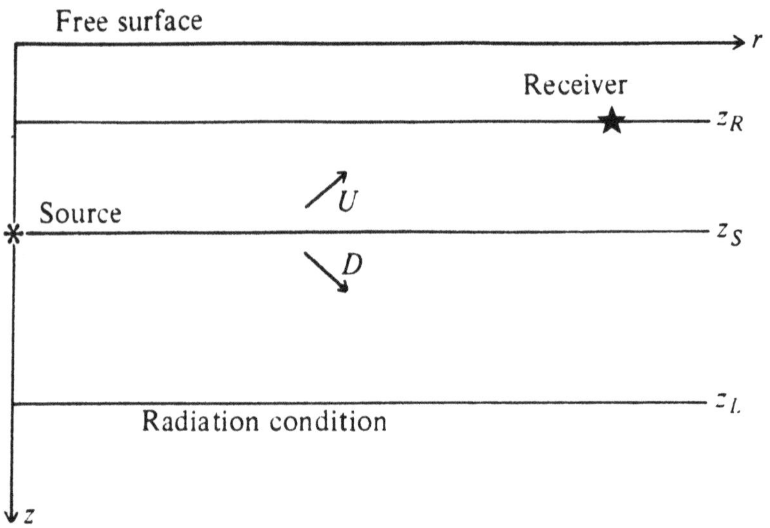

Figure 7.1. Configuration of elastic half space with a source at depth z_S and a receiver at depth z_R. Beneath z_L the medium has uniform properties so that a radiation condition has to be applied at this level. The conventions for up and downgoing waves are also indicated.

At the base of the stratification we can set up the radiation condition by making use of the eigenvector decomposition of the seismic field in $z > z_L$. Thus to exclude upgoing waves we take

$$\mathbf{b}(z_L) = \mathbf{D}(z_L)[0, \mathbf{c}_D]^T \tag{7.3}$$

in terms of a vector of downgoing wave elements $\mathbf{c}_D$ which will be subsequently specified in terms of the nature of the source and the properties of the half space. The choice of branch cut (3.8) for the radicals appearing in $\mathbf{D}(z_L)$ ensures that (7.3) has the correct character.

Starting with the form (7.3) we may now construct the stress-displacement vector $\mathbf{b}(z_S+)$ just below the source using the propagator matrix $\mathbf{P}(z_S, z_L)$,

$$\mathbf{b}(z_S+) = \mathbf{P}(z_S, z_L)\mathbf{b}(z_L). \tag{7.4}$$

With the jump in the $\mathbf{b}$ vector across z_S we find that just above the source.

$$\mathbf{b}(z_S-) = \mathbf{P}(z_S, z_L)\mathbf{b}(z_L) - \mathbf{S}(z_S). \tag{7.5}$$

We may now use the propagator $\mathbf{P}(0, z_S)$ acting on $\mathbf{b}(z_S-)$ to construct the surface displacement via

$$\begin{aligned}\mathbf{b}(0) &= \mathbf{P}(0, z_S)\{\mathbf{P}(z_S, z_L)\mathbf{b}(z_L) - \mathbf{S}(z_S)\},\\ &= \mathbf{P}(0, z_L)\mathbf{b}(z_L) - \mathbf{P}(0, z_S)\mathbf{S}(z_S),\end{aligned} \tag{7.6}$$

where we have used the propagator chain rule (2.89). The vector

$$\mathbf{S}(0) = \mathbf{P}(0, z_S)\mathbf{S}(z_S) = [\mathbf{S}_{W0}, \mathbf{S}_{T0}]^T, \tag{7.7}$$

represents an equivalent source at the surface which produces the same radiation in the half space as the original source, represented by the jump $\mathbf{S}(z_S)$, at depth.

The surface displacement may now be expressed in terms of the downgoing wavefield at z_L as

$$[\mathbf{w}_0, 0]^T = \mathbf{P}(0, z_L)\mathbf{D}(z_L)[0, \mathbf{c}_D]^T - \mathbf{S}(0). \tag{7.8}$$

Thus, whatever the depth of the source, the relation between $\mathbf{w}_0$ and $\mathbf{c}_D$ is controlled by

$$\mathbf{B}_{VL}(0) = \mathbf{P}(0, z_L)\mathbf{D}(z_L), \tag{7.9}$$

which we may recognise to be a fundamental stress-displacement matrix whose columns correspond to up and downgoing waves in the underlying half space. In terms of the partitions of $\mathbf{B}_{VL}(0)$ we write (7.8) as

$$\begin{bmatrix} \mathbf{w}_0 \\ 0 \end{bmatrix} = \begin{bmatrix} \mathbf{W}_{UL}(0) & \mathbf{T}_{UL}(0) \\ \mathbf{W}_{DL}(0) & \mathbf{T}_{DL}(0) \end{bmatrix} \begin{bmatrix} 0 \\ \mathbf{c}_D \end{bmatrix} - \begin{bmatrix} \mathbf{S}_{W0} \\ \mathbf{S}_{T0} \end{bmatrix}, \tag{7.10}$$

and from the surface conditions of vanishing traction the wavefield below z_L is specified by

$$\mathbf{c}_D = [\mathbf{T}_{DL}(0)]^{-1}\mathbf{S}_{T0}. \tag{7.11}$$

We may think of the source as equivalent to tractions $\mathbf{S}_{T0}$ which are neutralized by the reaction of the displacement field in order to satisfy the surface condition. The surface displacement

$$\mathbf{w}_0 = \mathbf{W}_{DL}(0)[\mathbf{T}_{DL}(0)]^{-1}\mathbf{S}_{T0} - \mathbf{S}_{W0}, \tag{7.12}$$

provided that the secular function for the half space, $\det\{\mathbf{T}_{DL}(0)\}$, does not vanish. The partitions $\mathbf{W}_{DL}(0)$, $\mathbf{T}_{DL}(0)$ are the displacement and traction components of the 'downgoing' vectors in the fundamental matrix $\mathbf{B}_{VL}$. The combination of partitions $\mathbf{W}_{DL}(0)[\mathbf{T}_{DL}(0)]^{-1}$ in (7.12) is analogous to those we have encountered in reflection problems (cf. 5.43). The elements of this matrix product for the *P-SV* wave case are therefore composed of ratios of 2×2 minors of the matrix $\mathbf{B}_{VL}(0)$.

The condition $\det\{\mathbf{T}_{DL}(0)\} = 0$, corresponds to the existence of a displacement field which satisfies both the free surface and the radiation conditions; this will only occur when $|p| > \beta_L^{-1}$, so that both *P* and *S* waves are evanescent in the underlying half space. In the p, ω domain these waves are associated with simple poles in the surface response (7.12). The poles with largest slowness will give rise to the surface wavetrain when the transforms are inverted (see Chapter 11).

The only other singularities in the full surface response $\mathbf{w}_0$ (7.12) are branch points at $|p| = \alpha_L^{-1}$, $|p| = \beta_L^{-1}$, where α_L, β_L are the elastic wavespeeds in the underlying half space. These branch points arise from the presence of $\mathbf{D}(z_L)$ in the fundamental matrix $\mathbf{B}_{VL}$, and the branch cuts will be specified by the choice (3.8)

$$\mathrm{Im}(\omega q_{\alpha L}) \geq 0, \quad \mathrm{Im}(\omega q_{\beta L}) \geq 0. \tag{7.13}$$

There are no branch points associated with the propagator matrices $\mathbf{P}(0, z_L)$, $\mathbf{P}(0, z_S)$. For a uniform layer $\mathbf{P}(z_{j-1}, z_j)$ is symmetric in $q_{\alpha j}$, $q_{\beta j}$ and so a continued product of propagators has no branch points. This property transfers to the continuous limit. The source vector $\mathbf{S}(z_S)$, (4.62), has no singularities.

Once we have found the surface displacement, the $\mathbf{b}$ vector at any other level may be found from

$$\begin{aligned} \mathbf{b}(z) &= \mathbf{P}(z, 0)[\mathsf{w}_0, 0]^T, && z < z_S, \\ &= \mathbf{P}(z, z_L)\mathbf{D}(z_L)[0, \mathsf{c}_D]^T, && z > z_S. \end{aligned} \tag{7.14}$$

The propagator solution enables us to get a complete formal specification of the seismic wavefield, but it is difficult to make any physical interpretation of the results.

If, however, we recall the representation of the displacement and traction matrices $\mathsf{W}_{DL}(0)$, $\mathsf{T}_{DL}(0)$ in terms of the reflection and transmission matrices $\mathbf{R}_D^{0L}$, $\mathbf{T}_D^{0L}$ (5.56), we have

$$\begin{aligned} \mathsf{W}_{DL}(0) &= (\mathsf{m}_{D0} + \mathsf{m}_{U0}\mathbf{R}_D^{0L})(\mathbf{T}_D^{0L})^{-1}, \\ \mathsf{T}_{DL}(0) &= (\mathsf{n}_{D0} + \mathsf{n}_{U0}\mathbf{R}_D^{0L})(\mathbf{T}_D^{0L})^{-1}, \end{aligned} \tag{7.15}$$

These expressions involve only the downward reflection and transmission matrices for the entire stratification beneath the free surface.

With the substitutions (7.15) the surface displacement (7.12) can be recast into the form

$$\mathsf{w}_0 = (\mathsf{m}_{D0} + \mathsf{m}_{U0}\mathbf{R}_D^{0L})(\mathsf{n}_{D0} + \mathsf{n}_{U0}\mathbf{R}_D^{0L})^{-1}\mathsf{S}_{T0} - \mathsf{S}_{W0}. \tag{7.16}$$

We have already seen that $\mathbf{R}_D^{0L}$ may be constructed without any numerical precision problems associated with growing exponential terms in evanescent regions, and so (7.16) provides a numerically stable representation of the half space response to surface sources. If the original form for the surface displacement (7.12) is used and the elements of $\mathsf{W}_{DL}(0)$, $\mathsf{T}_{DL}(0)$ are constructed via the propagator matrix, dramatic loss of precision problems occur at large slownesses for even moderate frequencies. An alternative procedure to the use of (7.16) is to set up the solution *ab initio* in terms of minors of the propagator and eigenvector matrices. Such schemes lead to very efficient computational algorithms (Woodhouse, 1981) but the physical content is suppressed.

From (7.16) the secular function for the half space is

$$\det\{\mathsf{T}_{DL}(0)\} = \det(\mathsf{n}_{D0} + \mathsf{n}_{U0}\mathbf{R}_D^{0L}) / \det \mathbf{T}_D^{0L}, \tag{7.17}$$

and this expression may be readily calculated by a single pass through the layering.

As it stands (7.16) still contains a propagator term $\mathbf{P}(0, z_S)$ in the definition of the surface source vector $\mathbf{S}(0)$. This causes very little difficulty for shallow sources, as for example in prospecting applications (Kennett, 1979a). For deep sources it is preferable to recast the entire response in terms of reflection and transmission matrices and this procedure is discussed in the following section.

At the free surface, the form of the upward reflection matrices arises from satisfying the condition of vanishing traction:

$$\mathbf{R}_F = -\mathsf{n}_{D0}^{-1}\mathsf{n}_{U0}. \tag{7.18}$$

Thus, by extracting a factor of n_{D0} from the matrix inverse in (7.16) we may generate an alternative form for the surface displacement response

$$w_0 = (\mathsf{m}_{D0} + \mathsf{m}_{U0}\mathbf{R}_D^{0L})[\mathbf{I} - \mathbf{R}_F\mathbf{R}_D^{0L}]^{-1}\mathsf{n}_{D0}^{-1}\mathsf{S}_{T0} - \mathsf{S}_{W0}. \tag{7.19}$$

The half space reverberation operator $[\mathbf{I} - \mathbf{R}_F\mathbf{R}_D^{0L}]^{-1}$ between the free surface and the stratification is now clearly displayed, and we can begin to see how the total surface displacement effects are generated.

7.2 A source at depth

We have just seen how the response of a half space can be found from an equivalent surface source and a displacement field which satisfies just the radiation condition at $z = z_L$.

An alternative scheme suggested by Kennett (1981) is to build the entire displacement field in the half space from elements which behave like up and downgoing waves at the source level - the displacement and traction partitions $\mathsf{W}_{US}, \mathsf{W}_{DS}$ and $\mathsf{T}_{US}, \mathsf{T}_{DS}$ of the fundamental matrix $\mathbf{B}_{VS}$.

Across the source plane we have discontinuities in displacement and traction (4.62)

$$\begin{aligned} \mathsf{W}(z_S+) - \mathsf{W}(z_S-) &= \mathsf{S}_W(z_S), \\ \mathsf{T}(z_S+) - \mathsf{T}(z_S-) &= \mathsf{S}_T(z_S). \end{aligned} \tag{7.20}$$

At the free surface $z = 0$ we require that there should be no traction, and at $z = z_L$ we wish to have only downgoing waves. We now construct displacement fields in $z < z_S$ and $z > z_S$ which satisfy the upper and lower boundary conditions respectively, but which have constant vector multipliers. These factors are then determined by imposing the source condition (7.20).

7.2.1 Treatment via free-surface reflection matrices

From our definition of the free-surface reflection matrix $\mathbf{R}_U^{fS}$ (5.76)-(5.78) the displacement matrix

$$\mathsf{W}_{1S}(z) = \mathsf{W}_{US}(z) + \mathsf{W}_{DS}(z)\mathbf{R}_U^{fS}, \tag{7.21}$$

has no associated traction at $z = 0$. In the region $z > z_S$, above the source, we can satisfy the free-surface boundary condition by taking a displacement field

$$\mathbf{W}(z) = \mathbf{W}_{1S}(z)\mathbf{v}_1, \quad z < z_S, \tag{7.22}$$

where $\mathbf{v}_1$ is a constant vector. For the region below the source the displacement matrix

$$\mathbf{W}_{2S}(z) = \mathbf{W}_{DS}(z) + \mathbf{W}_{US}\mathbf{R}_D^{SL} \tag{7.23}$$

will, from (5.90), satisfy the radiation condition; and so we choose a displacement field

$$\mathbf{W}(z) = \mathbf{W}_{2S}(z)\mathbf{v}_2, \quad z > z_S, \tag{7.24}$$

where $\mathbf{v}_2$ is a further constant vector. When we insert the representations (7.22) and (7.24) for the displacements above and below the source level into the source conditions (7.20), we obtain the following simultaneous equations in $\mathbf{v}_1$ and $\mathbf{v}_2$

$$\begin{aligned} \mathbf{W}_{2S}(z_S)\mathbf{v}_2 - \mathbf{W}_{1S}(z_S)\mathbf{v}_1 &= \mathbf{S}_W(z_S), \\ \mathbf{T}_{2S}(z_S)\mathbf{v}_2 - \mathbf{T}_{1S}(z_S)\mathbf{v}_1 &= \mathbf{S}_T(z_S). \end{aligned} \tag{7.25}$$

We may now eliminate variables between the displacement and traction equations by making use of the properties of matrix invariants, as in our treatment of reflection and transmission in (5.63)-(5.69). Thus we find

$$\begin{aligned} \mathbf{v}_1 &= \langle \mathbf{W}_{1S}, \mathbf{W}_{2S} \rangle^{-T}[\mathbf{W}_{2S}^T(z_S)\mathbf{S}_T(z_S) - \mathbf{T}_{2S}^T(z_S)\mathbf{S}_W(z_S)], \\ \mathbf{v}_2 &= \langle \mathbf{W}_{1S}, \mathbf{W}_{2S} \rangle^{-1}[\mathbf{W}_{1S}^T(z_S)\mathbf{S}_T(z_S) - \mathbf{T}_{1S}^T(z_S)\mathbf{S}_W(z_S)], \end{aligned} \tag{7.26}$$

since from (2.72)

$$\langle \mathbf{W}_{1S}, \mathbf{W}_{1S} \rangle = \langle \mathbf{W}_{2S}, \mathbf{W}_{2S} \rangle = 0. \tag{7.27}$$

The displacement field satisfying the source and boundary conditions is therefore: in $z < z_S$

$$\mathbf{W}(z) = \mathbf{W}_{1S}(z)\langle \mathbf{W}_{1S}, \mathbf{W}_{2S} \rangle^{-T}[\mathbf{W}_{2S}^T(z_S)\mathbf{S}_T(z_S) - \mathbf{T}_{2S}^T(z_S)\mathbf{S}_W(z_S)], \tag{7.28}$$

and for $z > z_S$

$$\mathbf{W}(z) = \mathbf{W}_{2S}(z)\langle \mathbf{W}_{1S}, \mathbf{W}_{2S} \rangle^{-1}[\mathbf{W}_{1S}^T(z_S)\mathbf{S}_T(z_S) - \mathbf{T}_{1S}^T(z_S)\mathbf{S}_W(z_S)], \tag{7.29}$$

and with our expressions for $\mathbf{W}_{1S}$, $\mathbf{W}_{2S}$ we can express (7.28) in terms of the reflection properties of the regions above and below the source.

We can calculate the invariant $\langle \mathbf{W}_{1S}, \mathbf{W}_{2S} \rangle$ most easily at the source level. From the definitions (7.21), (7.23)

$$\begin{aligned} \langle \mathbf{W}_{1S}, \mathbf{W}_{2S} \rangle = [\mathbf{R}_U^{fS}]^T \langle \mathbf{W}_{DS}, \mathbf{W}_{DS} \rangle + \langle \mathbf{W}_{US}, \mathbf{W}_{DS} \rangle \mathbf{R}_D^{SL} \\ + \langle \mathbf{W}_{US}, \mathbf{W}_{DS} \rangle + [\mathbf{R}_U^{fS}]^T \langle \mathbf{W}_{DS}, \mathbf{W}_{US} \rangle \mathbf{R}_D^{SL}, \end{aligned} \tag{7.30}$$

The terms which are linear in the reflection matrices vanish identically, and from (5.60) $<W_{US}, W_{DS}> = i\mathbf{I}$. Since the reflection matrices are symmetric we find

$$<W_{1S}, W_{2S}> = i[\mathbf{I} - \mathbf{R}_U^{fS}\mathbf{R}_D^{SL}] \tag{7.31}$$

and so the inverse invariant $<W_{1S}, W_{2S}>^{-1}$ appearing in the displacement solution (7.28) is the reverberation operator for the whole half space, including the effect of free-surface reflections.

The surface displacement can therefore be represented as

$$w_0 = -iW_{1S}(0)[\mathbf{I} - \mathbf{R}_D^{SL}\mathbf{R}_U^{fS}]^{-1}[\mathbf{W}_{2S}^T(z_S)\mathbf{S}_T(z_S) - \mathbf{T}_{2S}^T(z_S)\mathbf{S}_W(z_S)], \tag{7.32}$$

where W_{1S} is a displacement matrix satisfying the free-surface condition and W_{2S} satisfies the lower boundary condition. The expression (7.32) will prove to be very convenient when we come to discuss modal summation techniques in Chapter 11.

The contribution at the source level

$$\begin{aligned} W_{2S}^T(z_S)\mathbf{S}_T(z_S) - \mathbf{T}_{2S}^T(z_S)\mathbf{S}_W(z_S) \\ = \{\mathbf{m}_{DS}^T\mathbf{S}_T - \mathbf{n}_{DS}^T\mathbf{S}_W\} + \mathbf{R}_D^{SL}\{\mathbf{m}_{US}^T\mathbf{S}_T - \mathbf{n}_{US}^T\mathbf{S}_W\}, \end{aligned} \tag{7.33}$$

and the expressions in braces are just multiples of the quantities $\mathbf{\Sigma}_D(z_S), \mathbf{\Sigma}_U(z_S)$ introduced in Section 4.5, to describe the upward and downward radiation from a source. Thus

$$\begin{aligned} W_{2S}^T(z_S)\mathbf{S}_T(z_S) - \mathbf{T}_{2S}^T(z_S)\mathbf{S}_W(z_S) = -i[\mathbf{\Sigma}_U(z_S) + \mathbf{R}_D^{SL}\mathbf{\Sigma}_D(z_S)], \\ W_{1S}^T(z_S)\mathbf{S}_T(z_S) - \mathbf{T}_{1S}^T(z_S)\mathbf{S}_W(z_S) = -i[\mathbf{\Sigma}_D(z_S) + \mathbf{R}_U^{fS}\mathbf{\Sigma}_U(z_S)]. \end{aligned} \tag{7.34}$$

When we bring together the results (7.31), (7.34) we obtain a compact and useful representation of the displacement field at an arbitrary receiver level z_R:
for $z_R < z_S$,

$$W(z_R) = [W_{US}(z_R)+W_{DS}(z_R)\mathbf{R}_U^{fS}][\mathbf{I}-\mathbf{R}_D^{SL}\mathbf{R}_U^{fS}]^{-1}[\mathbf{\Sigma}_U(z_S)+\mathbf{R}_D^{SL}\mathbf{\Sigma}_D(z_S)], \tag{7.35}$$

and for $z_R > z_S$,

$$W(z_R) = [W_{DS}(z_R)+W_{US}(z_R)\mathbf{R}_D^{SL}][\mathbf{I}-\mathbf{R}_U^{fS}\mathbf{R}_D^{SL}]^{-1}[\mathbf{\Sigma}_D(z_S)+\mathbf{R}_U^{fS}\mathbf{\Sigma}_U(z_S)]. \tag{7.36}$$

This displacement representation separates into three contributions which we shall illustrate by considering a receiver above the source (cf. figure 7.2).

Firstly we have the source contribution

$$\mathbf{\Sigma}_U(z_S) + \mathbf{R}_D^{SL}\mathbf{\Sigma}_D(z_S), \tag{7.37}$$

which corresponds to the entire upward radiation associated with the source at the level $z = z_S$. This is produced in part by direct upward radiation $\{\mathbf{\Sigma}_U(z_S)\}$ and in part by waves which initially departed downwards, but which have been reflected back beneath the level of the source $\{\mathbf{R}_D^{SL}\mathbf{\Sigma}_D(z_S)\}$. This excitation vector, for an isotropic medium, will depend on azimuthal order m, whereas the reflection matrices are independent of m.

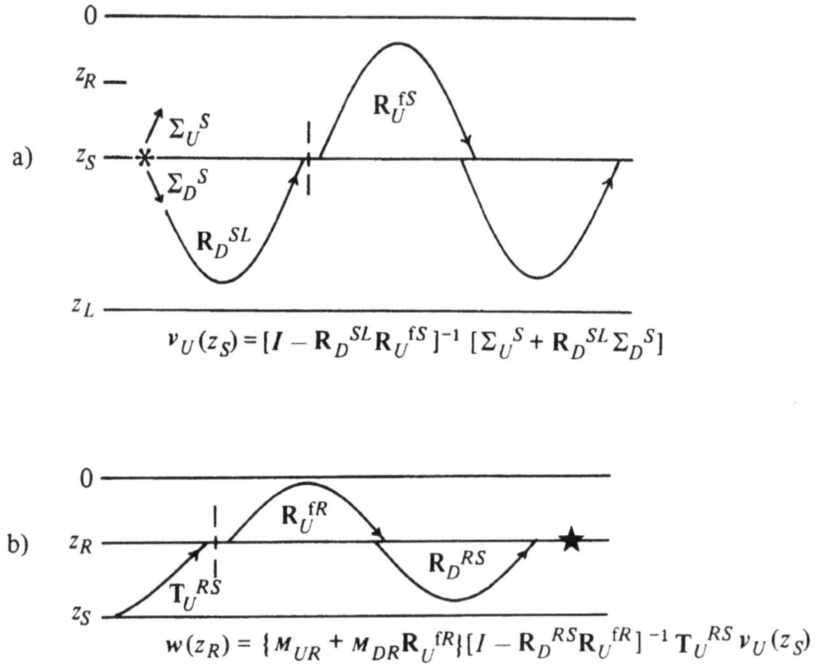

Figure 7.2. Schematic representation of the propagation elements for the response of buried source recorded at a buried receiver: a) formation of $\mathbf{v}_U(z_S)$; b) the receiver displacement $\mathbf{W}(z_R)$ in terms of $\mathbf{v}_U(z_S)$.

The second contribution arises from the inverse invariant $<\mathbf{W}_{1S}, \mathbf{W}_{2S}>^{-1}$ and is just a reverberation operator

$$[\mathbf{I} - \mathbf{R}_D^{SL}\mathbf{R}_U^{fS}]^{-1} \tag{7.38}$$

coupling the upper part of the half space, including the effect of free-surface reflections, to the lower part at the source level. The secular function for the half space is

$$\det[\mathbf{I} - \mathbf{R}_D^{SL}\mathbf{R}_U^{fS}] = 0. \tag{7.39}$$

The pole singularities in (7.35), which arise when (7.39) is satisfied, correspond to the existence of displacement fields which satisfy both upper and lower boundary conditions. For propagating waves we can make a series expansion of the reverberation operator as in (6.17)

$$[\mathbf{I} - \mathbf{R}_D^{SL}\mathbf{R}_U^{fS}]^{-1} = \mathbf{I} + \mathbf{R}_D^{SL}\mathbf{R}_U^{fS} + \mathbf{R}_D^{SL}\mathbf{R}_U^{fS}\mathbf{R}_D^{SL}\mathbf{R}_U^{fS} + \dots . \tag{7.40}$$

We may then recognise that the combined action of

$$[\mathbf{I} - \mathbf{R}_D^{SL}\mathbf{R}_U^{fS}]^{-1}[\mathbf{\Sigma}_U(z_S) + \mathbf{R}_D^{SL}\mathbf{\Sigma}_D(z_S)] = \mathbf{v}_U(z_S), \tag{7.41}$$

is to produce at the source level a sequence of upgoing wave groups corresponding to radiation from the source subjected to successively higher order multiple reverberations within the half space. This is represented schematically in figure 7.2a.

The portion of the response corresponding to the receiver location

$$W_{US}(z_R) + W_{DS}(z_R)\mathbf{R}_U^{fS} \tag{7.42}$$

needs to be recast in terms of reflection and transmission terms before its physical significance can be appreciated. We write the displacement matrices $W_{US}(z_R)$, $W_{DS}(z_R)$ in terms of up and downgoing fields at z_R, and the properties of the region 'RS' as in (6.7), and make use of the addition rule for free-surface reflections (6.10) to give

$$\begin{aligned} &W_{US}(z_R) + W_{DS}(z_R)\mathbf{R}_U^{fS} \\ &\quad = m_{UR}\mathbf{T}_U^{RS} + [m_{DR} + m_{UR}\mathbf{R}_D^{RS}]\mathbf{R}_U^{fR}[\mathbf{I} - \mathbf{R}_D^{RS}\mathbf{R}_U^{fR}]^{-1}\mathbf{T}_U^{RS}. \end{aligned} \tag{7.43}$$

This expression simplifies to the rather more compact form

$$\begin{aligned} &W_{US}(z_R) + W_{DS}(z_R)\mathbf{R}_U^{fS} \\ &\quad = \{m_{UR} + m_{DR}\mathbf{R}_U^{fR}\}[\mathbf{I} - \mathbf{R}_D^{RS}\mathbf{R}_U^{fR}]^{-1}\mathbf{T}_U^{RS}, \end{aligned} \tag{7.44}$$

which has a structure similar to (6.11).

When we construct the displacement at the receiver from (7.45) and (7.41) we have, for $z_R < z_S$

$$W(z_R) = \{m_{UR} + m_{DR}\mathbf{R}_U^{fR}\}[\mathbf{I} - \mathbf{R}_D^{RS}\mathbf{R}_U^{fR}]^{-1}\mathbf{T}_U^{RS}v_U(z_S), \tag{7.45}$$

which was first presented by Kennett & Kerry (1979) with a rather different derivation. Each of the wave groups in $v_U(z_S)$ is projected to the receiver level by the action of the transmission matrix $\mathbf{T}_U^{RS}$ (figure 7.2b). Reverberations near the receiver in the zone between the source and the surface are represented by the operator $[\mathbf{I} - \mathbf{R}_D^{RS}\mathbf{R}_U^{fR}]^{-1}$. The displacements are finally generated by adding the effect of the upgoing waves to the downgoing waves which have previously been reflected at the free surface, using the appropriate transformation matrices: $\{m_{UR} + m_{DR}\mathbf{R}_U^{fR}\}$.

Not only do we have a ready physical interpretation for the contributions to (7.35), but also they may all be represented in terms of reflection and transmission matrices which can be constructed without loss-of-precision problems. The representation (7.35) is therefore numerically stable for deep sources at high frequencies if we make use of (7.45) for the receiver term.

For a surface receiver (7.45) reduces to $\mathbf{W}_U^{fS}$ and so the surface displacement can be found from

$$w_0 = \mathbf{W}_F[\mathbf{I} - \mathbf{R}_D^{0S}\mathbf{R}_F]^{-1}\mathbf{T}_U^{0S}[\mathbf{I} - \mathbf{R}_D^{SL}\mathbf{R}_U^{fS}]^{-1}[\Sigma_U(z_S) + \mathbf{R}_D^{SL}\Sigma_D(z_S)] \tag{7.46}$$

where, as in (5.86), the surface amplification factor $\mathbf{W}_F = (\mathbf{m}_{U0} + \mathbf{m}_{D0}\mathbf{R}_F)$. This expression proves to be very convenient for the construction of the surface displacement from an arbitrary source at depth in the half space.

When the receiver lies beneath the source, the expression

$$\mathbf{v}_D(z_S) = [\mathbf{I} - \mathbf{R}_U^{fS}\mathbf{R}_D^{SL}]^{-1}[\boldsymbol{\Sigma}_D(z_S) + \mathbf{R}_U^{fS}\boldsymbol{\Sigma}_U(z_S)], \tag{7.47}$$

gives the net downward radiation from the source after reverberation through the whole half space. The receiver contribution can be evaluated in a similar way to that described above to give the displacement field for $z_R > z_S$ as

$$\mathbf{W}(z_R) = \{\mathbf{m}_{DR} + \mathbf{m}_{UR}\mathbf{R}_D^{RL}\}[\mathbf{I} - \mathbf{R}_U^{RS}\mathbf{R}_D^{RL}]^{-1}\mathbf{T}_D^{RS}\mathbf{v}_D(z_S), \tag{7.48}$$

which now allows for transmission and reverberation effects in the region below the source.

If we specialise to a surface source with a receiver just below the surface, then we can combine (7.47) and (7.48) to give

$$\mathbf{W}(0+) = (\mathbf{m}_{D0} + \mathbf{m}_{U0}\mathbf{R}_D^{0L})[\mathbf{I} - \mathbf{R}_F\mathbf{R}_D^{0L}]^{-1}[\boldsymbol{\Sigma}_D(0) + \mathbf{R}_F\boldsymbol{\Sigma}_U(0)]. \tag{7.49}$$

We recall that $\mathbf{R}_F = -\mathbf{n}_{D0}^{-1}\mathbf{n}_{U0}$, so that we may write

$$\mathbf{W}(0+) = (\mathbf{m}_{D0}+\mathbf{m}_{U0}\mathbf{R}_D^{0L})(\mathbf{n}_{D0}+\mathbf{n}_{U0}\mathbf{R}_D^{0L})^{-1}[\mathbf{n}_{D0}\boldsymbol{\Sigma}_D(0)-\mathbf{n}_{U0}\boldsymbol{\Sigma}_U(0)], \tag{7.50}$$

and from the definition of the up and downgoing source components (4.68) we can recognise $[\mathbf{n}_{D0}\boldsymbol{\Sigma}_D(0) - \mathbf{n}_{U0}\boldsymbol{\Sigma}_U(0)]$ as the traction components $\mathbf{S}_{T0}$ of the source vector $\mathbf{S}(0)$. The displacement at the surface includes the displacement components of $\mathbf{S}(0)$ and so

$$\mathbf{w}_0 = \mathbf{W}(0+) - \mathbf{S}_{W0}. \tag{7.51}$$

The surface displacement given by (7.50) and (7.51) has just the form (7.16) and so we see the equivalence of the initial value techniques used in Section 7.1 and the two-point boundary value method developed in this section.

The method we have just discussed for the construction of the displacement field is quite general and is not restricted to free-surface and downward radiation conditions (Kennett, 1981). More general cases can be constructed by replacing $\mathbf{R}_U^{fS}$ by $\mathbf{R}_1^S$, a reflection matrix appropriate to the new upper boundary condition at $z = 0$, and $\mathbf{R}_D^{SL}$ by $\mathbf{R}_2^S$ corresponding to the new lower boundary condition at $z = z_L$. The inverse invariant $\langle\mathbf{W}_{1S}, \mathbf{W}_{2S}\rangle^{-1}$ will appear in the displacement solutions, and in general

$$\langle\mathbf{W}_{1S}, \mathbf{W}_{2S}\rangle^{-1} = -i[\mathbf{I} - \mathbf{R}_1^S\mathbf{R}_2^S]^{-1}. \tag{7.52}$$

Thus the secular function for a particular problem can be represented as $\det[\mathbf{I} - \mathbf{R}_1^S\mathbf{R}_2^S]$ in terms of the appropriate reflection matrices $\mathbf{R}_1^S, \mathbf{R}_2^S$. With different source depths we obtain secular functions which differ only by a factor and have the same zeroes. The vanishing of the secular determinant represents a constructive interference condition for waves successively reflected above and below the level z_S.

For wave propagation in spherical stratification we can use the foregoing results to find the displacement field $\mathbf{W}(l, m, R, \omega)$. In this case we would choose the displacement matrix $\mathbf{W}_{1S}$ to satisfy the free-surface boundary condition at $R = r_e$, and $\mathbf{W}_{2S}$ to satisfy a regularity condition at the origin. The expressions (7.28) for the displacement field can be used directly, once we recall that z and R increase in opposite directions. We may also use the representation for the surface displacement in terms of reflection matrices (7.46) for the spherical case, provided we interpret $\mathbf{R}_D^{SL}$ as being the reflection matrix for the region $R < R_S$ with a regularity condition at the origin (cf. Section 5.2.4). The physical interpretation of the results is otherwise as for horizontal stratification and so approximations to the full response will run in parallel for the two cases.

7.2.2 Explicit representation of free-surface reflections

In the preceding treatment we chose to work in terms of displacement matrices which satisfied the upper and lower boundary conditions. If, however, we construct

$$\mathbf{W}_{3S}(z) = \mathbf{W}_{US}(z) + \mathbf{W}_{DS}(z)\mathbf{R}_U^{0S}, \tag{7.53}$$

this corresponds to a radiation boundary condition at $z = 0$. In order to satisfy the actual free-surface condition we choose a linear combination of the displacement matrices $\mathbf{W}_{3S}, \mathbf{W}_{2S}$ in $z < z_S$

$$\mathbf{W}(z) = \mathbf{W}_{3S}(z)\mathbf{u}_3 + \mathbf{W}_{2S}(z)\mathbf{u}_1, \tag{7.54}$$

and determine the relation between $\mathbf{u}_1$ and $\mathbf{u}_3$ by requiring vanishing traction at $z = 0$

$$\mathsf{T}_{3S}(0)\mathbf{u}_3 + \mathsf{T}_{2S}(0)\mathbf{u}_1 = 0. \tag{7.55}$$

Below the source we take

$$\mathbf{W}(z) = \mathbf{W}_{2S}(z)\mathbf{u}_2, \tag{7.56}$$

which satisfies the lower boundary condition. The source condition (7.20) now leads to simultaneous equations in $(\mathbf{u}_2 - \mathbf{u}_1)$ and $\mathbf{u}_3$:

$$\begin{aligned} \mathbf{W}_{2S}(z_S)(\mathbf{u}_2 - \mathbf{u}_1) - \mathbf{W}_{3S}(z_S)\mathbf{u}_3 &= \mathbf{S}_W(z_S), \\ \mathsf{T}_{2S}(z_S)(\mathbf{u}_2 - \mathbf{u}_1) - \mathsf{T}_{3S}(z_S)\mathbf{u}_3 &= \mathbf{S}_T(z_S). \end{aligned} \tag{7.57}$$

Since we will wish to concentrate on surface displacement we solve for $\mathbf{u}_3$,

$$\mathbf{u}_3 = \langle \mathbf{W}_{3S}, \mathbf{W}_{2S} \rangle^{-T} [\mathbf{W}_{2S}^T(z_S)\mathbf{S}_T(z_S) - \mathsf{T}_{2S}^T(z_S)\mathbf{S}_W(z_S)]. \tag{7.58}$$

The inverse invariant may be found by analogy with (7.31) and we have previously evaluated the source term in (7.34), so that

$$\mathbf{u}_3 = [\mathbf{I} - \mathbf{R}_D^{SL}\mathbf{R}_D^{0S}]^{-1}[\mathbf{\Sigma}_U(z_S) + \mathbf{R}_D^{SL}\mathbf{\Sigma}_D(z_S)]. \tag{7.59}$$

The displacement in the zone above the source is then given by

$$W(z) = \{W_{3S}(z) - W_{2S}(z)T_{2S}^{-1}(0)T_{3S}(0)\}u_3, \tag{7.60}$$

which has a particularly simple form at $z = 0$. At the surface by (5.57),

$$W_{3S}(0) = m_{U0}\mathbf{T}_U^{0S}, \quad T_{3S}(0) = n_{U0}\mathbf{T}_U^{0S}. \tag{7.61}$$

The displacement and traction matrices satisfying the lower boundary condition have a more complex form

$$\begin{aligned} W_{2S}(0) &= (m_{D0} + m_{U0}\mathbf{R}_D^{0L})(\mathbf{T}_D^{0S})^{-1}(\mathbf{I} - \mathbf{R}_U^{0S}\mathbf{R}_D^{SL}) \\ T_{2S}(0) &= (n_{D0} + n_{U0}\mathbf{R}_D^{0L})(\mathbf{T}_D^{0S})^{-1}(\mathbf{I} - \mathbf{R}_U^{0S}\mathbf{R}_D^{SL}) \end{aligned} \tag{7.62}$$

but $W_{2S}(0)T_{2S}^{-1}(0)$ is somewhat simpler

$$W_{2S}(0)T_{2S}^{-1}(0) = (m_{D0} + m_{U0}\mathbf{R}_D^{0L})(n_{D0} + n_{U0}\mathbf{R}_D^{0L})^{-1}. \tag{7.63}$$

This combination of terms has already appeared in our surface source representation (7.16).

The surface displacement now takes the form

$$w_0 = \{m_{U0} - (m_{D0} + m_{U0}\mathbf{R}_D^{0L})(n_{D0} + n_{U0}\mathbf{R}_D^{0L})^{-1}n_{U0}\}\sigma(z_S), \tag{7.64}$$

where we have introduced the upgoing wavevector

$$\sigma(z_S) = \mathbf{T}_U^{0S}[\mathbf{I} - \mathbf{R}_D^{SL}\mathbf{R}_U^{0S}]^{-1}[\Sigma_U(z_S) + \mathbf{R}_D^{SL}\Sigma_D(z_S)]. \tag{7.65}$$

The vector $\boldsymbol{\sigma}$ includes all interactions of the source with the stratified structure above and below the source, but unlike (7.41) does not allow for reflections generated at the free surface. In (7.64) such reflections are contained within the term in braces. This dependence may be emphasised by writing (7.64) in terms of the free-surface reflection matrix $\mathbf{R}_F$ and rearranging to give

$$w_0 = (m_{U0} + m_{D0}\mathbf{R}_F)[\mathbf{I} - \mathbf{R}_D^{0L}\mathbf{R}_F]^{-1}\sigma(z_S), \tag{7.66}$$

so that the reverberation operator for the half space is clearly displayed. When the source, is just at the surface $\sigma(z_S)$ takes a particularly simple form, since $\mathbf{T}_U^{0S}$ becomes the unit matrix and $\mathbf{R}_D^{0S}$ vanishes, so that

$$\sigma(0+) = [\Sigma_U(0+) + \mathbf{R}_D^{0L}\Sigma_D(0+)]. \tag{7.67}$$

7.3 Recovery of the response in space and time

We have just seen how we can construct the response of a stratified half space to excitation by a buried or surface source in the transform domain, as a function of frequency ω and slowness p. To get actual seismograms we must still invert the transforms. For the *P-SV* wave part of the seismograms we have the vector harmonic expansion (2.55),

$$\mathbf{u}_P(r,\phi,0,t) = \frac{1}{2\pi}\int_{-\infty}^{\infty} d\omega\, e^{-i\omega t}\int_0^{\infty} dk\, k\sum_m [U\mathbf{R}_k^m + V\mathbf{S}_k^m]. \tag{7.68}$$

We may recast this integral in terms of slowness p and the surface displacement vector w_0, by introducing the tensor field

$$\mathsf{T}_m(\omega pr) = [\mathbf{R}_k^m, \mathbf{S}_k^m]^T, \tag{7.69}$$

so that (7.68) may be expressed as

$$\mathbf{u}_P(r,\phi,0,t) = \frac{1}{2\pi}\int_{-\infty}^{\infty} d\omega e^{-i\omega t}\omega^2\int_0^{\infty} dp\, p\sum_m \mathbf{w}_0^T(p,m,\omega)\mathsf{T}_m(\omega pr). \tag{7.70}$$

We have a similar form for the *SH* response $\mathbf{u}_H$:

$$\mathbf{u}_H(r,\phi,0,t) = \frac{1}{2\pi}\int_{-\infty}^{\infty} d\omega\, e^{-i\omega t}\omega^2\int_0^{\infty} dp\, p\sum_m W_0^T(p,m,\omega)\mathbf{T}_m(\omega pr). \tag{7.71}$$

in terms of the harmonic $\mathbf{T}_k^m$ which we have rewritten in a form designed to display the dependence on frequency and slowness. The vector harmonics $\mathbf{R}_k^m, \mathbf{S}_k^m$ and $\mathbf{T}_k^m$ (2.56) can be cast entirely in terms of Bessel function entries by using the derivative property

$$J'_m(x) = J_{m-1}(x) - mJ_m(x)/x. \tag{7.72}$$

In terms of the orthogonal coordinate vectors $\mathbf{e}_z, \mathbf{e}_r, \mathbf{e}_\phi$ we have explicit forms for the harmonics.

$$\begin{aligned}
\mathbf{R}_m(\omega pr) &= \mathbf{e}_z J_m(\omega pr)e^{im\phi},\\
\mathbf{S}_m(\omega pr) &= [\mathbf{e}_r J_{m-1}(\omega pr) - (\mathbf{e}_r - i\mathbf{e}_\phi)mJ_m(\omega pr)/\omega pr]e^{im\phi},\\
\mathbf{T}_m(\omega pr) &= [-\mathbf{e}_\phi J_{m-1}(\omega pr) + (\mathbf{e}_\phi + i\mathbf{e}_r)mJ_m(\omega pr)/\omega pr]e^{im\phi}.
\end{aligned} \tag{7.73}$$

These forms are most convenient for $m \geq 0$, but the values for $m < 0$ are easily obtained from

$$J_{-m}(x) = (-1)^m J_m(x). \tag{7.74}$$

It is only on the horizontal components, for $|m| > 0$ that we get 'near-field' components depending on $mJ_m(\omega pr)/\omega pr$ which decay more rapidly than the contributions oriented along the coordinate vectors. These 'near-field' terms couple the radial and tangential components of motion so that there is no clear separation by component of *SV* and *SH* motion at small distances from the source.

When we represent an actual source by an equivalent point source consisting of force and dipole components, the azimuthal summation is restricted to angular orders $|m| < 2$. This sum presents no significant complication once the integrals over frequency and slowness have been performed for each m.

For the double integral (7.70) we have to choose the order in which the frequency and slowness integrals are undertaken. If the slowness integral is calculated first then the intermediate result is the complex frequency spectrum $\bar{\mathbf{u}}(r, \phi, 0, \omega)$ at a particular location. This approach may therefore be designated the *spectral method* and has been used in most attempts to calculate theoretical seismograms by numerical integration of the complete medium response (Kind, 1978; Kennett, 1980; Wang & Herrmann, 1980). When, alternatively, the frequency integral is evaluated first the intermediate result is a time response for each slowness p, corresponding to the illumination of the medium by a single slowness component. The final result is obtained by an integral over slowness and we follow Chapman (1978) by calling this approach the *slowness method.* Although this integral scheme can be used for the full response, most applications to date have been for approximate methods where the response is split up into generalized rays, e.g., Cagniard's method (Helmberger, 1968; Wiggins & Helmberger, 1974; Vered & BenMenahem, 1974), and a method due to Chapman suitable for smoothly stratified media (Dey-Sarkar & Chapman, 1978).

7.3.1 The spectral method

We construct the spectrum of the mth azimuthal contribution to a seismogram as a slowness integral,

$$\bar{\mathbf{u}}(r, m, 0, \omega) = \omega^2 \int_0^\infty dp\, p \boldsymbol{w}_0^T(p, m, \omega) \mathbf{T}_m(\omega p r). \tag{7.75}$$

The transform vector $\boldsymbol{w}_0(p, m, \omega)$ depends on the azimuthal order m through the source jump term $\mathbf{S}(z_S)$ (4.62), and a different slowness dependence is introduced for each order. The result is that $\boldsymbol{w}_0(p, m, \omega)$ is an odd function of p if m is even, and an even function of p if m is odd. We now recall that the elements of $\mathbf{T}_m(\omega p r)$ depend on $J_m(\omega p r) e^{im\phi}$ and so we may make a decomposition of this 'standing wave' form into a travelling wave representation in terms of the Hankel functions $H_m^{(1)}(\omega p r)$, $H_m^{(2)}(\omega p r)$. We write e.g. $\mathbf{T}_m^{(1)}(\omega p r)$ for the harmonics corresponding to outgoing waves from the origin. Now

$$\begin{aligned} J_m(\omega p r) &= \tfrac{1}{2}[H_m^{(1)}(\omega p r) + H_m^{(2)}(\omega p r)] \\ &= \tfrac{1}{2}[H_m^{(1)}(\omega p r) - e^{im\pi} H_m^{(1)}(-\omega p r)] \end{aligned} \tag{7.76}$$

and so when we use the symmetry properties of $\boldsymbol{w}_0$, we can express (7.75) as an integral along the entire slowness axis

$$\bar{\mathbf{u}}(r, m, 0, \omega) = \tfrac{1}{2}\omega|\omega| \int_{-\infty}^\infty dp\, p \boldsymbol{w}_0^T(p, m, \omega) \mathbf{T}_m^{(1)}(\omega p r), \tag{7.77}$$

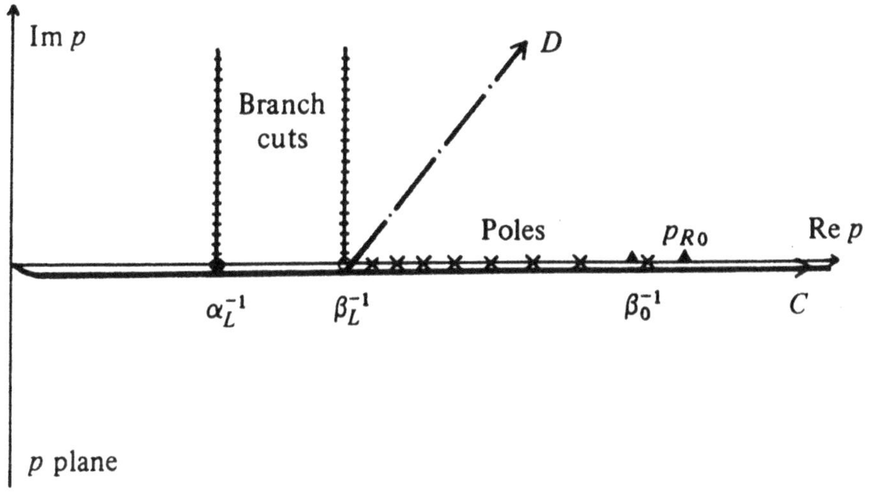

Figure 7.3. The singularities, branch cuts and integration contours for the full response of a half space.

where the contour of integration in the p plane is taken above the branch point for $H_m^{(1)}(\omega pr)$ at the origin. This form shows explicitly that we are only interested in waves which diverge from the source.

For large values of the argument, $H_m^{(1)}(\omega pr)$ may be replaced by its asymptotic form

$$H_m^{(1)}(\omega pr) \sim (2/\pi\omega pr)^{1/2} e^{\{i\omega pr - i(2m+1)\pi/4\}}, \tag{7.78}$$

and to the same approximation the vector harmonics take on the character of fields directed along the orthogonal coordinate vectors $\mathbf{e}_z, \mathbf{e}_r, \mathbf{e}_\phi$. Thus the tensor field $\mathbf{T}_m^{(1)}(\omega pr)$ is approximated by

$$\mathbf{T}_m^{(1)} \sim [\mathbf{e}_z, i\mathbf{e}_r]^T (2/\pi\omega pr)^{1/2} e^{\{i\omega pr - i(2m+1)\pi/4\}}, \tag{7.79}$$

and the tangential (*SH*) harmonic

$$\mathbf{T}_m^{(1)}(\omega pr) \sim -i\mathbf{e}_\phi (2/\pi\omega pr)^{1/2} e^{\{i\omega pr - i(2m+1)\pi/4\}}. \tag{7.80}$$

In this asymptotic limit we are faced with the same slowness and distance dependence in the integrand of (7.77) for all three components of displacement.

The symmetry properties we have described will be shared by approximations to the complete response w_0 and so we will always have the possibility of using a standing wave expression (7.75), or a travelling wave representation (7.77).

In the slowness plane, for the full surface response w_0 we have branch points at $p = \pm\alpha_L^{-1}$ for the *P-SV* case and $p = \pm\beta_L^{-1}$ in all cases. For both *P-SV* and *SH* wave contributions we have a sequence of poles in the region $\beta_L^{-1} < p < \beta_{min}^{-1}$, where β_{min} is the smallest shear wavespeed anywhere in the half space - this is normally attained at the surface. Over this slowness interval for *SH* waves, we have higher mode Love wave poles, which for a perfectly elastic medium lie on the real p axis. For *P-SV* waves we have higher mode Rayleigh poles whose locations are close to, but not identical to, the Love poles. In addition we have in $|p| > \beta_{min}^{-1}$ the fundamental Rayleigh mode which couples the evanescent *P* and *SV* waves in the half space; the limit point for this mode is p_{R0}, the Rayleigh waveslowness on a uniform half space with elastic properties at the surface. The distribution of poles depends strongly on frequency; at low frequency only a few poles occur in $|p| > \beta_L^{-1}$ whilst at high frequencies there are a great many (see Section 11.3).

The set of singularities for the *P-SV* wave case is sketched in figure 7.3, the contour of integration for $\omega > 0$ runs just below the singularities for $p > 0$ and just above for $p < 0$. This contour may be justified by allowing for slight attenuation of seismic waves within the half space (which we may well want on physical grounds) in which case the poles move into the first and third quadrants of the complex p plane. The line of the branch cuts from α_L^{-1}, β_L^{-1} is not critical provided that the conditions

$$\text{Im}(\omega q_{\alpha L}) \geq 0, \quad \text{Im}(\omega q_{\beta L}) \geq 0, \tag{7.81}$$

are maintained on the real p axis. We therefore follow Lamb (1904) by taking cuts parallel to the imaginary p axis.

The most direct approach to the evaluation of (7.75) or (7.77) is to perform a direct numerical integration along the real p axis, but for a perfectly elastic medium the presence of the poles on the contour is a major obstacle to such an approach.

However, if we deform the contour of integration in (7.77) into the upper half plane to D, we can pick up the polar residue contributions from all the poles to the right of β_L^{-1}. Convergence at infinity is ensured by the properties of $H_m^{(1)}(\omega pr)$. With this deformation the displacement spectrum is given as a sum of a contour integral and a residue series e.g.

$$\begin{aligned}\bar{\mathbf{u}}_P(r, m, 0, \omega) = \tfrac{1}{2}\omega|\omega|\{\int_D dp\, p\mathbf{w}_0^T(p, m, \omega)\mathbf{T}_m^{(1)}(\omega pr)\} \\ +\pi i\omega^2 \sum_{j=0}^{N(\omega)} p_j \text{Res}_j[\mathbf{w}_0^T\mathbf{T}_m^{(1)}].\end{aligned} \tag{7.82}$$

At even moderate frequencies the number $N(\omega)$ of modal residue contributions becomes very large indeed (cf., figure 11.3), and locating all the poles is a major computational problem. The poles with largest slowness give the major contribution to what would generally be regarded as the surface wavetrain, with relatively low group velocities. The summation of modes with smaller slownesses just synthesises *S* body wave phases by modal interference. With only a residue sum taken over a portion of the real p axis, good results can be obtained for the *S* wave coda (Kerry, 1981), and by forcing α_L, β_L to be very large even *P* waves can be synthesised (Harvey, 1981). We will consider such modal summation methods in more detail in Chapter 11.

The contour integration in (7.82) consists of a real axis slowness integral from $-\infty$ up to β_L^{-1}, and then a line segment off into the first quadrant where $H_m^{(1)}(\omega p r)$ is a decaying function of complex p. For large ranges r the contribution from negative slownesses is very small and has often been neglected.

This approach has been used by Wang & Herrmann (1980) who have deformed the contour D further to lie along the two sides of the branch cuts, taken along the imaginary p axis and along the real p axis to β_L^{-1}. They have used different numerical integration schemes along the real and imaginary axes.

An alternative to the contour deformation procedure we have just described is to arrange to move the poles in the response off the contour of integration. For an attenuative medium the poles will lie in the first and third quadrants away from the real p axis although their influence is strongly felt on the contour of integration. In most applications we anticipate that at least some loss will occur in seismic wave propagation and so introducing small loss factors is very reasonable.

Kind (1978) constructed the full response of an attenuative medium to excitation by a vertical point force, but has taken the asymptotic form of the Hankel function in (7.77) and thus excluded near-field effects; he also restricted his numerical integration over p to a band of positive slowness covering the main body and surface wave phases of interest.

At the origin the $J_m(\omega p r)$ Bessel functions remain well behaved and so when an attempt is made to calculate the complete response of the half space there are advantages in using the standing wave expression (7.75). If a fairly broad band of frequencies is required, for the shortest ranges and lowest frequencies $\omega p r$ can be quite small and so it is desirable to use a high accuracy approximation to $J_m(x)$ over the whole range of arguments, e.g., via Chebyshev polynomials as in Kennett (1980). For a broad-band signal the effects of velocity dispersion due to attenuation (1.19), (1.25) can become significant and should strictly be included when the response of the medium is calculated. When the loss factors are small (Q_α^{-1}, $Q_\beta^{-1} < 0.03$) and propagation distances are less than 500 km, the pulse distortion associated with neglect of dispersion is very slight; this is in agreement with the results of O'Neill & Hill (1979) who have shown that significant change in pulse form can occur for propagation of 600 km through a region with $Q_\alpha^{-1} > 0.01$.

With our reflection matrix representation (7.46), the construction of the theoretical seismograms for a general point source at $z = z_S$ proceeds in three stages. For a surface receiver we construct the matrix operator

$$\mathbf{Z}(p, \omega) = \mathbf{W}_U^{fS}[\mathbf{I} - \mathbf{R}_D^{SL}\mathbf{R}_U^{fS}]^{-1} \tag{7.83}$$

for the *P-SV* and *SH* parts of the response. $\mathbf{Z}(p, \omega)$ is independent of angular order m and source type and so needs to be formed only once for any source depth. The reflection matrix $\mathbf{R}_D^{SL}(p, \omega)$ for a stack of uniform layers or a piecewise smooth structure, can be constructed recursively by working up from z_L to the source level as discussed in Chapter 6. In a similar way the displacement matrix $\mathbf{W}_U^{fS}$ and free-surface reflection matrix $\mathbf{R}_U^{fS}$ can be calculated by working down from the free surface to the level z_S.

The calculation is simplified if we are able to assume that all elements in the source moment tensor M_{ij} have the same time dependence $M(t)$. We may then extract the corresponding spectrum $\bar{M}(\omega)$ from the source terms; at fixed slowness p, the source radiation terms $\mathbf{\Sigma}_U$ and $\mathbf{\Sigma}_D$ are then independent of frequency but depend on azimuthal order m through the character of the source. For each angular order m we may now construct the transform vector $\boldsymbol{w}_0$ at slowness p and frequency ω as

$$\boldsymbol{w}_0(p,m,\omega) = \mathbf{Z}(p,\omega)[\mathbf{R}_D^{SL}(p,\omega)\mathbf{\Sigma}_D(p,m) + \mathbf{\Sigma}_U(p,m)]\bar{M}(\omega). \qquad (7.84)$$

The change of variables from horizontal wavenumber k to slowness p gives an effective source spectrum of $\omega^2\bar{M}(\omega)$. We recall that the far-field displacement in an unbounded medium is controlled by the derivative of the moment time function and so it is often advantageous to specify the moment rate spectrum $i\omega\bar{M}(\omega)$. For excitation by a point force with time function $\mathcal{E}(t)$, we replace the moment spectrum in (7.84) by $(i\omega)^{-1}\bar{\mathcal{E}}(\omega)$.

For each azimuthal component we now have to perform a numerical integration over p to produce the spectrum of the three components of displacement at a range r. For a source specified by a general moment tensor we need five azimuthal orders and so we have 15 slowness integrations for each range when near-field terms are included by using the original forms of the vector harmonics (2.56). A final summation over the angular terms gives the three component seismograms for a given azimuth from the source. If the asymptotic forms of the harmonics (7.79) can be used we need perform only one integration for each displacement. Unfortunately, the circumstances in which it is appropriate to attempt to calculate complete synthetic seismograms, i.e. at moderate ranges so that the time separation between the fastest body waves and slowest surface waves is not too large, are just those in which the asymptotic approximation is barely adequate.

The slowness integrals in (7.75) have an infinite upper limit, but truncation is required for numerical integration. For frequencies around 2 Hz, Kennett (1980) integrated from the origin to a slowness $(0.85\,\beta_0)^{-1}$, for surface shear wavespeed β_0. This slowness is well beyond p_{R0} the high frequency asymptotic for the fundamental Rayleigh mode. At lower frequencies it is advantageous to extend the calculation to somewhat larger slownesses.

The integrand in (7.75) has a relatively unpleasant character. The vector harmonics are oscillatory through the presence of $J_m(\omega pr)$: $\boldsymbol{w}_0(p,m,\omega)$ is by no means smooth, particularly where the integration path passes over the shoulders of the poles displaced from the real axis by the inclusion of attenuation. The simplest approach to the numerical integration is to divide the slowness integral into sections and to use a trapezium rule in each section with panel spacing chosen to suit the character of the integrand; e.g., a finer sampling would be used for $p > \beta_L^{-1}$. An alternative is to modify Filon's (1928) method to Bessel function integrands and attempt a polynomial fit to $w_0(p, m, \omega)$ over each integration panel. The integral is then evaluated as a sum of contributions of the form $\int dx\, x^p J_m(kx)$ over the panels.

For large ωpr rather fine sampling in slowness is needed to give a good representation of the integrand. Without excessive computation, there is therefore an effective upper limit in frequency at given range, and a maximum range with a given frequency band when we seek to maintain a given accuracy.

At very low frequencies the fundamental Rayleigh mode is barely affected by the loss factors of the stratification and so very fine spacing in slowness is needed to cope with a near pole on the integration path. It is probably worthwhile to modify the contour of integration to pick up the fundamental Rayleigh mode pole explicitly. If we use (7.75) this would require two additional line segments starting to the left of the pole: one into the upper half p plane with $H_m^{(1)}(\omega pr)$ dependence and the other into the lower half plane depending on $H_m^{(2)}(\omega pr)$.

Cormier (1980) has used an equivalent representation to (7.77) with a piecewise smooth model. By deforming the contour of integration he has isolated the residue contribution from the fundamental Rayleigh mode and also taken a path into the upper half p plane to exclude very small slownesses and so avoid the singularity at the origin.

When the standing wave expression (7.75) is used we know that the entire response can be represented in terms of outgoing functions, the size of any apparently incoming waves provides a very useful check on the accuracy of any numerical integration.

It is desirable to construct the *P-SV* and *SH* wave parts of the seismogram at the same time, because the near-field contributions to the horizontal component seismograms can then be correctly calculated. Wang & Herrman (1980) have shown that neglect of the near-field terms from either *P-SV* or *SH* waves gives non-causal, non-propagating arrivals which cancel when both contributions are included. For long-period waves the near-field contributions can have a significant effect on the calculated waveform at moderate ranges. For example, with a simple crustal model, the surface wavetrains from a full calculation and one including only far-field terms, show visible differences out to 80 km range.

After the slowness integration and azimuthal summation we are left with a spectrum of the seismogram at a receiver location. The final integration to the time domain is commonly performed by using the Fast Fourier transform (Cooley & Tukey 1965) over a set of discrete frequencies. The finite bandwidth of practical recording equipment sets an upper limit on frequency, and this has to be taken below the Nyquist frequency for the transform. The time series obtained after transformation is of fixed length and is cyclic in nature. There is therefore the possibility of time 'aliasing', energy which would arrive after the end of the allotted time interval is wrapped back over the early part of the seismogram.

For complete seismograms we have a long duration of signal out to the end of the surface wavetrain; to generate such a long time interval we need very fine frequency spacing. In order to follow arrivals at varying ranges with a fixed time interval it is convenient to calculate $\mathbf{u}(r, t - p_{red}r)$ by multiplying the spectrum at r by $\exp(-i\omega p_{red}r)$ (Fuchs & Müller 1971). The reduction slowness p_{red} is chosen to confine the arrivals most conveniently. Thus, if at each frequency the slowness integral is split into a number of parts, a different reduction slowness may be used to compute a time series for each part. The final seismograms may then be obtained by superposition of the time series for the sections with appropriate time delays (Kennett 1980).

7.3.2 The slowness method

We now consider carrying out the integration over frequency for a particular combination of slowness p and range r. For the mth azimuthal component

$$\mathbf{u}_P(p, r, m, t) = \frac{1}{2\pi}\int_{-\infty}^{\infty} d\omega\, e^{-i\omega t}\omega^2 \mathbf{w}_0^T(p, m, \omega)\mathbf{T}_m(\omega pr) \tag{7.85}$$

and this transform of a product can be expressed as a convolution

$$\mathbf{u}_P(p, r, m, t) = -\partial_{tt}\{\check{\mathbf{W}}_0(p, m, t) * (1/pr)\check{\mathbf{T}}_m(t/pr)\} \tag{7.86}$$

where $\check{}$ indicates the inverse Fourier transform with respect to frequency. For the *SH* motion

$$u_H(p, r, m, t) = -\partial_{tt}\{\check{W}_0(p, m, t) * (1/pr)\check{\mathbf{T}}_m(t/pr)\}. \tag{7.87}$$

From (7.73) we see that the inverse transforms of the vector harmonics depend on being able to find the time transform of $J_m(\omega pr)$. We start with the integral representation

$$J_m(x) = \frac{i^{-m}}{\pi}\int_0^{\pi} d\theta\, e^{ix\cos\theta}\cos m\theta, \tag{7.88}$$

and now change variable to $t = \cos\theta$ to obtain

$$J_m(x) = \frac{i^{-m}}{\pi}\int_{-1}^{1} dt\, e^{ixt}\frac{T_m(t)}{\sqrt{1-t^2}}, \tag{7.89}$$

where $T_m(t)$ is a Chebyshev polynomial of the first kind with the property

$$T_m(\cos\theta) = \cos m\theta \tag{7.90}$$

so that $T_0(x) = 1$, $T_1(x) = x$, $T_2(x) = 2x^2 - 1$. Since (7.89) is in the form of a Fourier transform we can recognise the inverse transform of J_m as

$$\check{J}_m(t) = \frac{i^{-m}}{\pi}\frac{T_m(t)}{\sqrt{1-t^2}}\{H(t+1) - H(t-1)\} \tag{7.91}$$

which has integrable, square root, singularities at $t = \pm 1$.

We can find a comparable form for the near-field contributions, since

$$J_{m-1}(x) - J'_m(x) = mJ_m(x)/x, \tag{7.92}$$

$$= \frac{i^{-(m-1)}}{\pi} \int_0^{\pi} d\theta\, e^{ix\cos\theta} \sin m\theta \sin\theta, \tag{7.93}$$

and with the substitution $t = \cos\theta$, as before

$$L_m(x) = mJ_m(x)/x = \frac{i^{-(m-1)}}{\pi} \int_{-1}^{1} dt\, e^{ixt} U_{m-1}(t)\sqrt{1-t^2}. \tag{7.94}$$

Here $U_{m-1}(t)$ is now a Chebyshev polynomial of the second kind

$$U_{m-1}(\cos\theta) = \frac{1}{m} T'_m(\cos\theta) = \frac{\sin m\theta}{\sin\theta}, \tag{7.95}$$

and so $U_0(x) = 1, U_1(x) = 2x$. The inverse transform of the near-field term L_m is thus

$$\check{L}_m(t) = \frac{i^{-(m-1)}}{\pi} U_{m-1}(t)\sqrt{1-t^2}\{H(t+1) - H(t-1)\}, \tag{7.96}$$

and here only the derivative is singular at $t = \pm 1$.

The forms of the inverse transforms we have derived are only appropriate for $m \geq 0$, but once again we can derive the results for $m < 0$ from the symmetry relation

$$J^{-m}(t) = (-1)^m J_m(t). \tag{7.97}$$

With these results for the Bessel function transforms, we can find the time transforms of the vector harmonics which appear in (7.86). For $m = 0$ we have only far-field terms:

$$\begin{aligned}
\pi(1/pr)\check{\mathbf{R}}_0(t/pr) &= \mathbf{e}_z B(t, pr)(p^2r^2 - t^2)^{-1/2},\\
\pi(1/pr)\check{\mathbf{S}}_0(t/pr) &= -\mathbf{e}_r B(t, pr)(t/pr)(p^2r^2 - t^2)^{-1/2},\\
\pi(1/pr)\check{\mathbf{T}}_0(t/pr) &= \mathbf{e}_\phi B(t, pr)(t/pr)(p^2r^2 - t^2)^{-1/2},
\end{aligned} \tag{7.98}$$

where

$$B(t, pr) = \{H(t+pr) - H(t-pr)\}. \tag{7.99}$$

For $m > 0$, we also include near-field effects

$$\begin{aligned}
\pi(1/pr)\check{\mathbf{R}}_m(t/pr) &= i^{-m} B(t, pr)\mathbf{e}_z T_m(t/pr)(p^2r^2 - t^2)^{-1/2} e^{im\phi},\\
\pi(1/pr)\check{\mathbf{S}}_m(t/pr) &= i^{-(m-1)} B(t, pr)(p^2r^2 - t^2)^{-1/2} e^{im\phi}\\
&\qquad \{\mathbf{e}_r T_{m-1}(t/pr) - (\mathbf{e}_r - i\mathbf{e}_\phi)U_{m-1}(t/pr)(1 - t^2/p^2r^2)\},\\
\pi(1/pr)\check{\mathbf{T}}_m(t/pr) &= i^{-(m-1)} B(t, pr)(p^2r^2 - t^2)^{-1/2} e^{im\phi}\\
&\qquad \{-\mathbf{e}_\phi T_{m-1}(t/pr) + (\mathbf{e}_\phi + i\mathbf{e}_r)U_{m-1}(t/pr)(1 - t^2/p^2r^2)\}.
\end{aligned} \tag{7.100}$$

When we perform the convolutions in (7.86),(7.87) to calculate $\mathbf{u}(p, r, t)$, the far-field and near-field terms give rise to very different contributions to the final waveform. Near the singularities at $t = \pm pr$, $\check{J}_m(t)$ behaves like the time derivative of $\check{L}_m(t)$. The major contribution to $\boldsymbol{w}_0(p, m, t) * (1/pr)\check{T}_m(t/pr)$ will arise from the neighbourhood of these singularities, and so the far-field contribution will closely resemble the derivative of the near-field part. This behaviour, for a general stratified medium, is similar to our previous results for an unbounded medium (Section 4.3.2) where the far-field term had a time dependence which was the derivative of the nearer contributions.

The expressions we have just established for the time functions corresponding to the vector harmonics are valid for all slownesses. However, the inverse transform for the response vector

$$\check{\boldsymbol{W}}_0(p, m, t) = \frac{1}{2\pi}\int_{-\infty}^{\infty} d\omega\, e^{-i\omega t}\boldsymbol{w}_0(p, m, \omega), \tag{7.101}$$

depends strongly on slowness p.

Since we are interested in sources which start at $t = 0$, $\boldsymbol{w}_0(p, m, \omega)$ is analytic in the upper half plane Im $\omega > 0$. The exponential term $e^{-i\omega t}$ enables us to deform the contour, if necessary, into the lower half plane for $t > 0$.

The quantity $\boldsymbol{w}_0(p, m, t)$ can be thought of as the time response of the half space to irradiation by a single slowness component. In two dimensions this would correspond to a 'plane wave' seismogram. In the full half-space response for $0 < p < \beta_L^{-1}$, there are no pole singularities in $\boldsymbol{w}_0(p, m, \omega)$ as a function of ω, since we have the possibility of radiation loss into the underlying uniform half space. There will be a branch point at $\omega = 0$, and the branch cut can be conveniently taken along the negative imaginary ω axis. In this slowness range we get individual pulse-like arrivals corresponding to the major phases with a shape determined by the source time function (cf. figure 6.4) The pattern of arrivals across the band of slowness gets repeated in time with delays associated with multiple surface reflections. Because there is radiation leakage of *S* waves, at least, into the underlying uniform medium, each successive surface multiple set will be of smaller amplitude and this decay will be enhanced by the presence of attenuation in the medium. Nevertheless a long time series is needed to include all surface multiples and this can create difficulties when one tries to compute $\check{\boldsymbol{W}}_0(p, m, t)$ numerically.

When $p > \beta_L^{-1}$, both *P* and *S* waves are evanescent in the underlying half space and we have poles in $\boldsymbol{w}_0(p, m, \omega)$ which for perfectly elastic media lie on the real ω axis; the closest pole to the origin corresponds to fundamental mode surface waves. For an attenuative structure the poles move off the real axis into the lower half plane. Just at the branch point at β_L^{-1} we get the maximum density of poles along the ω axis (see figure 11.3) and the spacing expands as p increases to β_{min}^{-1}. For $p > \beta_{min}^{-1}$ we have only one pole for the *P-SV* case corresponding to the fundamental Rayleigh mode. Since the poles in ω are symmetrically disposed about the imaginary axis, the residue contribution to $\check{W}_0(p, m, t)$ takes the form (for $p < \beta_{min}^{-1}$)

$$\mathrm{Re}\left\{\sum_{k=0}^{\infty} e^{-i\omega_k t}\mathrm{Res}_k[\mathbf{w}_0^T(p, m, \omega_k)]\right\}, \tag{7.102}$$

which can often prove convenient in surface wave studies. There will in addition be a continuous spectrum contribution from the sides of the branch cut along the negative imaginary ω axis.

Once we have found the inverse transform of the medium response and performed the convolution (7.86) we need to carry out the slowness integral and summation over angular order

$$\mathbf{u}_P(r, \phi, 0, t) = \sum_m \int_0^{\infty} dp\, \mathbf{u}_P(p, r, m, t), \tag{7.103}$$

to generate the seismograms for a particular range. We illustrate the procedure by considering the mth azimuthal contribution to the radial component of motion (7.70):

$$u_{r0}(r, m, t) = \int_0^{\infty} dp\, p u_r(p, r, m, t). \tag{7.104}$$

When only the low frequency part of the seismic field is of interest it is probably most effective to calculate seismograms from (7.103) having previously evaluated the integral (7.86) by direct numerical integration.

The far-field contribution to (7.104) may be found from (7.88) and (7.101)

$${}_f u_{r0}(r, m, t) = -\frac{1}{\pi}\partial_{tt}\int_0^{\infty} dp\, p \int_{-pr}^{pr} ds\, i^{-(m-1)}T_{m-1}(s/pr)\frac{\check{V}_0(p, m, t-s)}{(p^2r^2 - s^2)^{1/2}} \tag{7.105}$$

using the explicit form for the convolution. The near-field contribution mixes both *P-SV* and *SH* elements

$${}_n u_{r0}(r, m, t) = -\frac{1}{\pi r}\partial_{tt}\int_0^{\infty} dp \int_{-pr}^{pr} ds\, i^{-(m-1)}U_m - 1(s/pr)$$
$$\times\{\check{V}_0(p, m, r-s) - i\check{W}_0(p, m, t-s)\}(p^2r^2 - s^2)^{1/2} \tag{7.106}$$

and, as we have noted in discussing the spectral method, both V_0 and W_0 need to be present in (7.106) or non-causal arrivals are generated. The same combination $\{\check{V}_0 - i\check{W}_0\}$ will also appear on the tangential component.

The total seismograms are now to be constructed by performing the summation over angular order

$$u_{r0}(r, t) = \sum_m \{{}_n u_{r0}(r, m, t) + {}_f u_{r0}(r, m, t)\}e^{im\phi}. \tag{7.107}$$

When pr is small we need to employ the expressions (7.105), (7.106) as they stand since the singularities at $s = \pm pr$ will be very close together. However, for large pr, the singularities become widely separated and so we may make an approximate development in terms of isolated singularities (Chapman 1978)

$$\frac{B(t,pr)T_m(t/pr)}{(p^2r^2-t^2)^{1/2}} = \frac{1}{(2pr)^{1/2}}\left\{\frac{H(pr-t)}{\sqrt{(pr-t)}} + (-1)^m\frac{H(t+pr)}{\sqrt{(pr+t)}}\right\}, \quad (7.108)$$

since

$$T_m(1) = 1, \quad T_m(-1) = (-1)^m. \quad (7.109)$$

We note that the contribution from $t = pr$ has a functional form with respect to t which is the Hilbert transform of that from $t = -pr$.

With this approximation the far-field displacement (7.105) becomes

$$\begin{aligned} {}_f u_{r0}(r,m,t) = &-\frac{1}{\pi(2r)^{1/2}}\partial_{tt}\int_0^\infty dp\, p^{1/2} i^{-(m-1)} \\ &\times\int_{-\infty}^{\infty} ds\, \check{V}_0(p,m,t-s)\left\{\frac{H(pr-s)}{\sqrt{(pr-s)}} + (-1)^m\frac{H(pr+s)}{\sqrt{(pr+s)}}\right\}, \end{aligned} \quad (7.110)$$

and by separating the singularities we lose the finite interval of integration. The integration domains for the two separated singularities stretch in opposite directions with respect to the time variable s. We can force a common time convolution operator for the two singularities when we make use of the properties of a convolution

$$\hat{f} * g = f * \hat{g}, \quad (7.111)$$

where ^ denotes a Hilbert transform. We transfer the Hilbert transform from the $s = pr$ singularity term to the response term, so that the s integral becomes

$$\int_{-\infty}^{\infty} ds\{\hat{\check{V}}_0(p,m,t-pr-s) + (-1)^m\check{V}_0(p,m,t+pr-s)\}\frac{H(s)}{s^{1/2}}. \quad (7.112)$$

When all the source elements have a common time dependence $M(t)$, the response terms can be written as a convolution

$$V_0(p,m,t) = M(t) * \check{v}_0(p,m,t). \quad (7.113)$$

In terms of this representation the double integral for the displacement may be written as

$$\begin{aligned} {}_f u_{r0}(r,m,t) = &-\frac{1}{\pi(2r)^{1/2}}\int_{-\infty}^{\infty} ds\, \mathcal{M}(t-s) \\ &\times\partial_s\int_0^\infty dp\, p^{1/2}\{\hat{\check{v}}_0(p,m,s-pr) + \check{v}_0(p,m,s+pr)\}, \end{aligned} \quad (7.114)$$

where we have introduced an 'effective' source function $\mathcal{M}(t)$ (Chapman 1978)

$$\mathcal{M}(t) = \int_0^t dl\, \partial_l M(l) H(t-l)/(t-l)^{1/2}. \tag{7.115}$$

This convolution with $H(t)/t^{1/2}$ may be well approximated by a finite length recursive operator (Wiggins, 1976) which facilitates numerical evaluation of the effective source. We chose to work with $\partial_t M$, since this will correspond to the far-field displacement time function and so may often be well estimated from observations.

In general, the main contribution to the far-field displacement (7.114) comes from $\hat{v}_0(p, m, t - pr)$ which corresponds to outgoing waves, and the incoming part $\check{v}_0(p, m, t + pr)$ can be neglected. The process of separating the singularities (7.108) and then retaining only outgoing waves is equivalent to taking the asymptotic form (7.79) for the Hankel function and then restricting attention to positive slownesses. As we shall see in Chapter 10, the representation (7.114) becomes particularly convenient when $V_0(p, m, \omega)$ is represented as a sum of contributions for which the inversion to the time domain can be performed analytically. However, such 'generalized-ray' representations are not very suitable for synthesising surface wavetrains.

In general we may construct $\partial_t \hat{\mathbf{W}}_0(p, m, t)$ by numerical inversion of a Fourier transform (Fryer, 1980). The Hilbert transform corresponds to a multiplier of $-i\,\mathrm{sgn}\,\omega$ in the frequency domain, and differentiation with respect to time to a further factor of $-i\omega$. Thus we take the inverse Fourier transform of $-|\omega|\mathbf{w}_0(p, m, \omega)$ to construct the quantities we need. This transform can be performed numerically on the full half space response with a fast Fourier transform and a very long time series for $p < \beta_L^{-1}$. For larger slownesses it will be more effective to use a residue summation as in (7.102), over the frequency band of interest.

Once we have constructed the slowness-time response (cf., figures 6.4, 6.5) we can perform the p-integration along linear trajectories in p, t depending on range r to form

$$v_0(r, m, t) = \int_0^{p_{max}} dp\, p^{1/2} \hat{v}_0(p, m, t - pr). \tag{7.116}$$

The convolution with the effective source can then be performed at leisure to give

$$_f u_r(r, m, t) = -\frac{1}{\pi(2r)^{1/2}} \int_0^\infty ds\, \mathcal{M}(t-s) v_0(r, m, s), \tag{7.117}$$

for the far-field radial displacement at range r.

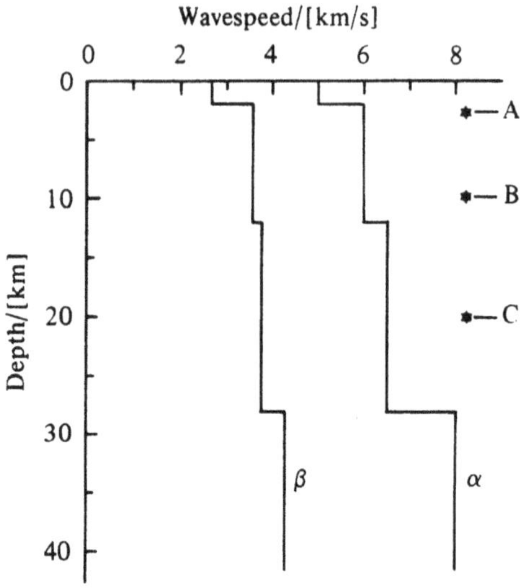

Figure 7.4. Crustal structure used in calculations of complete theoretical seismograms. The three focal depths illustrated in figure 7.5 are indicated A - 2.5 km, B - 10 km, C - 20 km. A 45° dip-slip event is used in all cases, recorded along an azimuth of 10°

7.3.3 Examples of complete theoretical seismograms

As an illustration of the variety of seismic wave phenomena which can occur when we calculate the full seismic wavetrain, we consider a source with a fixed mechanism at different focal depths in a simple model.

The calculations were carried out using the spectral approach described in Section 7.3.1, for a simple attenuative crustal model (figure 7.4) with $Q_\alpha^{-1} = 0.001$, $Q_\beta^{-1} = 0.002$. The source was chosen to be a 45° dip-slip dislocation source with a moment tensor

$$M_{ij} = M(t)\mathrm{diag}[1, 0, -1]. \tag{7.118}$$

This particular type of source excites only the angular orders $m = 0, \pm 2$. For sources at 2.5 km, 10 km and 20 km depth we present record sections of the three components of displacement as a function of range, along an azimuth of 10° in figure 7.5. The moment rate function was a delta function (corresponding to $M(t)$ being a step function), and the calculation was performed for a frequency band from 0.04 to 4.0 Hz with a simple half-cycle sinusoidal filter response.

The displays in figures 7.5a,b,c consist of composite record sections for all three displacement components. For each distance we present a triad of seismograms in the order vertical (Z), radial (R) and tangential (T). The radial seismograms are plotted at the correct ranges and the vertical and tangential seismograms at constant offset. The net effect is thus to give three interleaved record sections with the same time distance relations. A scaling factor of $1.0 + 0.1\,r$ is applied to all seismograms at a range r.

a)

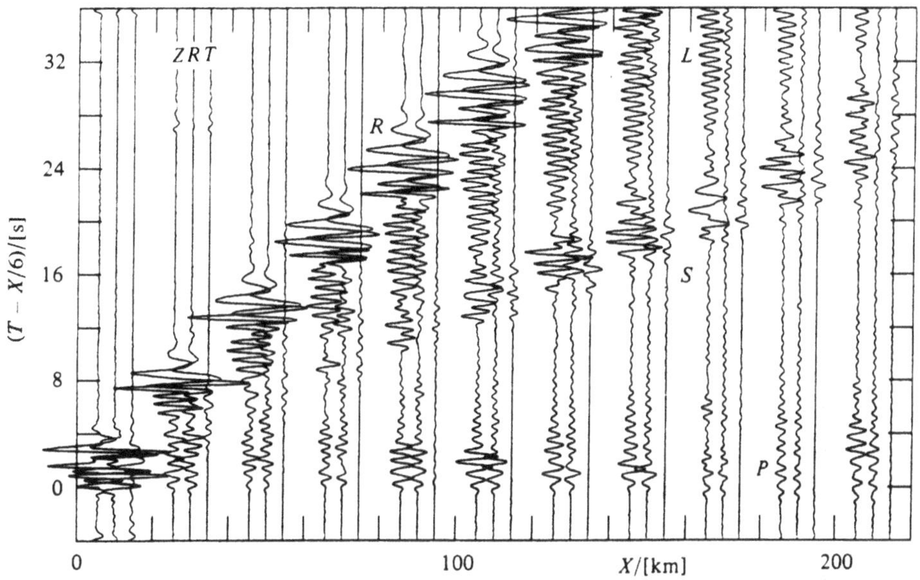

Figure 7.5. Complete theoretical seismograms calculated for the crustal structure illustrated in figure 7.4, records are shown for all three components of displacement: a) 2.5 km deep source

The seismograms from the shallowest source (case A - 2.5 km depth) are displayed in figure 7.5a. We note immediately that we have well developed *P*, *S* and Rayleigh wavetrains of quite complex form influenced strongly by reverberations in the surface channel with reduced velocity. The propagation times for both *P* and *S* waves from the source to the surface are sufficiently short that there is no clear separation of the surface reflected phases (*pP*, *sS*) from direct propagated phases (*P*, *S*). The source lies just below the surface channel and we get a pronounced Airy phase for the Rayleigh waves with a group velocity close to 2.7 km/s (the surface *S* wave velocity). The main phase is preceded by a rather oscillatory higher mode train. There is noticeable velocity dispersion with frequency and we can see an indication of much lower frequency Rayleigh waves emerging from the tail of the seismograms. For this azimuth of observation the radiation pattern of the source gives rather weak excitation of the tangential component, particularly for the surface waves. The higher mode Love waves on the tangential component arrive along with the higher mode Rayleigh waves, but the group velocity of the fundamental mode Love wave is somewhat faster than that for the Rayleigh wave.

The final time series were generated by adding together the results for different slowness intervals. For phase velocities greater than 3.57 km/s, i.e. slowness $p < 0.28$ s/km, 40 s of time series were computed with a reduction slowness of 0.16 s/km, and this contribution includes most of the *P* and *S* wavetrains. In order to

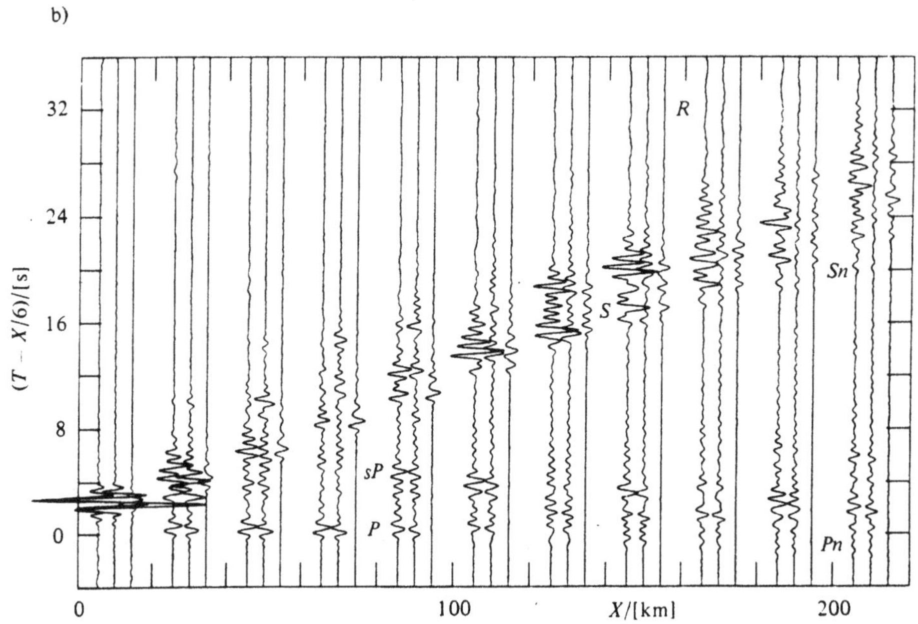

Figure 7.5. Complete theoretical seismograms calculated for the crustal structure illustrated in figure 7.4, records are shown for all three components of displacement: b) 10 km deep source.

achieve a good representation of the surface train a double length (80 s) time series was used for phase velocities between 3.57 and 2.3 km/s (i.e. $0.28 < p < 0.434$), with a reduction slowness of 0.28 s/km - since no significant *P* waves should be present in this slowness interval. When we add this portion to the single length time series for higher phase velocities we get incomplete cancellation of the numerical arrivals associated with splitting the slowness integration at $p = 0.28$ s/km. The resulting arrival (apparently incoming) may be seen at large reduced times (≈ 30 s) on the first two sets of seismograms in figure 7.5a, but decays rapidly with distance and gives little contamination of the response.

The seismograms for the midcrustal source (case B - 10 km depth) are illustrated in figure 7.5b. We immediately notice that, as expected, the surface wave excitation is very much reduced and a low frequency Rayleigh wave is just visible beyond 100 km emerging from the tail of the *S* wavetrain. The time differential between surface reflected phases and the direct phases is now more significant and we are beginning to get a clearer separation at shorter ranges. At the larger ranges we see the emergence of *Pn* and *Sn* phases refracted along the crust-mantle interface. As expected, there is a change in polarity of the *P* wave between 70 and 90 km range corresponding to the switch between the upper and lower lobes of the *P* wave radiation pattern from our dip-slip source.

In the seismograms for the deepest source (case C - 20 km depth) shown in figure

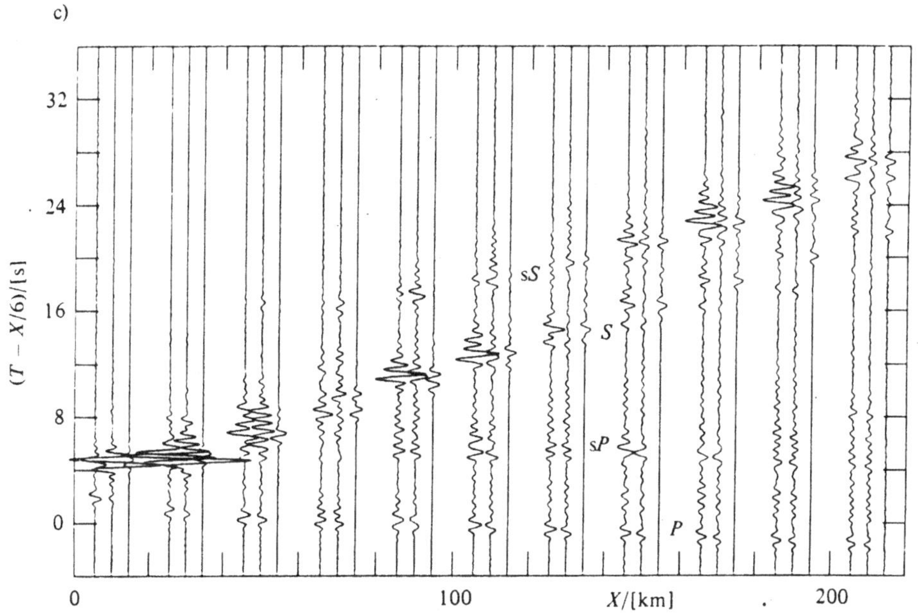

Figure 7.5. Complete theoretical seismograms calculated for the crustal structure illustrated in figure 7.4, records are shown for all three components of displacement: c) 20 km deep source.

7.5c, we have no significant surface waves and the surface reflected phases are now very prominent. Multiple reflections within the whole crustal channel are also just beginning to influence the seismograms.

These calculations for a point source in a simple structure show that we can give a good representation of the general character of local events. The calculations can be extended to higher frequencies and greater ranges at the cost of considerable computer time. However, for regional and teleseismic ranges (r >300 km), attention is usually focussed on more limited portions of the seismic records, and then it is often more convenient to make an approximation to the full response and model the features of interest (see Chapter 9).

Chapter 8

The Seismic Wavefield

So far in this book we have shown how we may calculate complete theoretical seismograms for a horizontally layered medium. Such calculations are most useful when the total time span of the seismic wavetrain is fairly short and there is no clear separation between different types of wave propagation processes.

As the distance between source and receiver increases, the wavetrain becomes longer and waves which have travelled mostly as P waves arrive much earlier than those which propagate mostly as S. Also the surface waves which are principally sensitive to shallow S wavespeed structure separate out from the S body waves which are returned from the higher wavespeeds at depth. Once the seismic wavetrain begins to resemble a sequence of isolated phases it becomes worthwhile to develop approximate techniques designed to synthesise a particular phase. However, in order that such approximations can be made efficiently, with a due regard for the nature of the propagation process, we need to have a good idea of the character of the seismic wavefield.

In this chapter we will therefore survey the character of the seismograms which are recorded at different epicentral ranges.

8.1 Controlled source seismology

In the application of seismic techniques to the determination of geological structure, the source of seismic radiation is usually man-made, such as an explosive charge. In this case the origin time is known with precision and with high frequency recording the fine detail in the seismograms can be retained. Experiments of this type can be loosely divided into two classes characterised by the maximum range at which recordings are made and the density of recording points.

In *reflection* seismic studies attention is concentrated on P waves reflected at depth and returned at small offsets from the source. The propagation paths are then close to the vertical, particularly for reflections from deep structure. The major use of the reflection method has been in prospecting for minerals and hydrocarbons. Here the features of interest usually lie shallower than 5 km and the array of geophones at the surface rarely extends to more than 5 km from the source. Many receivers are used for each source (typically 48 or 96 in current practice), and the source point is then moved slightly and the recording repeated. The multiplicity of subsurface coverage can then be exploited to enhance the weak reflections from depth. Reflection methods are now also being used for investigations of the deep crust (Schilt et al., 1979) and here longer recording arrays are often used.

In *refraction* studies the receivers extend to horizontal ranges which are eight to ten times the depths of interest, and the density of observations is often fairly low. At the largest ranges the main features in the seismograms are refracted phases or wide-angle reflections from major structural boundaries. These can sometimes be traced back into reflections at steeper angles if adequate coverage is available at small ranges.

Early work in seismic refraction in Western countries used very limited numbers of receivers, but latterly the benefits of very much denser recording have been appreciated (see, e.g., Bamford et al., 1976). In Eastern Europe and the USSR, reflection and refraction techniques have been combined in 'Deep Seismic Sounding' (Kosminskaya, 1971), which has closely spaced receivers along profiles which can be hundreds of kilometres long. Such an arrangement gives a very detailed description of the seismic wavefield.

8.1.1 Reflection studies

For many years the major source of seismic radiation used in reflection work was small explosive charges, both at sea and on land. Now, however, a large proportion of the work on land uses an array of surface vibrators to generate the seismic waves. This avoids drilling shot holes and allows more control over the frequency content of the signal transmitted into the ground. For marine work the commonest source is now an array of airguns, which generate P energy in water by the sudden release of high pressure air.

All these energy sources are at, or close to, the surface and so tend to excite significant amplitude arrivals travelling in the low wavespeed zone at the surface (water or weathered rock). In marine records these slowly travelling waves are mostly direct propagation in the water and multiple bottom reflections, but in very shallow water there may also be effects from the weak sediments at the bottom giving strong arrivals known as 'mud-roll'. On land a surface vibrator is a very efficient generator of fundamental mode Rayleigh waves and such 'ground-roll' phases show up very strongly when single geophone recording is used (figure 1.3). When explosive charges are used they are usually fired beneath the weathered zone, this reduces the excitation of the ground-roll, but it can still have large amplitude.

The weak reflections from depth have very small apparent slownesses on a surface array and tend to be obscured in part by the shallow propagating phases. In order to remove the ground-roll and water phases, the seismic records are normally obtained not from a single sensor, but from an array of sensors. If such an array is chosen to span a wavelength of the ground-roll at the dominant frequency, the

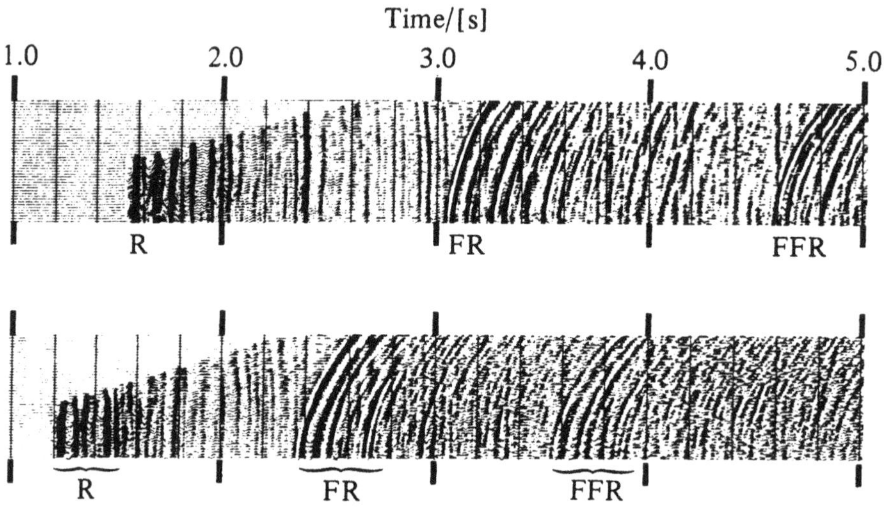

Figure 8.1. Seismic reflection gather showing prominent surface-reflected multiples indicated by markers. This set of traces has been selected to have a common midpoint between source and receiver and then has been corrected for the time shifts associated with source-receiver offset on reflection at depth, with the result that primary reflections appear nearly flat. (Courtesy of Western Geophysical Company).

resulting summed amplitude is substantially reduced. The residual ground-roll can be removed by exploiting the separation in slowness from the reflections and so designing a filter, e.g., in the frequency-slowness domain, to leave only the reflections with small slowness.

The use of arrays of sources and receivers makes it difficult to produce good theoretical models of the source radiation. For airgun arrays the guns are usually so close together that very complex interference effects occur. With surface vibrators the ground coupling can be highly variable and several vibrators in a similar location can have very different seismic efficiency.

The reflections from depth are weak compared with the early refracted arrivals from the shallow structure. These refractions are often forcibly removed from reflection records ('muting') and an attempt is then made to equalize the amplitude of the traces in time and distance to compensate for losses in propagation. As a result a set of reflection records will normally appear to become more ragged with increasing time, since noise is amplified along with the coherent signal.

In most reflection situations the reflection coefficient at the surface is larger than any of the reflection coefficients in the subsurface, particularly at near-normal incidence (cf. figure 5.4). Waves which have been reflected back from below the source can be reflected at the free surface, and then reflected again by the structure. A good example is shown in figure 8.1 where a prominent reflector (R) is mirrored at twice the time by its free-surface multiple (FR). This surface multiple obscures genuine reflections from greater depth. When there is a strong contrast in properties at the base of the low wavespeed zone, multiple reverberations within this zone associated with each significant reflection lead to a very complex set of records. Often predictive deconvolution (Peacock & Treitel, 1969) will give help in 'cleaning' the records to leave only primary reflections. Sometimes internal multiples between major reflectors can also give significant interference with deeper reflections.

Most reflection recordings on land have been made with vertical component geophones and so P waves are preferentially recorded. However, waves which have undergone conversion to S at some stage of their path can sometimes be seen at the largest offsets on a single shot gather. Converted phases can occur at small offsets in the presence of strongly dipping reflectors. At sea, pressure sensors are used and the recording array is not usually long enough to pick up effects due to conversion. The standard data processing techniques, particularly stacking of traces with an estimated wavespeed distribution to attempt to simulate a normally incident wave, tend to suppress conversions.

Although the object of seismic reflection work is to delineate the lateral variations in subsurface structure, studies of wave propagation in stratified models can help in understanding the nature of the records, as for example in multiple and conversion problems.

8.1.2 Refraction studies

Whereas the object of a reflection experiment is to delineate the fine detail in geological structure, in refraction work resolution is sacrificed to penetration in depth. As a result the interpretation of a refraction profile, which does not cross any major vertical discontinuities such as deep faults, will give only the broad outline of the lateral variations in structure. The principal phases which can be correlated from record to record are refracted arrivals (quite often interference head waves, see Section 9.2.2) and wide angle reflections from major horizontal boundaries. As a result the portion of the wavefield which is studied is most sensitive to the wavespeed distribution near interfaces and strong gradient zones and reveals little information about the rest of the structure.

The source of seismic radiation for refraction work is normally an explosive charge recorded at an array of receivers. After a shot the receiver array is moved to a new location and a further shot fired. In this way a detailed profile can be built up to considerable range with only a limited number of recording stations. Since studies of deep structure require a detailed knowledge of the shallower regions, multiple shots at a variety of ranges are often fired into the same receiver array so that detailed results can be built up at both large and small ranges (Bamford et al., 1976). Such a procedure also allows a test of the degree of lateral homogeneity along a refraction profile. In work on land there is often considerable variability in amplitudes between nearby recorders, and in order to reduce local variations

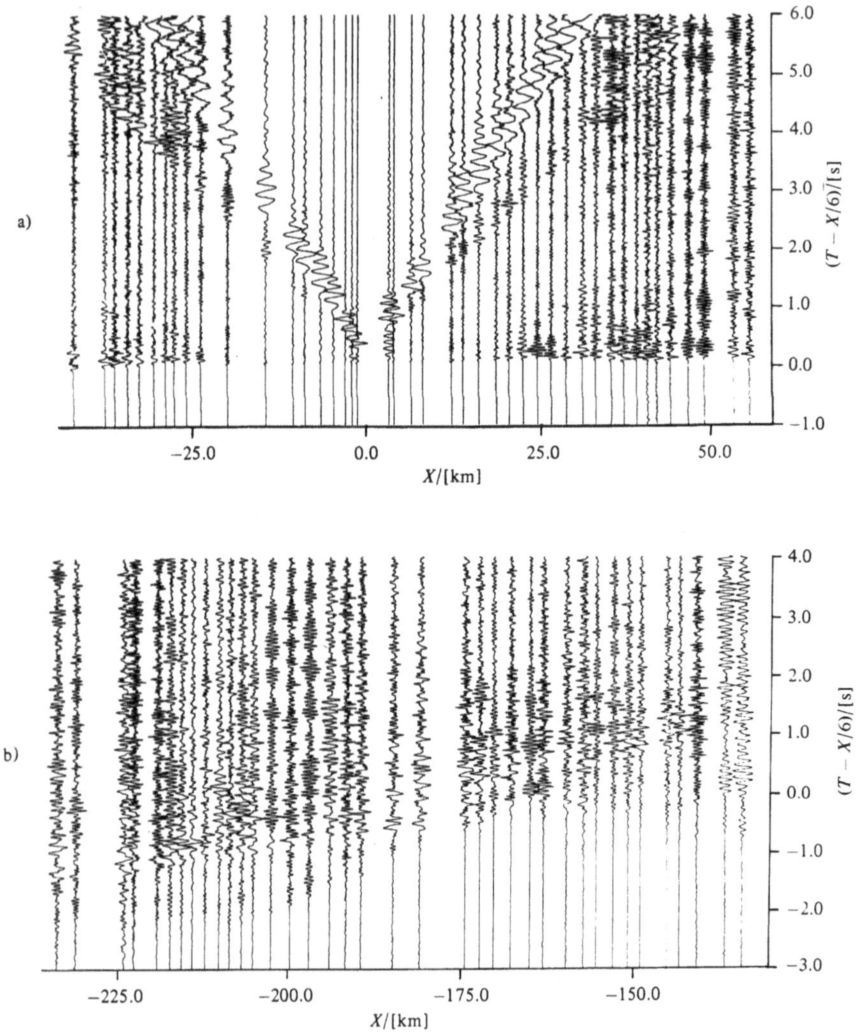

Figure 8.2. Seismic refraction records from the Arabian Penisula (Courtesy of U.S. Geological Survey): a) Ranges out to 150 km showing *P*, *S* and Rayleigh waves; b) Detail of *P* wavetrain at larger ranges.

observations are often low-pass filtered before interpretation (Fuchs & Müller, 1971).

The depth of the shot on land must be such as to contain the explosion. The resulting seismic wavefield is rich in *P* waves, and some *S* waves are generated by reflection at the free surface. These deep sources are not very efficient generators of Rayleigh waves, but a surface wavetrain is often seen late on refraction records. In figure 8.2a we show the close range seismograms from an experiment conducted by the United States Geological Survey in Saudi Arabia (Healy et al., 1981), which show clearly the three major wave contributions. We display the beginning of the *P* wavetrain at larger ranges in figure 8.2b, and we can see considerable detail in the phases with closely spaced receivers. At large ranges (300 km - 600 km) detailed refraction experiments have revealed character within the first arriving *P* waves normally referred to as *Pn* (Hirn et al., 1973) which could not have been observed with the coarse spacing available with permanent stations. These observations suggest the presence of fine structure in the uppermost part of the mantle which could not be resolved in previous studies.

Seismic refraction work at sea has significant differences from work on land, since it is difficult to use more than a few receivers and so a suite of observations is built up using multiple shots. At short ranges ($<$ 20 km) airguns have been used to give high data density and provide continuous coverage from near-vertical incidence out to wide angle reflections (see, e.g., White, 1979). This data gives good control on the structure of the uppermost part of the oceanic crust and the longer range refractions fill in the picture at depth. For refraction experiments which have been shot along isochrons in the oceans, lateral variations in structure are not too severe, and stratified models give a good representation of the structure.

The interpretation of refraction records was initially based on the times of arrival of the main *P* phases, but latterly this has been supplemented with amplitude information. Frequently the amplitude modelling has been done by computing theoretical seismograms for an assumed model, and then refining the model so that observations and theoretical predictions are brought into reasonable agreement. This approach has spurred on many of the developments in calculating theoretical seismograms both by generalized ray methods (Helmberger, 1968 - see Chapter 10) and reflectivity techniques (Fuchs & Müller, 1971 - see Section 9.3.1). The use of amplitude information has resulted in more detail in the postulated wavespeed distributions with depth. In particular this has led to a considerable change in our picture of the oceanic crust (cf. Kennett, 1977; Spudich & Orcutt, 1980). Braile & Smith (1975) have made a very useful compilation of theoretical seismograms for the continental crust, illustrating the effect of a variety of features in the wavespeed distribution.

In refraction work most attention is given to the *P* arrivals, but on occasion very clear effects due to *S* waves or conversion can be seen and these may be used to infer the *S* wavespeed distribution. Even when such waves are not seen the *S* wavespeed distribution can have significant influence on the character of the *P* wavefield (White & Stephen, 1980).

8.2 Ranges less than 1500 km

For epicentral distances out to 1500 km the properties of the seismic wavefield are dominated by the wavespeed distribution in the crust and uppermost mantle.

8.2.1 Strong ground-motion

In the immediate neighbourhood of a large earthquake the Earth's surface suffers very large displacements which are sufficient to overload many seismic instruments. However, specially emplaced accelerometer systems can record these very large motions. These systems are normally triggered by the *P* wavetrain and so the accelerograms consist almost entirely of *S* waves and surface waves. Normally the strong-motion instruments are somewhat haphazardly distributed relative to the fault trace since they are placed in major buildings. However, in the Imperial Valley in California, an array of accelerometers was installed across the trace of the Imperial fault. These stations recorded the major earthquake of 1979 October 15 with some accelerometers lying almost on top of the fault (Archuleta & Spudich, 1981).

In figure 8.3 we show the three components of velocity for a group of stations close to the fault obtained by numerical integration of the accelerograms. The earthquake rupture started at depth on the southern portion of the fault and propagated to the north-west. The major features of the velocity records are associated with the progression of the rupture, and show strong excitation of higher mode surface waves on the horizontal components. The early high frequency arrivals on the vertical component are probably multiple *P* phases. The oscillatory tails to the records arise from surface waves trapped in the sediments.

In order to understand such strong ground motion records we have to be able to calculate complete theoretical seismograms as in Section 7.3 and, in addition, need to simulate the effect of large-scale fault rupture. In a stratified medium this can be achieved by setting up a mesh of point sources on the fault plane with suitable weighting and time delays, and then summing the response at each receiver location. Such a representation will fail at the highest frequencies because the wavelengths will be smaller than the mesh spacing, but will describe the main character of the event. Point-source models are still useful since they allow the study of the effects of wave propagation in the crustal structure rather than source processes.

The aftershocks of major events are often quite small and these may be modelled quite well with equivalent point sources. In many areas the surface motion is strongly affected by the sedimentary cover, particularly where this is underlain by high wavespeed material. A detailed study of such amplification effects has been made by Johnson & Silva (1981) using an array of accelerometers at depth in a borehole.

8.2.2 Local events

Most seismic areas now have a fair density of short-period seismic stations which have been installed to allow detailed mapping of seismicity patterns and so have a high frequency response. A common features of seismograms at such sites are

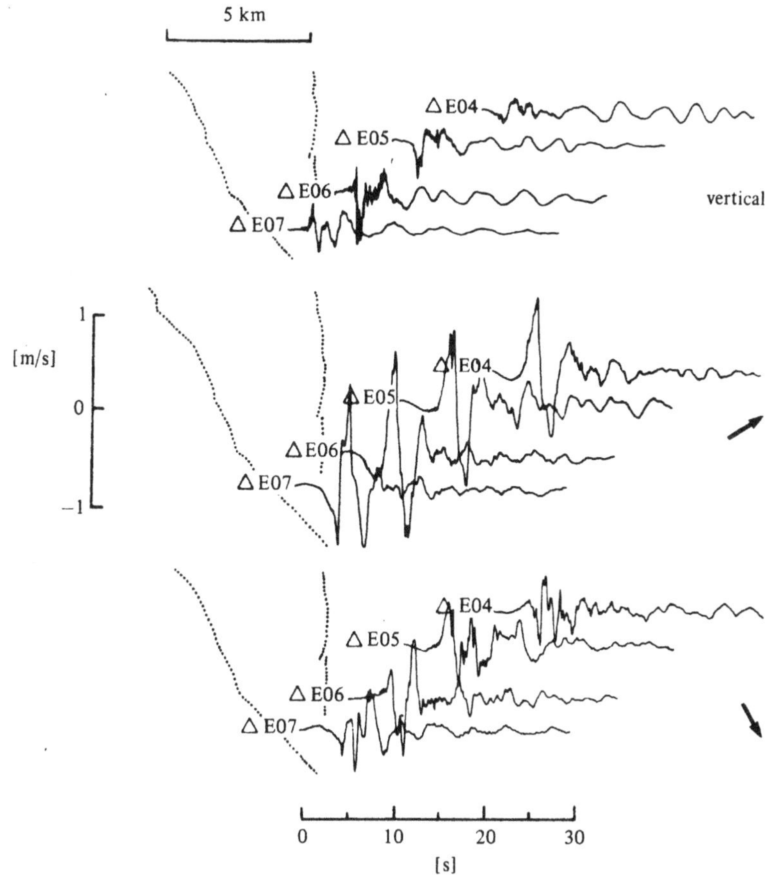

Figure 8.3. Three-component velocity records for the 1979 earthquake in the Imperial Valley, California. The epicentre lay about 20 km to the south-east of the group of stations shown in map view. The surface fault break is also marked. (Courtesy of U.S. Geological Survey)

short bursts of energy associated with local earthquakes, less than 200 km or so from the station (see figures 8.4 and 8.5). These records are dominated by *P* and *S* body waves, although at larger ranges there are sometimes hints of surface waves which increase in importance on broad-band records.

In figure 8.4 we show seismograms recorded on the North Anatolian fault zone (Crampin et al., 1980), for an earthquake at 13 km depth at a hypocentral distance of 18 km. The *P* waveform is quite simple and is followed by very clear *S* onsets with a lower frequency content indicating significant attenuation in the fault zone. Close recordings of small aftershocks also show such a pattern of arrivals but the details of the waveform can be strongly influenced by near-surface structure, such as sediments.

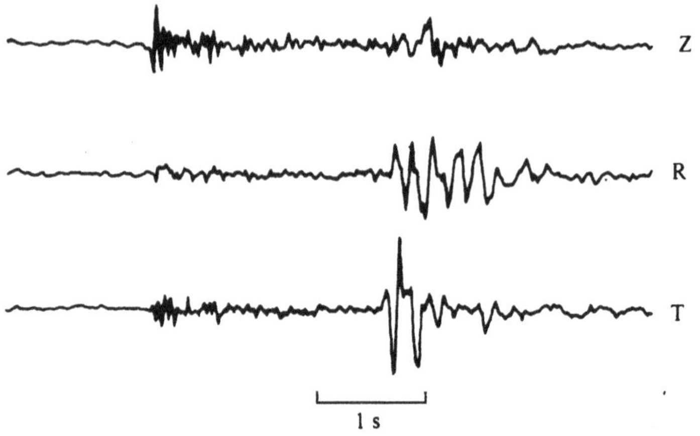

Figure 8.4. Rotated three-component seismograms for a small earthquake on the North Anatolian fault zone, hypocentral distance 18 km.

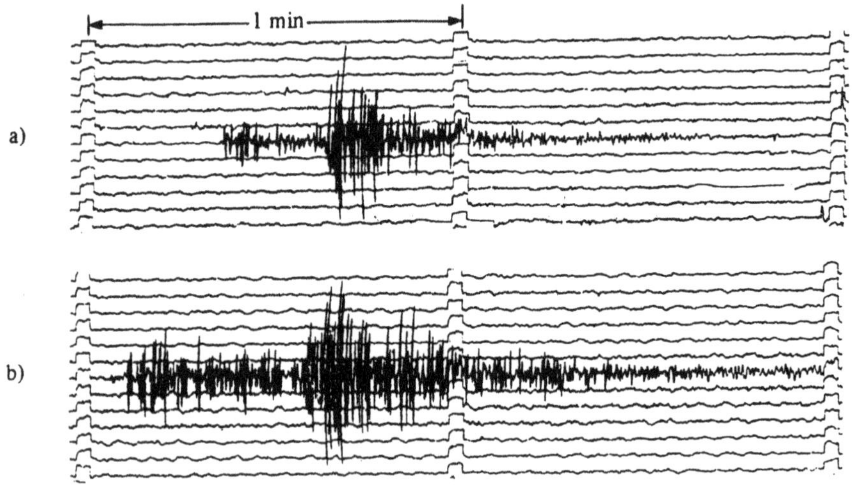

Figure 8.5. Vertical component short-period seismograms recorded at JAS, showing the effect of increasing epicentral distance.

At most permanent seismic stations with visual recording, it is difficult to separate the *P* and *S* wave arrivals from close events and so the records show strong excursion followed by swift decay. As the epicentral distance increases, the time separation between *P* and *S* waves is such that distinct phases are seen. In figure 8.5 we show vertical component records for two local events recorded at Jamestown in Northern California. The closer event (figure 8.5a) is about 100 km away from the station, and shows a clear crustal guided *Pg* group which begins to die away before the *Sg* waves which carry most of the energy. The *S* wave coda has a generally

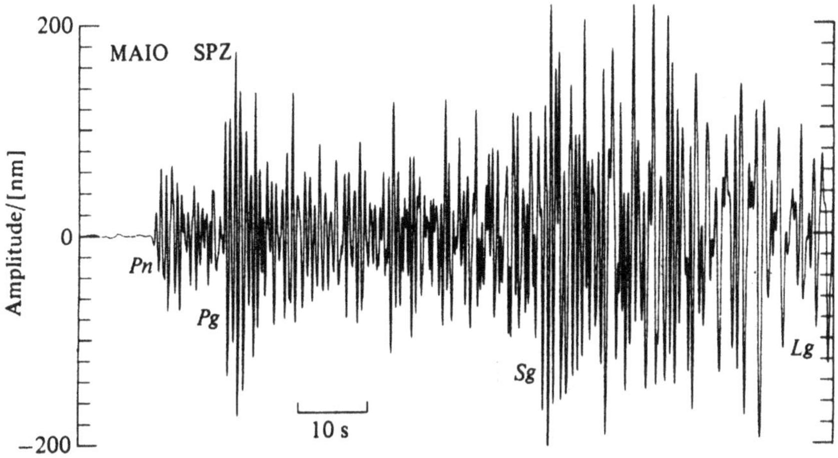

Figure 8.6. Short period SRO record from Mashad, Iran showing distinct *Pn*, *Pg*, *Sg* and *Lg* phases.

exponential envelope and the later arrivals may well be due to scattering in the neighbourhood of the recording station. The more distant event (figure 8.5b) is 200 km away from the station and now *Pn* waves propagating in the uppermost mantle have separated from the front of the *Pg* group. The corresponding *Sn* phase is often difficult to discern because of the amplitude of the *P* coda, but there is a hint of its presence before the large amplitude *Sg* group in figure 8.5b.

8.2.3 Regional events

In the distance range from 200-1500 km from the epicentre of an earthquake the pattern of behaviour varies noticeably from region to region. In the western United States there is a rapid drop in short period amplitude with distance with a strong minimum around 700 km (Helmberger, 1973) which seems to be associated with a significant wavespeed inversion in the upper mantle. Such a pattern is not seen as clearly in other regions, and for shield areas there is little evidence for an inversion. In general, however, the coverage of seismic stations in this distance interval is somewhat sparse and the structure of the top 200 km of the mantle is still imperfectly known.

At moderate ranges the character of the wavetrain is still similar to the behaviour we have seen for local events. In figure 8.6 we show a plot of the digital short-period channel of the SRO station at Mashad, Iran for a small earthquake at a range of 390 km. A very clear *Pn* phase is seen preceding the *Pg* phase, but once again it is difficult to pick the onset of the *Sn* phase, though there is the beginning of an apparent interference effect near the expected arrival time. The *Sg* waves grade at later times into a longer period disturbance composed of higher mode surface waves, the *Lg* phase. There is no clear distinction

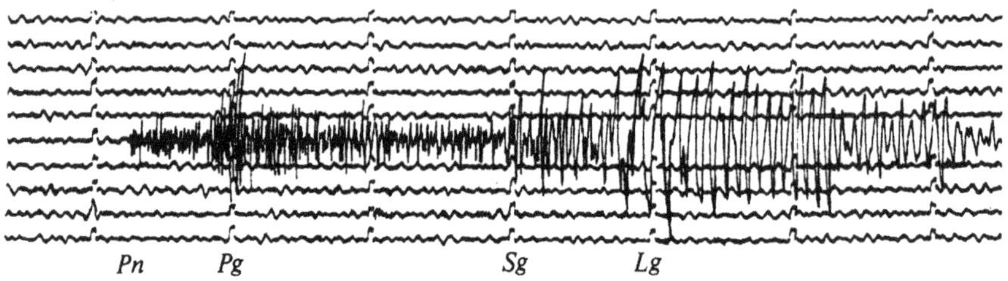

Figure 8.7. Broad-band seismic recording at Boulder, Colorado from a nuclear explosion at the Nevada test site, epicentral distance 980 km.

between the *Sg* and *Lg* phases and so, to synthesise seismograms for this distance range, we again need to include as much of the response as possible. For the *S* waves and their coda, this may be achieved by the modal summation techniques discussed in Chapter 11, in addition to the integration methods discussed in Chapter 7.

At longer ranges, the higher mode surface waves are only rarely seen on short-period records, but show up clearly with broad-band instruments. Figure 8.7 illustrates such a record at 980 km from a nuclear test in Nevada. The explosive source gives stronger *P* waves than in the previous earthquake examples. The *P* waves show high frequency effects modulating the long-period behaviour (*PL*) which has been studied by Helmberger & Engen (1980). Although *Sg* is not very strong, a well developed *Lg* wavetrain is seen. At the same range long-period earthquake records are dominated by fundamental mode Love and Rayleigh waves, which in areas with thick sedimentary sequences can have significant energy at very low group velocities.

8.3 Body waves and surface waves

Beyond about 1500 km from the source the *P* and *S* body waves are sufficiently well separated in time that we can study them individually. Out to 9000 km the earliest arriving waves are reflected back from the mantle, beyond this range the effect of the core is very significant. Reflected waves from the core *PcP*, *ScS* arrive just behind *P* and *S* between 8000 and 9000 km. For *P* waves the core generates a shadow zone and there is a delay before *PKP* is returned. The *P* wavespeed in the fluid core is higher than the *S* wave speed in the mantle and so, beyond 9200 km, the *SKS* phase penetrating into the core overtakes *S*. Simplified travel-time curves for the major phases seen on seismograms are illustrated in figure 8.8.

With increasing range the surface reflected phases such as *PP*, *PPP* separate from the *P* coda to become distinct arrivals. With each surface reflection the waves have passed through a caustic, and so the waveform is the Hilbert transform of the previous surface reflection. For *S*, such multiple reflections constitute a fair part

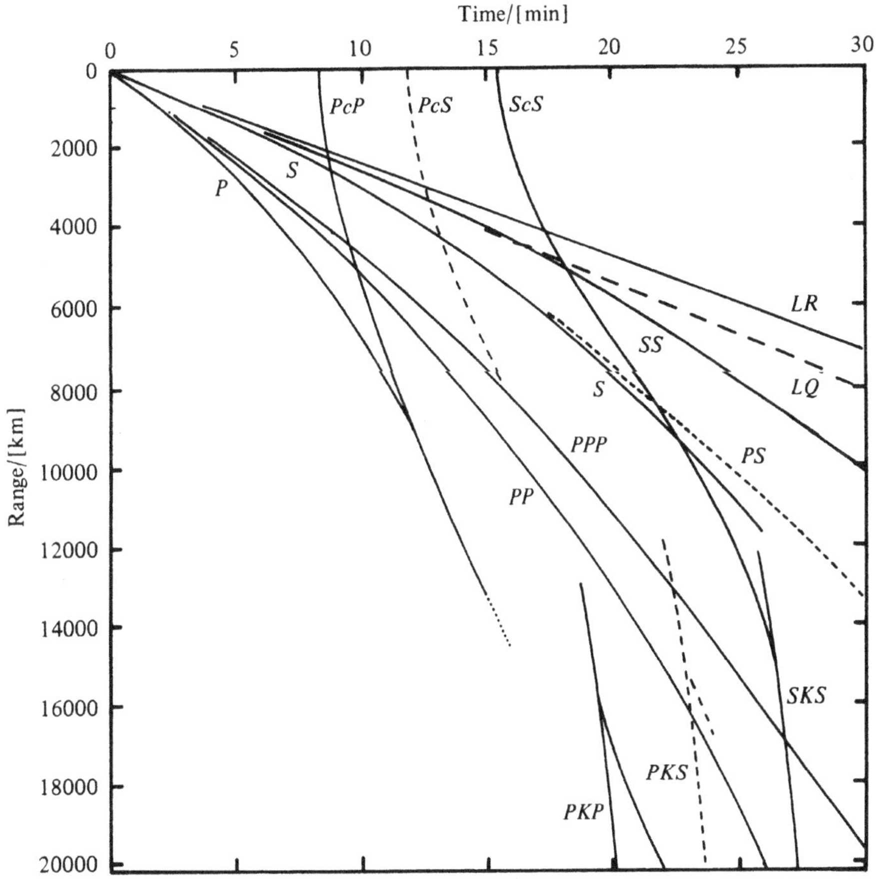

Figure 8.8. Simplified travel-time curves for the major seismic phases.

of what is commonly characterised as the surface wavetrain and so the *SS* and *SSS* phases appear to emerge from the travel-time curve for the surface waves.

8.3.1 Body waves

For epicentral ranges between 1500 and 3500 km the *P* and *S* waves are returned from the major transition zone in the upper mantle which occupies the depth interval from 300-800 km. In this region there are substantial wave speed gradients and near 400 km and 670 km very rapid changes in wavespeeds which act as discontinuities for large wavelengths. This complicated wavespeed distribution leads to travel-time curves for the *P* and *S* phases which consist of a number of overlapping branches, associated with variable amplitudes. As a result of interference phenomena the waveforms in this distance range are rather complex

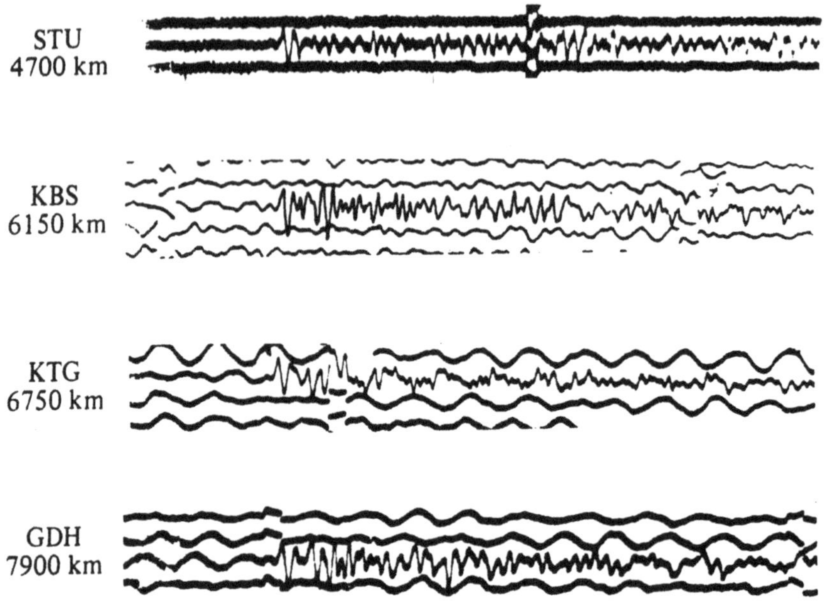

Figure 8.9. Short period WWSSN records at distant stations from an earthquake in Iran, focal depth 31 km.

and have mostly been studied in an attempt to elucidate the wavespeed distribution in the upper mantle (see, e.g., Burdick & Helmberger, 1978).

Beyond 3500 km the *P* and *S* waves pass steeply through the upper mantle transition zone and very little complication is introduced until the turning levels approach the core-mantle boundary. The resulting window from 3500-9000 km enables us to use *P* and *S* waveforms to study the characteristics of the source. For *P* waves, the character of the beginning of the wavetrain is determined by the interference of the direct *P* wave with the surface-reflected phases *pP* and *sP*. The relative amplitude of these reflected phases varies with the take-off angle from the source and the nature of the source mechanism.

In figure 8.9 we show short-period records from a number of WWSSN stations for a shallow event in Iran. The stations lie in a narrow range of azimuths and allow us to see the stability of the direct *P* wave shape over a considerable distance range. The ISC estimate of the focal depth of this event is 31 km, but the time interval between *P* and *pP* suggest a slightly smaller depth. The later parts of the seismograms are associated with crustal reverberations near source and receiver. The relative simplicity of mantle propagation illustrated by these records can be exploited to produce a scheme for calculating theoretical seismograms for teleseismic *P* and *S* phases discussed in Section 9.3.3. For teleseismic *S* waves, a *P* wave precursor can be generated by conversion at the base of the crust and on a vertical component record this can easily be misread as *S*.

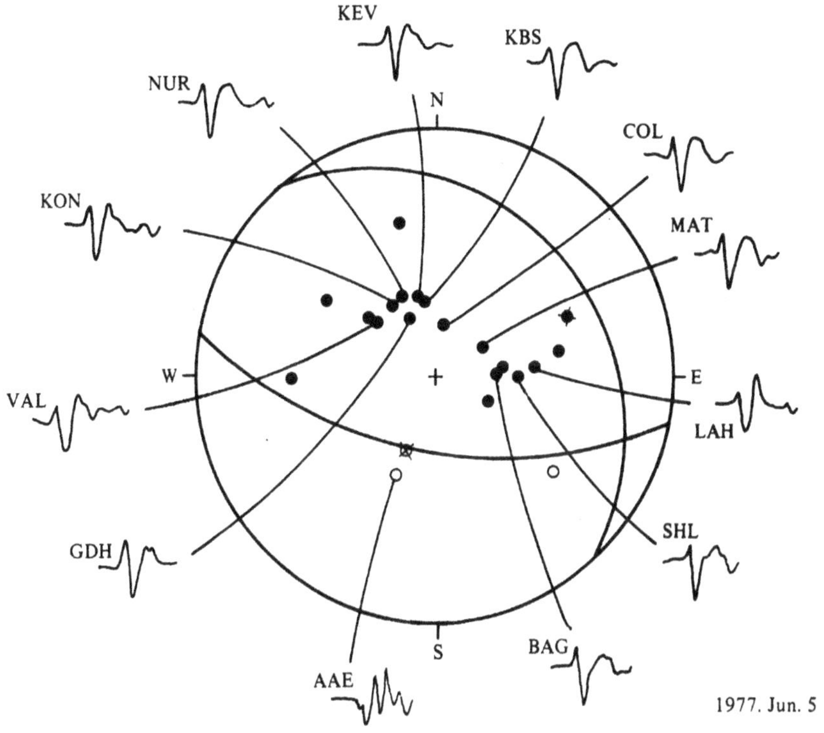

Figure 8.10. Long period WWSSN records in relation to their position on the focal sphere for an Iranian event.

On teleseismic long-period records the time resolution is normally insufficient to allow separation of the direct *P* from the surface reflections *pP*, *sP* for shallow events. However, the appearance of the onset of the *P* wavetrain provides a strong constraint on the depth of source particularly when many stations at different distances are available (Langston & Helmberger, 1975). This procedure relies on a very simple construction scheme for long-period records which is discussed in Section 9.3.3. To get depth estimates from these long-period records we need a model of the source time function. For small to moderate size events the far-field radiation can be modelled by a trapezoid in time. Figure 8.10 shows the long-period records from WWSSN stations for an Iranian event as a function of their position on the focal sphere. The simplicity of these long-period waveforms enables the sense of initial motion to be determined very reliably (Sykes, 1967) and so improves the estimate of the focal mechanism.

Once the earthquake focus lies well below the crust the surface reflections are seen as distinct phases, particularly for deep events. Surface reflections can also be returned as core reflections so that phases like *pPcP*, *sPcP* can often be found for intermediate or deep events. In figure 8.11 we show a vertical component broad band recording at Boulder, Colorado from an intermediate depth event (100 km) in

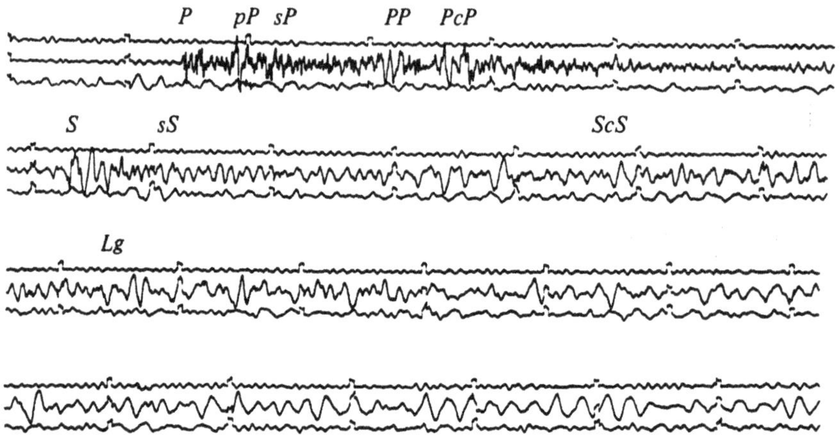

Figure 8.11. Broad-band record at Boulder, Colorado for an intermediate depth event, focal depth 100 km, epicentral distance 4200 km.

the Mona passage. The epicentral distance is 4200 km and now we see clear *pP* and *sP* phases. About a minute later *PP* arrives, with a longer period since most of its path has been spent in the attenuative regions of the upper mantle. This is followed by *PcP* and its surface reflections. The *P* wave coda then dies down and a clear *S* wave arrival is seen with a small *P* precursor. On a standard short-period instrument this *S* wave would not be seen at all clearly because of the strong roll-off in the instrumental response to suppress the microseism band (figure 1.2). The *S* coda grades into a weak *Lg* train rich in moderate frequencies but very few long period surface waves are seen. Such an intermediate depth event will be more successful in exciting higher mode Rayleigh waves than the fundamental (see Chapter 11).

8.3.2 Surface waves

For all but deep earthquakes (focal depths $>$ 300 km) the largest arrivals on long-period records occur after the *P* and *S* body waves. These surface waves have travelled with their energy confined to the crust and upper mantle and so have not suffered as much wavefront spreading as the body waves.

On the horizontal component oriented transverse to the path between the epicentre and the station, just behind the *S* body phases a very long disturbance (*G*) appears which at later times is replaced by short-period oscillations often denoted *LQ*. These two features arise from the fundamental Love mode for which the group slowness normally increases with frequency. As a result the apparent frequency of the record increases with time (see figure 8.12). Superimposed on this wavetrain

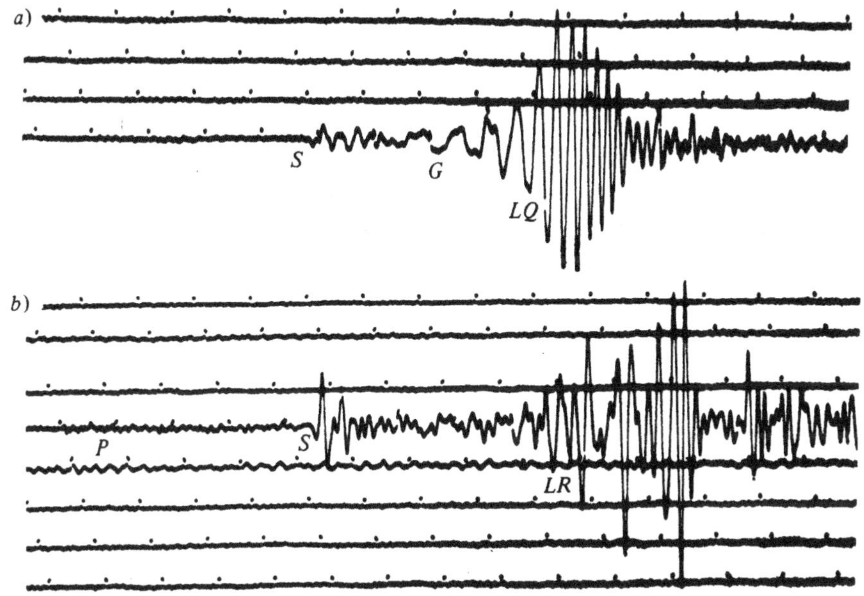

Figure 8.12. WWSSN long-period records for Baja California events at Atlanta, Georgia, epicentral distance 3000 km: a) transverse component; b) radial component.

are smaller high frequency waves with small group slowness. These are higher mode Love waves whose excitation increases with increasing depth of focus.

On the vertical component and the horizontal component oriented along the path to the source, the principal disturbance *LR* occurs some time after the commencement of the *LQ* waves and arises from the fundamental Rayleigh mode. This is preceded by very long-period waves, arriving after the *G* waves and with much lower amplitude. The wavetrain leads up to an abrupt diminuation of amplitude, followed by smaller late arrivals with high frequency. This Airy phase phenomena is associated with a maximum in the group slowness for the fundamental Rayleigh mode at 0.06 Hz (figure 11.10). Waves with frequencies both higher and lower then 0.06 Hz will arrive earlier than those for 0.06 Hz, but the frequency response of long-period instruments reduces the effect of the high frequency branch. The late arrivals following the Airy phase *Rg* arise from scattering and sedimentary effects.

In figure 8.12 we illustrate long-period seismograms from the WWSSN station at Atlanta, Georgia for shallow events (focal depth 25 km) in the northern part of Baja California, with an epicentral distance of 3000 km. The path is such that the North-South component (figure 8.12a) is almost perfectly transverse and so displays only Love waves, whereas the East-West component is radial and shows only the Rayleigh wave contribution (figure 8.12b). The vertical component is illustrated in figure 11.9. There is a clear contrast between the Love and Rayleigh

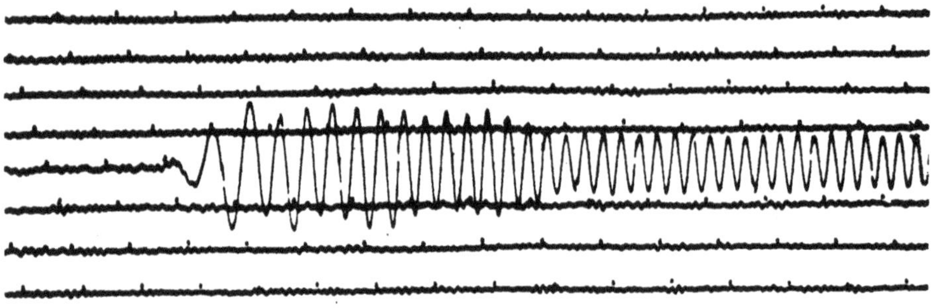

Figure 8.13. Example of Rayleigh wave dispersion for an oceanic path (WWSSN long-period record at Atlanta, Georgia).

waves; the Love waves show lower group slownesses but do not have the distinctive Airy phase. Higher frequency higher mode Rayleigh waves can often be seen in front of the main *LR* group and a good example can be seen in the broad-band Galitsin record shown in figure 1.1.

In oceanic regions, Rayleigh waves are affected by the presence of the low wavespeed material at the surface. The group slowness of the Rayleigh waves increases at higher frequencies when the wavelength is short enough to be influenced by the presence of the water layer. As a result the group slowness curve is very steep between frequencies of 0.05 and 0.1 Hz. This leads to a wavetrain with a clear long-period commencement followed by a long tail with nearly sinusoidal oscillations which is well displayed in figure 8.13.

In Chapter 11 we will discuss the dispersion of Love and Rayleigh waves and the way in which these influence the nature of the surface wave contribution to the seismograms. The dispersion of surface wave modes is controlled by the wavespeed structure along their path, whilst the excitation of the modes as a function of frequency and azimuth depends on the source mechanism. We are therefore able to achieve a partial separation of the problems of estimating the source properties and the structure of the Earth. The dispersion information can be inverted to give an estimate of the wavespeed distribution with depth and then with this information we have a linear inverse problem for the source moment tensor components.

8.4 Long range propagation

Beyond 10000 km *P* waves are diffracted along the core-mantle boundary and so their amplitude drops off with distance; this effect is particularly rapid at high frequencies. This leaves *PP* and core phases as the most prominent features on the early part of the seismogram.

The pattern of seismic phases is well illustrated by figure 8.14, a compilation of long-period WWSSN and CSN records made by Müller & Kind (1976). This event off the coast of Sumatra (1967 August 21) has a focal depth of 40 km. The focal

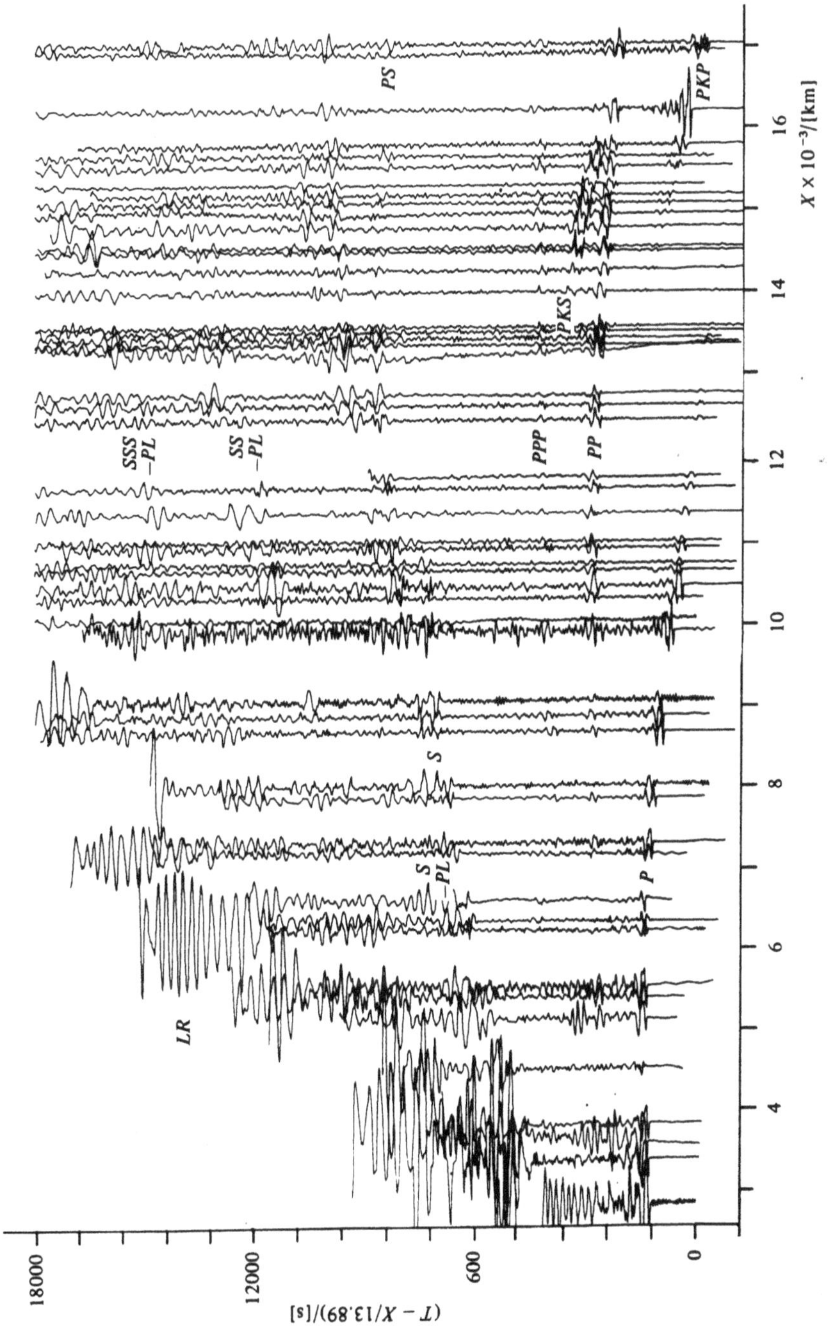

Figure 8.14. Vertical component seismogram section of an earthquake near Sumatra, as recorded by long period WWSSN and CSN stations. The amplitude scale of all traces is the same (after Müller & Kind, 1976).

mechanism has one *P* wave node nearly vertical and the other nearly horizontal; this leads to strong *S* wave radiation horizontally and vertically.

The phases on the vertical component section in figure 8.14 mirror the travel-time curves in figure 8.8. The radiation pattern is not very favourable for the excitation of *PcP* at short ranges, and near 9000 km there is insufficient time resolution to separate *P* and *PcP*. The core reflection *ScS* is also obscured, but now because of the train of long-period waves following *S*. This shear-coupled *PL* phase arises when a *SV* wave is incident at the base of the crust in the neighbourhood of the recording station at a slowness close to the *P* wave speed in the uppermost mantle. Long-period *P* disturbances excited by conversion at the crust-mantle interface then reverberate in the crust, losing energy only slowly by radiation loss into *S* waves in the mantle. Such *PL* waves are associated with *S* and its multiple reflections, and as in figure 8.14 can be the largest body wave phases on the record. No such effect occurs for *SH* waves and so *S* and *ScS* can be separated on the transverse component.

Around 9200 km *SKS* begins to arrive before *S* but can only just be discerned on figure 8.14. However, the converted phase *PKS* is quite strong and appears just after *PP*. The shadow zone caused by the core gives a couple of minutes delay between diffracted *P* and *PKP*. Near 15600 km there is a caustic for the *PKP* phase associated with very large amplitudes and this shows up very clearly on figure 8.14.

Chapter 9

Approximations to the Response of the Stratification

We have already seen how we may generate the complete response of a stratified medium, and we now turn our attention to the systematic construction of approximations to this response, with the object of understanding, and modelling, the features we have seen on the seismograms in Chapter 8. We will develop these approximations by exploiting the physical character of the solution and a very valuable tool will be the partial expansion of reverberation operators. The identity

$$[\mathbf{I}-\mathbf{R}_{\mathrm{U}}^{\mathrm{AB}}\mathbf{R}_{\mathrm{D}}^{\mathrm{BC}}]^{-1}=\mathbf{I}+\mathbf{R}_{\mathrm{U}}^{\mathrm{AB}}\mathbf{R}_{\mathrm{D}}^{\mathrm{BC}}+\mathbf{R}_{\mathrm{U}}^{\mathrm{AB}}\mathbf{R}_{\mathrm{D}}^{\mathrm{BC}}\mathbf{R}_{\mathrm{U}}^{\mathrm{AB}}\mathbf{R}_{\mathrm{D}}^{\mathrm{BC}}[\mathbf{I}-\mathbf{R}_{\mathrm{U}}^{\mathrm{AB}}\mathbf{R}_{\mathrm{D}}^{\mathrm{BC}}]^{-1}, \quad (9.1)$$

enables us to recognise the first internal multiple in 'AC', cf. (6.18), whilst retaining an exact expression for the effect of the second and all higher multiples. Higher order partial expansions may be obtained by recursive application of (9.1).

With an expansion of a representation of the full response we can identify the major constituents of the wavefield and so gain physical insight into the character of the propagation process. We can also devise techniques for extracting certain portions of the response and use the remainder terms as indicators of the conditions under which we may make such approximations. For *P-SV* wave problems we also have to consider the extent to which we can decouple the *P* and *SV* wave propagation in generating approximations to the response. Generally strong *P-SV* coupling occurs at the free surface and the core-mantle interface but significant effects can be introduced by discontinuities or very rapid changes in elastic parameters, e.g., at the Moho.

9.1 Surface reflections

We start by considering the expression (7.66) for the surface displacement generated by a buried source, in which the free-surface reverberation effects are represented explicitly

$$w_0=\mathbf{W}_{\mathrm{F}}[\mathbf{I}-\mathbf{R}_{\mathrm{D}}^{0\mathrm{L}}\mathbf{R}_{\mathrm{F}}]^{-1}\boldsymbol{\sigma}(z_{\mathrm{S}}), \quad (9.2)$$

with

$$\boldsymbol{\sigma}(z_{\mathrm{S}})=\mathbf{T}_{\mathrm{U}}^{0\mathrm{S}}[\mathbf{I}-\mathbf{R}_{\mathrm{D}}^{\mathrm{SL}}\mathbf{R}_{\mathrm{U}}^{0\mathrm{S}}]^{-1}[\boldsymbol{\Sigma}_{\mathrm{U}}(z_{\mathrm{S}})+\mathbf{R}_{\mathrm{D}}^{\mathrm{SL}}\boldsymbol{\Sigma}_{\mathrm{D}}(z_{\mathrm{S}})]. \quad (9.3)$$

We now make a partial expansion of the surface reflection operator out to the first surface reflection to give

$$w_0 = \mathbf{W}_F \left\{ \mathbf{I} + \mathbf{R}_D^{0L}\mathbf{R}_F + \mathbf{R}_D^{0L}\mathbf{R}_F\mathbf{R}_D^{0L}\mathbf{R}_F[\mathbf{I} - \mathbf{R}_D^{0L}\mathbf{R}_F]^{-1} \right\} \sigma(z_S), \tag{9.4}$$

and examine the contributions separately.

The portion of the field which has undergone no surface reflection is then

$${}^0w_0 = \mathbf{W}_F\mathbf{T}_U^{0S}[\mathbf{I} - \mathbf{R}_D^{SL}\mathbf{R}_U^{0S}]^{-1}[\Sigma_U(z_S) + \mathbf{R}_D^{SL}\Sigma_D(z_S)], \tag{9.5}$$

and this is just the displacement field we would expect if we had a uniform half space lying in $z < 0$, with a correction $\mathbf{W}_F = \mathbf{m}_{U0} + \mathbf{m}_{D0}\mathbf{R}_F$ to allow for free-surface magnification effects.

Once we remove the free surface boundary condition we make a radical change in the behaviour of the displacement representation as a function of p and ω. The poles associated with the secular function $\det(\mathbf{I} - \mathbf{R}_D^{0L}\mathbf{R}_F) = 0$ are no longer present and there are now branch points at $|p| = \alpha_0^{-1}$, $|p| = \beta_0^{-1}$ associated with the outward radiation condition at $z = 0$. The new secular function $\det(\mathbf{I} - \mathbf{R}_D^{SL}\mathbf{R}_U^{0S}) = 0$ will only have roots on the top Riemann sheet when the structure contains a significant wavespeed inversion. The poles will lie in the slowness range $\beta_0^{-1} < p < \beta_{min}^{-1}$ (assuming $\beta_L > \beta_0$), where β_{min} is the minimum shear wavespeed in the stratification.

The reverberation operator $[\mathbf{I} - \mathbf{R}_D^{SL}\mathbf{R}_U^{0S}]^{-1}$ includes all internal multiples purely within the stratification. If, therefore, we concentrate on the earliest arriving energy and neglect any such delayed reverberatory effects between the regions above and below the source we would take

$${}^0w_0 \approx \mathbf{W}_F\mathbf{T}_U^{0S}[\mathbf{R}_D^{SL}\Sigma_D(z_S) + \mathbf{\Sigma}_U(z_S)], \tag{9.6}$$

allowing for reflection beneath the source level. This portion of the response was used by Kennett & Simons (1976) to calculate the onset of seismograms for a source model of the 650 km deep earthquake of 1970 July 30 in Columbia.

When we consider the part of the response including free-surface reflections we often wish to distinguish between waves which suffer their first reflection in the neighbourhood of the source and those which have undergone reflection by the structure beneath the source level before undergoing reflection at the surface. If we allow for up to a single surface reflection, the surface displacement is given by

$${}^1w_0 = \mathbf{W}_F(\mathbf{I} + \mathbf{R}_D^{0L}\mathbf{R}_F)\sigma(z_S). \tag{9.7}$$

The wavevector $\boldsymbol{\sigma}$ is already partitioned at the source level and we may separate out internal multiples in the stratification by writing

$$\sigma(z_S) = \mathbf{T}_U^{0S} \left\{ \mathbf{I} + \mathbf{R}_D^{SL}\mathbf{R}_U^{0S}[\mathbf{I} - \mathbf{R}_D^{SL}\mathbf{R}_U^{0S}]^{-1} \right\} (\mathbf{R}_D^{SL}\mathbf{\Sigma}_D^S + \mathbf{\Sigma}_U^S) \tag{9.8}$$

where for brevity we have written e.g. $\mathbf{\Sigma}_D^S$ for $\mathbf{\Sigma}_D(z_S)$. We can also split the reflection matrix at the source level by using the addition rule (6.3) and then make a comparable expansion to (9.8) to give

$$\mathbf{R}_D^{0L} = \mathbf{R}_D^{0S} + \mathbf{T}_U^{0S}\mathbf{R}_D^{SL}\mathbf{T}_D^{0S} + \mathbf{T}_U^{0S}\mathbf{R}_D^{SL}\mathbf{R}_U^{0S}[\mathbf{I} - \mathbf{R}_D^{SL}\mathbf{R}_U^{0S}]^{-1}\mathbf{R}_D^{SL}\mathbf{T}_D^{0S}. \qquad (9.9)$$

We now insert (9.8), (9.9) into the approximate surface displacement representation (9.7) and concentrate on those parts of the wavefield which have not undergone any internal multiples in the stratification. Thus

$$\begin{aligned} {}^1w_0 = \mathbf{W}_F \Big\{ & \mathbf{T}_U^{0S}\mathbf{\Sigma}_U^S + \mathbf{T}_U^{0S}\mathbf{R}_D^{SL}(\mathbf{\Sigma}_D^S + \mathbf{T}_D^{0S}\mathbf{R}_F\mathbf{T}_U^{0S}\mathbf{\Sigma}_U^S) \\ & + \mathbf{T}_U^{0S}\mathbf{R}_D^{SL}\mathbf{T}_D^{0S}\mathbf{R}_F\mathbf{T}_U^{0S}\mathbf{R}_D^{SL}\mathbf{\Sigma}_D^S \\ & + \mathbf{R}_D^{0S}\mathbf{R}_F\mathbf{T}_U^{0S}(\mathbf{R}_D^{SL}\mathbf{\Sigma}_D^S + \mathbf{\Sigma}_U^S) + \text{Remainder} \Big\}. \end{aligned} \qquad (9.10)$$

The remainder takes account of all contributions with internal multiples of the type $\mathbf{R}_D^{SL}\mathbf{R}_U^{0S}$. The nature of the entries in (9.9) is indicated schematically in figure 9.1.

Direct upward propagation is represented by $\mathbf{T}_U^{0S}\mathbf{\Sigma}_U^S$ and that part of the energy which initially departed downward from the source, but which has been reflected back from beneath the source level, appears in $\mathbf{T}_U^{0S}\mathbf{R}_D^{SL}\mathbf{\Sigma}_D^S$. The combination $\mathbf{T}_U^{0S}(\mathbf{R}_D^{SL}\mathbf{\Sigma}_D^S + \mathbf{\Sigma}_U^S)$ will therefore represent the main *P* and *S* wave arrivals and will allow for the possibility of conversions beneath the source level through $\mathbf{R}_D^{SL}$.

The term

$$\mathbf{T}_U^{0S}\mathbf{R}_D^{SL}\mathbf{T}_D^{0S}\mathbf{R}_F\mathbf{T}_U^{0S}\mathbf{\Sigma}_U^S \qquad (9.11)$$

represents energy which initially was radiated upwards, but which has been reflected at the free surface before reflection beneath the source. In teleseismic work this term represents the surface reflected phases *pP*, *sP* and *sS*, *pS* generated near the source (figure 9.1a). As we have noted above, *P* to *S* conversion can be quite efficient at the free surface so that the off-diagonal terms in $\mathbf{R}_F$ are often important. The converted phases *sP*, *pS* can be quite large and have a significant influence on the character of the seismograms (cf. figure 8.11). The combined term

$$\mathbf{T}_U^{0S}\mathbf{R}_D^{SL}(\mathbf{\Sigma}_D^S + \mathbf{T}_D^{0S}\mathbf{R}_F\mathbf{T}_U^{0S}\mathbf{\Sigma}_U^S) \qquad (9.12)$$

represents all the energy which has been returned once from beneath the level of the source. The combination $(\mathbf{\Sigma}_D^S + \mathbf{T}_D^{0S}\mathbf{R}_F\mathbf{T}_U^{0S}\mathbf{\Sigma}_U^S)$ will appear as an equivalent source term for downward radiation and this forms the basis of the approximate technique for long-range propagation described by Langston & Helmberger (1975).

In reflection seismic work the *pP* reflection appears as the surface 'ghost' reflection associated with a shallowly buried source. The effective source waveform for reflection from deep horizons is provided by the interference of the downward radiation and the surface reflection in $(\mathbf{\Sigma}_D^S + \mathbf{T}_D^{0S}\mathbf{R}_F\mathbf{T}_U^{0S}\mathbf{\Sigma}_U^S)$. It is frequently assumed that the effective waveform may be evaluated at $p = 0$ and does not vary

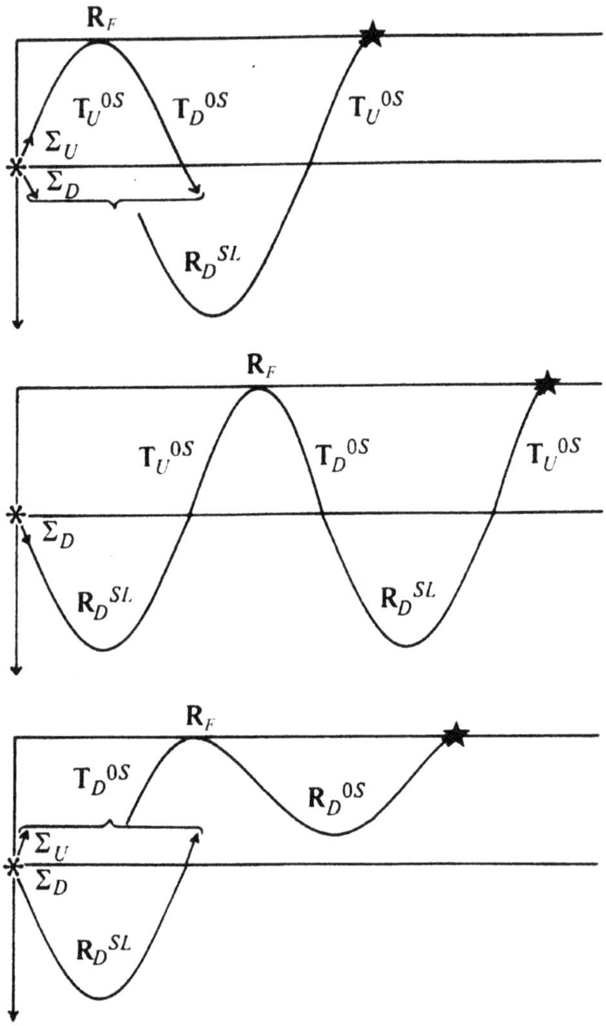

Figure 9.1. Illustration of the contributions to the response appearing in equation (9.9): a) Direct *P,S* and surface reflected phases *pP, sP, pS, sS*; b) Double reflection beneath the source *PP, PS, SP, SS*; c) Reverberations near the receiver.

with the incident angle on a reflector. A more accurate representation is obtained by allowing for the full slowness dependence of the interference.

A further class of surface reflections arises from waves which departed downwards from the source;

$$\mathbf{T}_U^{0S}\mathbf{R}_D^{SL}\mathbf{T}_D^{0S}\mathbf{R}_F\mathbf{T}_U^{0S}\mathbf{R}_D^{SL}\mathbf{\Sigma}_D^{S} \qquad (9.13)$$

represents energy which has been reflected back twice from below the level of the source (figure 9.1b). The first part is similar to straight *P* or *S* propagation, and then on reflection we get *PP*, *SS* and with conversion *PS*, *SP*. In long-range propagation the major contribution to the response will arise from nearly the same level on the two sides of the surface reflection point.

In reflection seismology we are interested in the total contribution to surface-generated multiple reflections and so have to include the effect of the 'ghost' reflection as well. The first surface multiples are therefore generated by

$$\mathbf{T}_U^{0S}\mathbf{R}_D^{SL}\mathbf{T}_D^{0S}\mathbf{R}_F\mathbf{T}_U^{0S}\mathbf{R}_D^{SL}(\Sigma_D^S + \mathbf{T}_D^{0S}\mathbf{R}_F\mathbf{T}_U^{0S}\Sigma_U^S), \tag{9.14}$$

and now there is rarely any symmetry in the reflection process. Surface-generated multiples are of particular importance in areas with shallow water cover or low velocity material near the surface because they tend to obscure the reflected arrivals of interest (see figure 8.2). Considerable effort has therefore been devoted to methods designed to eliminate such reflections from observed records and in following sections we will discuss the theoretical basis of such methods.

The last class of contributions which appear in (9.10) are governed by $\mathbf{R}_D^{0S}\mathbf{R}_F\mathbf{T}_U^{0S}(\mathbf{R}_D^{SL}\Sigma_D^S + \Sigma_U^S)$, and arise from the beginning of a reverberation sequence, near the receiver, between the free surface and the layering above the source (figure 9.1c). The higher terms arise from the reverberation operator $[\mathbf{I} - \mathbf{R}_D^{0S}\mathbf{R}_F]^{-1}$. This is most easily seen from the expressions (7.41), (7.46) for the surface displacement; the matrix $\mathbf{W}_U^{fS}$ which generates displacement at the surface from an upgoing wave at the source level includes this shallow reverberation operator.

9.1.1 Surface multiples

At moderate to large ranges from a source, free-surface reflections play an important role in determining the shape of the *P* and *S* waveforms due to the interference of the direct and reflected phases. The surface waves also owe their existence to the interaction of waves which have been reflected many times from the surface.

At short ranges the presence of the highly reflecting free surface is of major importance in determining the character of the entire seismogram. Waves which have been reflected at the surface and back from the stratification are important for both recordings of near earthquakes and reflection seismology. The profound effect of such multiples is well illustrated by theoretical seismograms for the full response (9.2) of an elastic model and approximate calculations neglecting any free-surface reflections using (9.5). The elastic wavespeed distributions and density for the very simple model we have used are illustrated in figure 9.2. We have taken $Q_\alpha^{-1} = 0.001$ and $Q_\beta^{-1} = 0.002$ in all layers, these values are too small to be realistic but help to avoid aliasing problems in time. For both calculations the source was an explosion at 10 m depth and a frequency band from 5 to 75 Hz has been constructed using a spectral approach (as in Section 7.3.3) and we present the vertical component traces in figure 9.3.

In the absence of surface reflections (figure 9.3a) the principal events are *P* wave

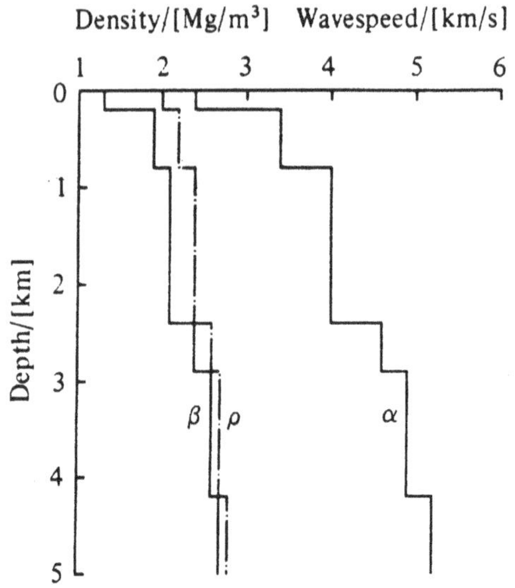

Figure 9.2. Sedimentary structure used for the calculations in figure 9.3.

reflections from the interfaces and the reflection times at vertical incidence are indicated by solid triangles. The model parameters are such that *P* to *S* conversion is quite efficient at the first interface and there are a number of reflected events associated with such conversions for which the amplitude increases away from the closest traces. The most important conversions correspond to *S* wave reflections from the second or deeper interfaces with either *P* or *S* upward legs in the shallowest layer. Even though there are no extreme contrasts in elastic parameters across the interfaces in this model, internal multiples between interfaces are important and complicate the reflection pattern. For the shallowest interface the array of receivers extends beyond the critical distance and a clear head wave is seen to emerge from the first interface reflection and to overtake the direct wave.

When all surface reflections are included (figure 9.3b) we obtain the previous set of reflections, and in addition all their surface multiples, which leads to a rather complex pattern. The first surface multiples of the *P* reflections are very clear and their vertical reflection times are indicated by open triangles. The apparent waveform of the primary reflection is now modified by the interference of initially downward propagating waves with the surface reflected 'ghosts' and is of longer duration than before. A new feature is the prominent ground-roll with high group slowness which cuts across the reflected phases. This is preceded by a weak *S* wave generated by surface conversion of the direct *P* wave and *S* reflection at the first interface. The shallow explosion gives significant excitation of fundamental mode Rayleigh waves which are the main component of the ground-roll; there is much weaker excitation of higher modes. Normal field procedures are designed to

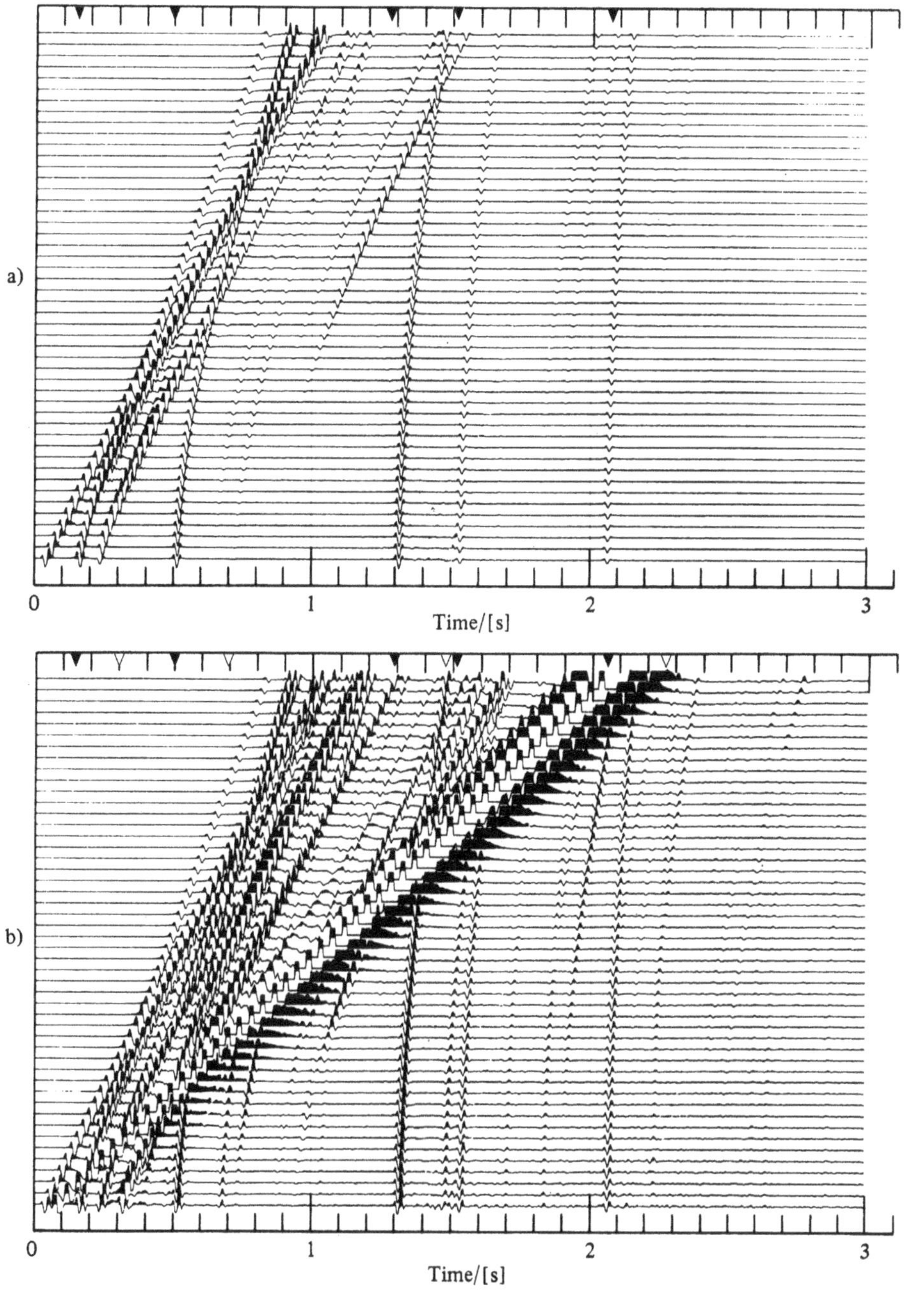

Figure 9.3. Theoretical seismograms for the sedimentary structure illustrated in figure 9.2. The vertical component of velocity is illustrated and a linear gain in time and distance has been applied: a) no free surface reflections, but all internal multiples and conversions; b) all free surface multiples and surface waves.

eliminate the ground-roll by using arrays of geophones at each receiver point. The surface waves are, however, often seen on noise spreads or where single geophone recording has been used (figure 1.3). The dispersion of the surface wave modes is most sensitive to the shallow S wave structure (see Chapter 11) and can itself provide valuable structural information.

For interpretation of seismic reflection records it is desirable that only simple P wave returns from each reflector should be displayed and considerable efforts have therefore been expended on trying to remove all other wave phenomena. The most important of these is, as we have seen, the effect of the free surface, and it is interesting to see how surface multiples could be removed with ideal data.

We will consider surface receivers and specialise to a surface source. In this case the full response of the half space in the frequency-slowness domain may be expressed as

$$w_0 = \mathbf{W}_F[\mathbf{I} - \mathbf{R}_D^{0L}\mathbf{R}_F]^{-1}(\boldsymbol{\Sigma}_U^0 + \mathbf{R}_D^{0L}\boldsymbol{\Sigma}_D^0), \tag{9.15}$$

where $\boldsymbol{\Sigma}_U^0$, $\boldsymbol{\Sigma}_D^0$ are the up and downgoing wave components which would be produced by the source embedded in a uniform medium with the surface properties. We can recast (9.15) in a form which accentuates the portion of the field which is reflected back by the stratification:

$$w_0 = \mathbf{W}_F[\mathbf{I} - \mathbf{R}_D^{0L}\mathbf{R}_F]^{-1}\mathbf{R}_D^{0L}\{\boldsymbol{\Sigma}_D^0 + \mathbf{R}_F\boldsymbol{\Sigma}_U^0\} + \mathbf{W}_F\boldsymbol{\Sigma}_U^0. \tag{9.16}$$

Here $\{\boldsymbol{\Sigma}_D^0 + \mathbf{R}_F\boldsymbol{\Sigma}_U^0\}$ represents the effective downward radiation from the source in the presence of the free surface.

The part of the seismic field which has not undergone any surface reflections is

$${}^0w_0 = \mathbf{W}_F\mathbf{R}_D^{0L}\{\boldsymbol{\Sigma}_D^0 + \mathbf{R}_F\boldsymbol{\Sigma}_U^0\} + \mathbf{W}_F\boldsymbol{\Sigma}_U^0. \tag{9.17}$$

The portion which is of interest in reflection work is therefore, for a specified source, dependent on the matrix $\mathbf{W}_F\mathbf{R}_D^{0L}$. We see from figure 9.3a that for small offsets the dominant contribution will come from the PP component of $\mathbf{R}_D^{0L}$. If therefore we can recover $\mathbf{R}_D^{0L}$ from the full response w_0 we could remove all surface reflections by constructing (9.12).

In terms of the operator

$$\mathbf{Y} = [\mathbf{I} - \mathbf{R}_D^{0L}\mathbf{R}_F]^{-1}\mathbf{R}_D^{0L}, \tag{9.18}$$

the full response w_0 takes the form,

$$w_0 = \mathbf{W}_F\mathbf{Y}\{\boldsymbol{\Sigma}_D^0 + \mathbf{R}_F\boldsymbol{\Sigma}_U^0\} + \mathbf{W}_F\boldsymbol{\Sigma}_U^0; \tag{9.19}$$

and the downward reflection matrix $\mathbf{R}_D^{0L}$ can be recovered from

$$\mathbf{R}_D^{0L} = \mathbf{Y}[\mathbf{I} + \mathbf{R}_F\mathbf{Y}]^{-1}. \tag{9.20}$$

Is it therefore possible to recover $\mathbf{Y}$ from (9.13)? Since $\mathbf{Y}$ is a 2×2 matrix we would need two distinct vector equations of the form (9.19) to find $\mathbf{Y}$, i.e. two experiments with recordings of both vertical and horizontal components for different sources.

This result parallels the work of Kennett (1979b) who worked with expressions based on the equivalent stress and displacement source elements at the surface (7.16). From vertical component seismograms we may estimate the PP component of $\mathbf{Y}$, allowing for the effect of shear at the surface through $\mathbf{W}_F^{UP}$ but neglecting conversion at depth. The PP element of $\mathbf{R}_D^{0L}$ is then given approximately by

$$R_D^{PP} \approx Y^{PP}/[1 + R_F^{PP} Y^{PP}]. \tag{9.21}$$

Kennett (1979b) has shown that such a scheme gives good results for theoretical records at small offsets from the source. With field recordings a major difficulty is to get an adequate representation of the seismograms in the frequency-slowness domain (see, e.g., Henry, Orcutt & Parker, 1980).

We may extend this treatment to buried sources by making use of *equivalent* sources. In (7.7) we have introduced an equivalent stress-displacement jump at the surface via a propagator from the source depth to the surface. We can do the same with the wavevector jump with the aid of the wave-propagator

$$\boldsymbol{\Sigma}^0 = \mathbf{Q}(0, z_S)\boldsymbol{\Sigma}^S. \tag{9.22}$$

From (5.45) we can express the wave-propagator in terms of the reflection and transmission properties of the region between the source and the surface, and so, using (4.66),

$$\begin{aligned} \boldsymbol{\Sigma}_D^0 &= (\mathbf{T}_D^{0S})^{-1}(\boldsymbol{\Sigma}_D^S + \mathbf{R}_U^{0S}\boldsymbol{\Sigma}_U^S), \\ \boldsymbol{\Sigma}_U^0 &= \mathbf{T}_U^{0S}\boldsymbol{\Sigma}_U^S + \mathbf{R}_D^{0S}\boldsymbol{\Sigma}_D^0, \end{aligned} \tag{9.23}$$

and if we construct the combination of equivalent source terms $\boldsymbol{\sigma}(0+)$ we have

$$\mathbf{R}_D^{0L}\boldsymbol{\Sigma}_D^0 + \boldsymbol{\Sigma}_U^0 = \mathbf{T}_U^{0S}[\mathbf{I} - \mathbf{R}_D^{SL}\mathbf{R}_U^{0S}]^{-1}[\boldsymbol{\Sigma}_U^S + \mathbf{R}_D^{SL}\boldsymbol{\Sigma}_D^S], \tag{9.24}$$

which is just the original $\boldsymbol{\sigma}(z_S)$ and so the surface displacement is the same. If we know the elastic properties down to the source level z_S we can find $\boldsymbol{\Sigma}_U^0$, $\boldsymbol{\Sigma}_D^0$ from (9.23) and then proceed as before. The term $(\mathbf{T}_D^{0S})^{-1}$ in (9.23) represents the advance in time and amplitude gain needed to be able to describe downgoing waves starting at the true origin time at the source level z_S, via a surface source.

9.1.2 Alternative approach for surface reflections

We have so far made use of representations of the seismic response which display the free surface reflection matrix $\mathbf{R}_F$ explicitly (7.66), but as we have seen in Chapter 7 we have the alternative form (7.41), (7.46)

$$w_0 = \mathbf{W}_U^{fS}[\mathbf{I} - \mathbf{R}_D^{SL}\mathbf{R}_U^{fS}]^{-1}(\boldsymbol{\Sigma}_U^S + \mathbf{R}_D^{SL}\boldsymbol{\Sigma}_D^S). \tag{9.25}$$

The operator $\mathbf{W}_U^{fS}$ generates surface displacement from upgoing waves at the source level z_S (6.11)

$$\mathbf{W}_U^{fS} = \mathbf{W}_F[\mathbf{I} - \mathbf{R}_D^{0S}\mathbf{R}_F]^{-1}\mathbf{T}_U^{0S}. \tag{9.26}$$

We may now rearrange (9.25) to isolate the effect of the direct upgoing waves ($\mathbf{W}_{U}^{fS}\mathbf{\Sigma}_{U}^{S}$), and to emphasise the role of reflection above the source level

$$w_0 = \mathbf{W}_{U}^{fS}[\mathbf{I} - \mathbf{R}_{D}^{SL}\mathbf{R}_{U}^{fS}]^{-1}\mathbf{R}_{D}^{SL}\{\mathbf{\Sigma}_{D}^{S} + \mathbf{R}_{U}^{fS}\mathbf{\Sigma}_{U}^{S}\} + \mathbf{W}_{U}^{fS}\mathbf{\Sigma}_{U}^{S}. \tag{9.27}$$

The combination $\{\mathbf{\Sigma}_{D}^{S} + \mathbf{R}_{U}^{fS}\mathbf{\Sigma}_{U}^{S}\}$ models the direct downward radiation from the source and all waves turned back by the structure or the free surface above the source. In the previous discussion of the *pP* phase we have only allowed for transmission in the passage of waves between the source and the surface. When, however, there is significant structure above the source $\mathbf{R}_{U}^{fS}\mathbf{\Sigma}_{U}^{S}$ provides a better representation of the reflected phases. An important case is for earthquakes which occur just below the Moho, for which reflections from the Moho immediately follow the direct *P* and *S* waves. For suboceanic earthquakes the reflection from the water/rock interface (sometimes designated *pwP*) can approach the size of the reflection from the sea surface.

9.2 Split stratification

As we have already noted it is very common for there to be a near-surface zone of low wavespeeds, and this feature occurs on a wide range of scales. In reflection work this region would be the weathered zone, and in crustal studies the sedimentary layers. For seismic studies of the ocean floor, the ocean itself acts as a low velocity waveguide. When attention is focussed on seismic wave propagation in the mantle or core, the entire crust appears to be a zone of low wavespeeds.

Often we would like to separate wave propagation effects in this shallow region from the waves which penetrate more deeply. In order to do this we introduce a level z_J at which we split the stratification, and this will normally be taken close to the base of the surface zone.

We can obtain very useful results for the displacement field by making use of the device of *equivalent* sources. We move the source effects from the true source level z_S, to the separation level z_J and then use the equivalent source terms in our previous expressions for the seismic response.

The equivalent stress-displacement jump $\mathbf{S}(z_J)$ at $z = z_J$ is obtained from $\mathbf{S}(z_S)$, the representation of the original source, by the action of the propagator from z_S to z_J.

$$\mathbf{S}(z_j) = \mathbf{P}(z_J, z_S)\mathbf{S}(z_S). \tag{9.28}$$

This equivalent source will give the same effective seismic radiation as the original. As in the surface case, we can also introduce an equivalent wavevector jump $\mathbf{\Sigma}^J$ at z_J which is related to $\mathbf{\Sigma}^S$ by the wave-propagator from z_S to z_J

$$\mathbf{\Sigma}^J = \mathbf{Q}(z_J, z_S)\mathbf{\Sigma}^S. \tag{9.29}$$

We will discuss first the case when the source lies in the low wave speed zone i.e. $z_S < z_J$. The wave-propagator can then be related to the reflection and transmission properties of the region between z_S and z_J using (5.46). The upward and downward radiation components at z_J are therefore

$$\begin{aligned}\mathbf{\Sigma}_U^J &= [\mathbf{T}_U^{SJ}]^{-1}(\mathbf{\Sigma}_U^S + \mathbf{R}_D^{SJ}\mathbf{\Sigma}_D^S),\\ \mathbf{\Sigma}_D^J &= \mathbf{T}_D^{SJ}\mathbf{\Sigma}_D^S + \mathbf{R}_U^{SJ}[\mathbf{T}_U^{SJ}]^{-1}(\mathbf{\Sigma}_U^S + \mathbf{R}_D^{SJ}\mathbf{\Sigma}_D^S).\end{aligned} \tag{9.30}$$

The contribution $[\mathbf{T}_U^{SJ}]^{-1}$ arises because we are attempting to move upgoing waves back along their propagation path. However, the expression $\mathbf{C}_D^J = \{\mathbf{\Sigma}_D^J + \mathbf{R}_U^{fJ}\mathbf{\Sigma}_U^J\}$ corresponding to the net downward radiation at the level z_J is free of such terms,

$$\begin{aligned}\mathbf{C}_D^J &= \mathbf{\Sigma}_D^J + \mathbf{R}_U^{fJ}\mathbf{\Sigma}_U^J,\\ &= \mathbf{T}_D^{SJ}[\mathbf{I} - \mathbf{R}_U^{fS}\mathbf{R}_D^{SJ}]^{-1}\{\mathbf{\Sigma}_D^S + \mathbf{R}_U^{fS}\mathbf{\Sigma}_U^S\}.\end{aligned} \tag{9.31}$$

The right hand side of (9.31) can be recognised as the total downward radiation at the level z_J produced by a source at the level z_S; this expression allows for reverberation between the surface and z_J, in the neighbourhood of the source.

We will now use these equivalent source expressions in the representation (9.27) for the seismic response

$$w_0 = \mathbf{W}_U^{fJ}[\mathbf{I} - \mathbf{R}_D^{JL}\mathbf{R}_U^{fJ}]^{-1}\mathbf{R}_D^{JL}\{\mathbf{\Sigma}_D^J + \mathbf{R}_U^{fJ}\mathbf{\Sigma}_U^J\} + \mathbf{W}_U^{fJ}\mathbf{\Sigma}_U^J. \tag{9.32}$$

The 'upward radiation' term

$$\mathbf{W}_U^{fJ}\mathbf{\Sigma}_U^J = \mathbf{W}_U^{fS}[\mathbf{I} - \mathbf{R}_D^{SJ}\mathbf{R}_U^{fS}]^{-1}\mathbf{T}_U^{SJ}[\mathbf{T}_U^{SJ}]^{-1}(\mathbf{\Sigma}_U^S + \mathbf{R}_D^{SJ}\mathbf{\Sigma}_D^S), \tag{9.33}$$

where we have used the expansion (6.11) for $\mathbf{W}_U^{fJ}$ and so

$$\mathbf{W}_U^{fJ}\mathbf{\Sigma}_U^J = \mathbf{W}_U^{fS}[\mathbf{I} - \mathbf{R}_D^{SJ}\mathbf{R}_U^{fS}]^{-1}(\mathbf{\Sigma}_U^S + \mathbf{R}_D^{SJ}\mathbf{\Sigma}_D^S). \tag{9.34}$$

We recognise this to be just the response of the stratified half space, truncated at the level z_J, to a source at z_S. The expression (9.34) therefore describes waves whose propagation is confined to the region of low wavespeeds.

The remaining contribution to the displacement takes the form

$$\mathbf{W}_U^{fJ}[\mathbf{I} - \mathbf{R}_D^{JL}\mathbf{R}_U^{fJ}]^{-1}\mathbf{R}_D^{JL}\mathbf{C}_D^J, \tag{9.35}$$

and includes all propagation effects below the separation level z_J through the reflection matrix $\mathbf{R}_D^{JL}$. The whole half space reverberation operator appears in (9.35) and near-receiver interactions in the low wavespeed zone are contained in $\mathbf{W}_U^{fJ}$.

The expression (9.32) for the seismic displacements is valid for an arbitrary level z_J below the source level z_S, and so we may use it to understand the approximations involved when we consider only a limited portion of an Earth model. For a structure terminated at z_J the entire response is contained in $\mathbf{W}_U^{fJ}\mathbf{\Sigma}_U^J$ (9.34) and the contribution (9.35) can be thought of as the error term corresponding to the neglect of deeper parts of the model. This remainder contains the secular operator for the shallow part through the source and receiver terms $\mathbf{C}_D^J$, $\mathbf{W}_U^{fJ}$. The neglected reflections from depth appear in $\mathbf{R}_D^{JL}$ and these couple into the shallower structure through the reverberation operator from the surface down to z_L. In general the truncated structure will provide an adequate approximation for small ranges, and at long-range may be used to describe certain features of the full response.

9.2.1 Propagation in the upper zone

We have just seen that by choosing a separation level z_J in the half space we can isolate the portion of the response which is confined to the region between the surface and z_J.

The choice of z_J is therefore crucial in defining the particular seismic phases which are well represented by $\mathbf{W}_U^{fJ}\Sigma_U^J$ (9.34). This contribution

$$^{sh}w_0 = \mathbf{W}_U^{fS}[\mathbf{I} - \mathbf{R}_D^{SJ}\mathbf{R}_U^{fS}]^{-1}(\Sigma_U^S + \mathbf{R}_D^{SJ}\Sigma_D^S), \qquad (9.36)$$

gives a full representation of all reverberations in the region between z_J and the free surface.

For seismic studies of the ocean floor, a choice of z_J just below the surface of the seismic basement will give an excellent description of the water wavetrain. In this case $^{sh}w_0$ will describe propagation in the water column itself, the effects of the marine sediments and the main sub-bottom reflection from the sediment/basement transition. The uniform half space below this interface will allow for radiation loss from the acoustic field in the water into both *P* and *S* waves. Within the water we have only *P* waves, but in the sediments and below we must allow for shear wave conversion effects. The reflection matrix for the composite region 'SJ', including both fluid and solid zones, can be constructed by the usual recursion scheme, as discussed in the appendix to Chapter 6. The waves returned by the oceanic crust itself will be described by the deeper contribution (9.35).

For continental areas with extensive sediment cover we would choose z_J to lie just below the base of the sediments and then we may isolate all phases which are confined to the sediments in $^{sh}w_0$ or are reflected from it.

In studies of the earth's mantle, the choice of separation level is less obvious. It is tempting to take z_J just below the crust-mantle interface and so model phases which are trapped in the crust alone, principally *Pg*, *Lg* and *PL* (see figures 8.4–8.7). This is basically the approach taken by Helmberger & Engen (1980) who have modelled long-period *P* wave-propagation in a simple crust out to regional distances (1500 km). In this case the Pn arrival is represented by a head wave on the crust-mantle interface and multiple reverberations in the crust give a complex coda in which the proportion of *SV* wave motion increases with time. However, this choice of z_J leads to neglect of sub-Moho velocity gradients on the *Pn* arrivals and also of the influence of any mantle wavespeed inversion. At long-periods the wavelengths are sufficiently large that a *P* wave with an apparent turning point some way above the inversion will lose energy by tunnelling through the intervening evanescent region (figure 9.4).

For a full description of crustal and subcrustal phases a convenient location for z_J is at about 200 km; which for most continental models, at least, is below any wavespeed inversion and about the level of a change in the velocity gradients in the mantle. This zone includes nearly all the reverberative features in the entire Earth model and allows interactions between the crust and the wavespeed inversion to be modelled. As we shall see in Chapter 11, such effects have a strong influence on moderate frequency (~ 0.2 Hz) surface wave propagation. For high frequency waves the immediate sub-Moho wavespeed gradients determine the character of the *Pn* and *Sn* arrivals. Even a slight positive gradient induced by sphericity is sufficient to produce an 'interference' head wave (Červený & Ravindra, 1975) in which multiple reflections from the gradients constructively interfere to give a larger arrival than can occur with a uniform half space (figure 9.5).

This type of propagation has been investigated by Menke & Richards (1980) who term the components of the interference head wave 'whispering gallery' phases. In the reflection matrix representation, if conversion below the Moho can be neglected, these contributions arise from the infinite expansion of the *P* wave reflection coefficient just above the Moho:

$$\begin{aligned} \mathbf{R}_D(z_M-) &= \mathbf{R}_D^M + \mathbf{T}_U^M\mathbf{R}_D^{MJ}[\mathbf{I} - \mathbf{R}_U^M\mathbf{R}_D^{MJ}]^{-1}\mathbf{T}_D^M, \\ &= \mathbf{R}_D^M + \mathbf{T}_U^M\mathbf{R}_D^{MJ}\mathbf{T}_D^M + \mathbf{T}_U^M\mathbf{R}_D^{MJ}\mathbf{R}_U^M\mathbf{R}_D^{MJ}\mathbf{T}_D^M + \dots. \end{aligned} \tag{9.37}$$

where $\mathbf{R}_D^M$ etc. are the interface matrices at the Moho, and $\mathbf{R}_D^{MJ}$ is the reflection matrix for the sub-Moho zone. The principal contribution will normally come from $\mathbf{T}_U^M\mathbf{R}_D^{MJ}\mathbf{T}_D^M$. At short ranges the geometrical picture for the *P* wave portion is a ray

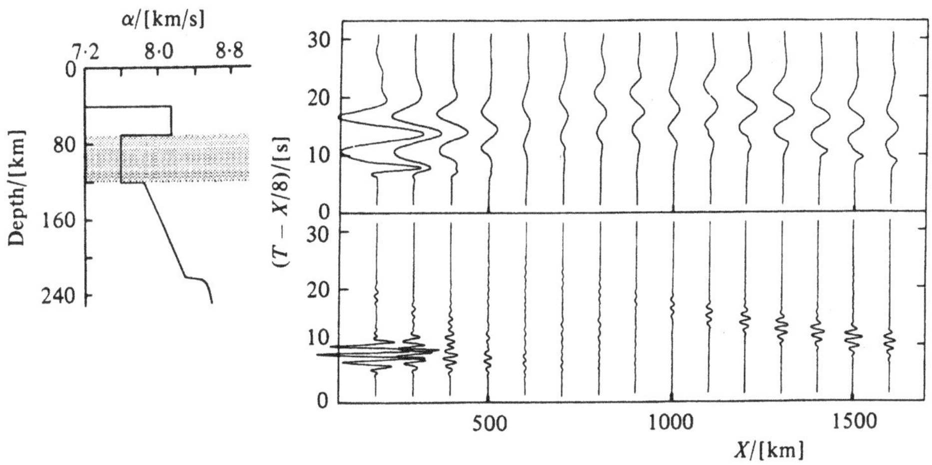

Figure 9.4. Illustration of the effect of a thin lid over a velocity inversion: low frequency energy tunnels through the inversion and is turned back by the deeper structure, whilst the high frequencies are reflected by the lid or propagate within it.

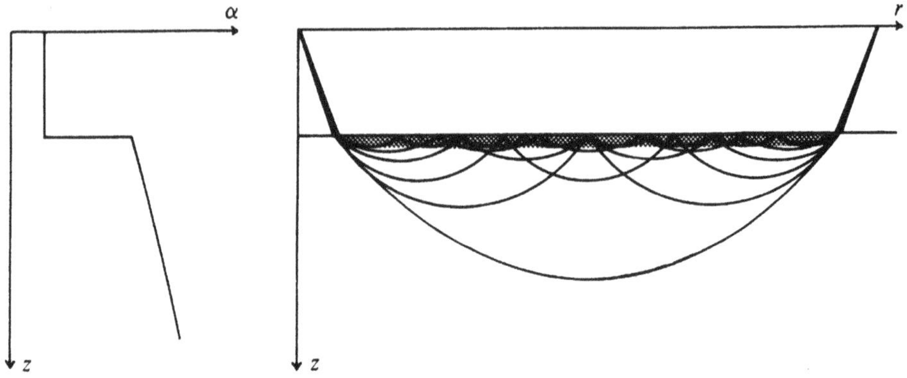

Figure 9.5. Illustration of the formation of an interference head wave.

turned back by the sub-Moho velocity gradients; this is the behaviour shown by the high frequencies, but there is a progressive loss of high frequencies by tunnelling into the wavespeed inversion. At large ranges the *P* turning point for $\mathbf{T}_U^M\mathbf{R}_D^{MJ}\mathbf{T}_D^M$ drops to below the wavespeed inversion and this contribution to *Pn* is lost. Low wavespeed gradients favour the development of the interference head wave to large ranges.

The reverberation operator representation in (9.37) is much more convenient than the expansion for computational purposes since the full interference effects are retained.

The sub-Moho structure has a significant effect on short-period crustal reverberations. Menke & Richards have pointed out that crustal reverberations are enhanced at large ranges for models with high sub-Moho gradients.

A similar system can be excited at large ranges by an incident *SV* wave at the base of the crust, arising from the presence of the contribution $\mathbf{W}_U^{fJ}\mathbf{R}_D^{JL}\mathbf{T}_D^{SJ}\Sigma_D^S$ in the full response. For slownesses such that *S* waves propagate in the crust and mantle, but *P* waves have turning points in the sub-Moho gradients, we can get an interference head wave for *P* generated by conversion at the Moho. This wave will be coupled into both *P* and *S* wave reverberations in the crust to give the shear-coupled PL waves (figure 8.14) following the main *SV* arrivals (see, e.g., Poupinet & Wright 1972). The *SV* pulse will be preceded by a conversion to *P* on transmission through the Moho.

9.2.2 Deeper propagation

For a source in the upper zone above the the separation level z_J the portion of the displacement response which involves propagation in the region below z_J is given by (9.27) and with the expanded forms for $\mathbf{C}^{\mathrm{J}}_{\mathrm{D}}$ (9.24) and $\mathbf{W}^{\mathrm{fJ}}_{\mathrm{U}}$ (6.12) we have

$$\begin{aligned} {}^{\mathrm{d}}w_0 = \mathbf{W}_{\mathrm{F}}[\mathbf{I} - \mathbf{R}^{0\mathrm{J}}_{\mathrm{D}}\mathbf{R}_{\mathrm{F}}]^{-1}\mathbf{T}^{0\mathrm{J}}_{\mathrm{U}} \\ \times[\mathbf{I} - \mathbf{R}^{\mathrm{JL}}_{\mathrm{D}}\mathbf{R}^{\mathrm{fJ}}_{\mathrm{U}}]^{-1}\mathbf{R}^{\mathrm{JL}}_{\mathrm{D}} \\ \times\mathbf{T}^{\mathrm{SJ}}_{\mathrm{D}}[\mathbf{I} - \mathbf{R}^{\mathrm{fS}}_{\mathrm{U}}\mathbf{R}^{\mathrm{SJ}}_{\mathrm{D}}]^{-1}\{\boldsymbol{\Sigma}^{\mathrm{S}}_{\mathrm{D}} + \mathbf{R}^{\mathrm{fS}}_{\mathrm{U}}\boldsymbol{\Sigma}^{\mathrm{S}}_{\mathrm{U}}\}. \end{aligned} \tag{9.38}$$

This representation allows for reverberations in the upper zone near the source through $[\mathbf{I}-\mathbf{R}^{\mathrm{fS}}_{\mathrm{U}}\mathbf{R}^{\mathrm{SJ}}_{\mathrm{D}}]^{-1}$ and similar effects near the receiver through $[\mathbf{I}-\mathbf{R}^{0\mathrm{J}}_{\mathrm{D}}\mathbf{R}_{\mathrm{F}}]^{-1}$. For a surface source these reverberation operators take on a symmetric form.

In reflection studies, multiples of deep seated reflections (in $\mathbf{R}^{\mathrm{JL}}_{\mathrm{D}}$) generated by reflection between the free surface and high contrast interfaces near the surface are of considerable importance. Such multiples can be particularly strong for reflection profiles conducted in shallow shelf seas. The strength of these effects depends on the reflectivity of the sea bed which can be as high as 0.4 at normal incidence. The automatic gain applied to most reflection displays acts to compensate for the multiple reflection losses and so multiples dominate giving a ‘singing’ record. Some success has been achieved with predictive convolution operators (Peacock & Treitel, 1969), which rely on the relative statistical randomness of the reflection elements in $\mathbf{R}^{\mathrm{JL}}_{\mathrm{D}}$ (with z_{J} just below sea bed), compared to the organised multiple train. If a good model exists for the sea bed structure, or if the reflection from the sea bed can be isolated, the reflection coefficient $\mathbf{R}^{0\mathrm{J}}_{\mathrm{D}}(p,\omega)$ can be estimated. Then, in favourable circumstances, it is possible to use the operator $[\mathbf{I}+\mathbf{R}^{0\mathrm{J}}_{\mathrm{D}}(p,\omega)]$ twice on the acoustic wavefield in the slowness-frequency domain, to suppress most of the near-surface multiples.

In regions of deep weathering on land, multiples in the weathered zone can also be important but here with only vertical component records it is more difficult to achieve multiple suppression.

Even when the shallow multiples are eliminated, the deep reflections are entangled in the combination

$$[\mathbf{I} - \mathbf{R}^{\mathrm{JL}}_{\mathrm{D}}\mathbf{R}^{\mathrm{fJ}}_{\mathrm{U}}]^{-1}\mathbf{R}^{\mathrm{JL}}_{\mathrm{D}}, \tag{9.39}$$

which represents the surface-generated multiples of the deep reflections. This is in the same form as the quantity $\mathbf{Y}$ introduced in (9.18) and the analysis is very similar to the previous case.

The expression ${}^{\mathrm{d}}w_0$ (9.35) for the displacement includes the reverberation operator for the whole half space through the contribution (9.39) but often we wish to restrict attention to a single reflection from beneath z_{J}. We may do this by making a partial expansion of (9.39)

$$\mathbf{R}^{\mathrm{JL}}_{\mathrm{D}} + [\mathbf{I} - \mathbf{R}^{\mathrm{JL}}_{\mathrm{D}}\mathbf{R}^{\mathrm{fJ}}_{\mathrm{U}}]^{-1}\mathbf{R}^{\mathrm{JL}}_{\mathrm{D}}\mathbf{R}^{\mathrm{fJ}}_{\mathrm{U}}\mathbf{R}^{\mathrm{JL}}_{\mathrm{D}}, \tag{9.40}$$

and then retain only the first term. The resulting approximation for the displacement response is

$${}^{\mathrm{de}}w_0 = \mathbf{W}^{\mathrm{fJ}}_{\mathrm{U}}\mathbf{R}^{\mathrm{JL}}_{\mathrm{D}}\mathbf{C}^{\mathrm{J}}_{\mathrm{D}}; \tag{9.41}$$

which separates very neatly into near-source and near-receiver contributions and the return from depth in $\mathbf{R}_\mathrm{D}^\mathrm{JL}$. The remaining part of ${}^\mathrm{d}w_0$,

$$\mathbf{W}_\mathrm{U}^\mathrm{fJ}[\mathbf{I} - \mathbf{R}_\mathrm{D}^\mathrm{JL}\mathbf{R}_\mathrm{U}^\mathrm{fJ}]^{-1}\mathbf{R}_\mathrm{D}^\mathrm{JL}\mathbf{R}_\mathrm{U}^\mathrm{fJ}\mathbf{R}_\mathrm{D}^\mathrm{JL}\mathbf{C}_\mathrm{D}^\mathrm{J}, \tag{9.42}$$

represents that portion of the wavefield which has undergone one or more reflections in the region between z_J and the free surface as well as reflection below z_J.

The approximation ${}^\mathrm{de}w_0$ may be employed in a wide variety of problems where the elastic wavespeeds increase rapidly with depth so that there is a large time separation between deeply penetrating waves and those confined to the upper zone and also any surface reflections of deep phases. This form with various further approximations for $\mathbf{W}_\mathrm{U}^\mathrm{fJ}, \mathbf{C}_\mathrm{D}^\mathrm{J}, \mathbf{R}_\mathrm{D}^\mathrm{JL}$ is the basis of most work on relatively long-range seismic wave propagation.

The use of ${}^\mathrm{de}w_0$ requires an adroit choice of the level z_J at which the half space is separated, so that the mutual interaction is kept to a minimum. In some cases a natural break point in the structure occurs, as we have seen in our discussion of shallow propagation, but usually some compromises have to be made. For mantle studies, as we have noted, a good choice is at about 200 km. However, for many teleseismic studies where attention has been focussed on crustal effects at source and receiver, z_J is taken just below Moho.

For slownesses such that *P* or *S* waves have their turning points above 800 km or near the core-mantle boundary, the complications in the elastic parameter distribution are such that it is worthwhile trying to get a good representation for $\mathbf{R}_\mathrm{D}^\mathrm{JL}$ or some of its elements. This has been the goal of many authors (e.g., Helmberger & Wiggins, 1971; Cormier & Choy, 1981; Choy, 1977; Müller, 1973) who have used a variety of techniques to generate theoretical seismograms which will be discussed in subsequent sections. Very simple approximations have usually been used for $\mathbf{W}_\mathrm{U}^\mathrm{fJ}, \mathbf{C}_\mathrm{D}^\mathrm{J}$. Frequently the response is constructed from just the direct transmission term $\mathbf{W}_\mathrm{F}\mathbf{T}_\mathrm{U}^\mathrm{0J}$ at the receiver, and at the source only the downward radiation term, including surface reflections, to give

$$\mathbf{W}_\mathrm{F}\mathbf{T}_\mathrm{U}^\mathrm{0J}\mathbf{R}_\mathrm{D}^\mathrm{JL}(\boldsymbol{\Sigma}_\mathrm{D}^\mathrm{S} + \mathbf{R}_\mathrm{U}^\mathrm{fS}\boldsymbol{\Sigma}_\mathrm{U}^\mathrm{S}) \tag{9.43}$$

as an approximation to ${}^\mathrm{de}w_0$.

On the other hand for turning points in the lower mantle below 800 km, but well away from the core-mantle boundary, the effect of mantle propagation is relatively simple and crustal effects at source and receiver become rather more important for frequencies around 1 Hz. For example Douglas, Hudson & Blamey (1973) have used the approximation ${}^\mathrm{de}w_0$ with full calculation of source and receiver crustal reverberations, to calculate body waves from shallow sources. They have made a simple allowance for lateral variations in crustal structure, by taking different crustal parameters on the source side and the receiver side. This scheme provides a good model for *P* wave arrivals in the range 3500 to 9500 km, for which a good approximation to $\mathbf{R}_\mathrm{D}^\mathrm{0L}$ may be obtained with asymptotic ray theory results.

Improved representations of $\mathbf{R}_D^{JL}$ to allow for upper mantle structure, using a piecewise smooth medium as in Section 6.3, enable the approximation ${}^{de}w_0$ to be used from epicentral distances of 1200 km outwards.

9.2.3 Sources at depth

For a source which lies *below* the separation level z_J which we have imposed on the stratification, we may once again make use of the device of equivalent sources at the level z_J to produce a convenient form for the seismic response. Since z_J now lies above z_S we use the partitioned form (5.45) for the wave-propagator in the relation (9.22) which defines the source jump $\mathbf{\Sigma}^J$. The upward and downward radiation components at z_J are thus

$$\begin{aligned}\mathbf{\Sigma}_D^J &= (\mathbf{T}_D^{JS})^{-1}(\mathbf{\Sigma}_D^S + \mathbf{R}_D^{JS}\mathbf{\Sigma}_U^S),\\ \mathbf{\Sigma}_U^J &= \mathbf{T}_U^{JS}\mathbf{\Sigma}_U^S + \mathbf{R}_U^{JS}\mathbf{\Sigma}_D^J.\end{aligned} \tag{9.44}$$

The net upward radiation $\mathbf{C}_U^J$ at z_J, allowing for reflection from below z_J, now has the equivalent representations

$$\begin{aligned}\mathbf{C}_U^J &= \mathbf{\Sigma}_U^J + \mathbf{R}_D^{JL}\mathbf{\Sigma}_D^J,\\ &= \mathbf{T}_U^{JS}[\mathbf{I} - \mathbf{R}_D^{SL}\mathbf{R}_U^{JS}]^{-1}(\mathbf{\Sigma}_U^S + \mathbf{R}_D^{SL}\mathbf{\Sigma}_D^S).\end{aligned} \tag{9.45}$$

The latter expression includes the full interactions of the waves from the source with the region below the separation level z_J. We may emphasise reflections at this level by writing

$$\mathbf{C}_U^J = \mathbf{T}_U^{JS}\{[\mathbf{I} - \mathbf{R}_D^{SL}\mathbf{R}_U^{JS}]^{-1}\mathbf{R}_D^{SL}(\mathbf{\Sigma}_D^S + \mathbf{R}_U^{JS}\mathbf{\Sigma}_U^S) + \mathbf{\Sigma}_U^S\}. \tag{9.46}$$

The surface displacement field in this case can be found from (9.25) as

$$w_0 = \mathbf{W}_U^{fJ}[\mathbf{I} - \mathbf{R}_D^{JL}\mathbf{R}_U^{fJ}]^{-1}\mathbf{C}_U^J. \tag{9.47}$$

and there is no simple separation of a shallow propagation term in this case.

The reverberative effects of the upper zone $0 < z < z_J$ are contained within the receiver operator $\mathbf{W}_U^{fJ}$. Energy is brought into the surface channel by waves travelling upward from the true source level, either directly or after reflection from beneath the source. The most important contributions for body waves are represented by

$$\mathbf{W}_F[\mathbf{I} - \mathbf{R}_D^{0J}\mathbf{R}_F]^{-1}\mathbf{T}_U^{0J}\mathbf{T}_U^{JS}\{\mathbf{\Sigma}_U^S + \mathbf{R}_D^{SL}[\mathbf{\Sigma}_D^S + \mathbf{R}_U^{JS}\mathbf{\Sigma}_U^S]\}, \tag{9.48}$$

where we have allowed for at most one reflection from the region between z_S and z_J in our approximation to $\mathbf{C}_U^J$ (9.40). When we are interested in propagation effects in the upper zone we choose the separation level z_J in the way we have previously discussed.

For a suboceanic earthquake with z_J taken at the basement, the main tsunamigenic effect of the event is described by $\mathbf{W}_U^{fJ}\mathbf{T}_U^{JS}\mathbf{\Sigma}_U^S$ which represents the upward radiation into the water column; there may also be some contribution from waves reflected back from just below the source level contained within $\mathbf{W}_{\mathrm{U}}^{\mathrm{fJ}}\mathbf{T}_{\mathrm{U}}^{\mathrm{JS}}\mathbf{R}_{\mathrm{D}}^{\mathrm{SL}}\boldsymbol{\Sigma}_{\mathrm{D}}^{\mathrm{S}}$.

Crustal propagation effects can be important for earthquakes which occur just below the Moho. $\mathbf{R}_{\mathrm{U}}^{\mathrm{JS}}$ will then mostly arise from reflections at the Moho. High angle reverberations within the crust will be possible, but as the slowness increases to the inverse wavespeeds at the source level, the crustal field for that wave type will become evanescent in the crust, and as a result reverberations will be damped. The most noticeable effects will therefore occur close to the source and become less pronounced as the range increases. On a smaller scale there will be a similar pattern of behaviour for events occurring below sedimentary cover.

For propagation deep into the stratification we can emphasise the physical character of the solution by taking a slightly different choice of separation level z_{J}. The surface displacement field (9.47)

$$w_0 = \mathbf{W}_{\mathrm{U}}^{\mathrm{fJ}}(\mathbf{I} - \mathbf{R}_{\mathrm{D}}^{\mathrm{JL}}\mathbf{R}_{\mathrm{U}}^{\mathrm{fJ}})^{-1}\mathbf{C}_{\mathrm{U}}^{\mathrm{J}} \tag{9.49}$$

allows for both shallow and deep reverberations. Now the displacement operator

$$\mathbf{W}_{\mathrm{U}}^{\mathrm{fJ}} = \mathbf{W}_{\mathrm{F}}(\mathbf{I} - \mathbf{R}_{\mathrm{D}}^{0\mathrm{J}}\mathbf{R}_{\mathrm{F}})^{-1}\mathbf{T}_{\mathrm{U}}^{0\mathrm{J}}, \tag{9.50}$$

and with an expansion of $\mathbf{R}_{\mathrm{U}}^{\mathrm{fJ}}$ to emphasise surface reflections

$$\mathbf{R}_{\mathrm{D}}^{\mathrm{JL}}\mathbf{R}_{\mathrm{U}}^{\mathrm{fJ}} = \mathbf{R}_{\mathrm{D}}^{\mathrm{JL}}\{\mathbf{R}_{\mathrm{U}}^{0\mathrm{J}} + \mathbf{T}_{\mathrm{D}}^{0\mathrm{J}}\mathbf{R}_{\mathrm{F}}(\mathbf{I} - \mathbf{R}_{\mathrm{D}}^{0\mathrm{J}}\mathbf{R}_{\mathrm{F}})^{-1}\mathbf{T}_{\mathrm{U}}^{0\mathrm{J}}. \tag{9.51}$$

If, therefore, we can choose the separation level z_{J} such that $\mathbf{R}_{\mathrm{D}}^{0\mathrm{J}}, \mathbf{R}_{\mathrm{U}}^{0\mathrm{J}}$ are small, we can get a good approximation by allowing for only transmission effects in '0J':

$$\begin{aligned} \mathbf{W}_{\mathrm{U}}^{\mathrm{fJ}} &\approx \mathbf{W}_{\mathrm{F}}\mathbf{T}_{\mathrm{U}}^{0\mathrm{J}}, \\ (\mathbf{I} - \mathbf{R}_{\mathrm{D}}^{\mathrm{JL}}\mathbf{R}_{\mathrm{U}}^{\mathrm{fJ}})^{-1} &\approx (\mathbf{I} - \mathbf{R}_{\mathrm{D}}^{\mathrm{JL}}\mathbf{T}_{\mathrm{D}}^{0\mathrm{J}}\mathbf{R}_{\mathrm{F}}\mathbf{T}_{\mathrm{U}}^{0\mathrm{J}})^{-1}. \end{aligned} \tag{9.52}$$

Such a split is possible in oceanic regions if we now take z_{J} to lie just above the sea floor. For slownesses appropriate to seismic wave-propagation in the sub-basement rocks, the reflection from the water column will be very slight and the sound speed structure can be approximated by a uniform medium with P wavespeed α_0 and thickness h_0. The secular function for the oceanic case is therefore, from (9.52),

$$1 + \mathbf{R}_{\mathrm{D}}^{\mathrm{JL}}|_{\mathrm{PP}} e^{2i\omega q_{\alpha 0} h_0} = 0, \tag{9.53}$$

which represents a constructive interference condition between waves reflected back from the sea surface and the rocks beneath the sea floor. We may use the addition rule to express $\mathbf{R}_{\mathrm{D}}^{\mathrm{JL}}$ in terms of the reflection properties of the sea bed $\mathbf{R}_{\mathrm{D}}^{\mathrm{B}}$ etc. and the deeper reflection response $\mathbf{R}_{\mathrm{D}}^{\mathrm{BL}}$ as

$$\mathbf{R}_{\mathrm{D}}^{\mathrm{JL}} = \mathbf{R}_{\mathrm{D}}^{\mathrm{B}} + \mathbf{T}_{\mathrm{U}}^{\mathrm{B}}\mathbf{R}_{\mathrm{D}}^{\mathrm{BL}}[\mathbf{I} - \mathbf{R}_{\mathrm{U}}^{\mathrm{B}}\mathbf{R}_{\mathrm{D}}^{\mathrm{BL}}]^{-1}\mathbf{T}_{\mathrm{D}}^{\mathrm{B}}, \tag{9.54}$$

where, as discussed in the appendix to Chapter 6, we work with 2×2 matrices throughout, but $\mathbf{R}_{\mathrm{D}}^{\mathrm{JL}}$ will have only a PP entry.

The effect of the low wavespeed water layer is, in this approximation, to produce just a phase delay and no amplitude change. However a further consequence of this choice of z_J is that, seen from below there will be strong reflections back from the region between the source and z_J. Thus the elements of $\mathbf{R}_U^{JS}$ will be significant, especially for *S* waves for which there will be only small radiation loss into the water by conversion to *P* near the sea bed. For long-range propagation through the oceanic crust and mantle an important role is therefore played by the contribution

$$[\mathbf{I} - \mathbf{R}_D^{SL}\mathbf{R}_U^{JS}]^{-1}\mathbf{R}_D^{SL}\{\mathbf{\Sigma}_D^S + \mathbf{R}_U^{JS}\mathbf{\Sigma}_U^S\}, \tag{9.55}$$

within $\mathbf{C}_U^J$ (9.47). This represents a guided wave system controlled by the oceanic crust and mantle structure with further reinforcement by multiple reflections within the water column. With slow lateral changes in structure, such a system can explain the persistence of high frequency oceanic *Pn* and *Sn* to large ranges. The excitation of these phases depends strongly on source depth and the details of the crustal and mantle structure.

For sources in the mantle we may make a similar split just above the Moho, and for low frequencies the crust will appear relatively transparent. We therefore have the approximate secular function

$$\det[\mathbf{I} - \mathbf{R}_D(z_M-)\mathbf{T}_D^{0M}\mathbf{R}_F\mathbf{T}_U^{0M}], \tag{9.56}$$

and so the dispersion characteristics are dominated by the mantle structure at low frequencies (see Chapter 11). The interference head wave system we have already mentioned is an analogue of the wave channelling in the oceanic case, and other sub-Moho reflections will contribute to the *P* and *S* wave codas.

9.3 Approximate integration techniques

In the two previous sections we have established a number of useful approximations to the displacement response of a stratified half space in the slowness-frequency domain. We now examine ways in which these approximations can be used to generate theoretical seismograms for specific portions of the body wave response.

The methods which will be discussed here aim to produce as complete a representation as possible for a seismic phase or group of phases, and so involve numerical integration over frequency and slowness with the integrand specified in terms of reflection and transmission elements for parts of the stratification. The approximation to the response or the interval of slowness integration will be such that we encounter no difficulties from surface wave poles. The synthesis of surface wavetrains and their relationship to body wave pulses is presented in Chapter 11.

An alternative approximate approach for body waves is presented in Chapter 10, where we will discuss *generalized ray* methods. These rely on casting a portion of the seismic response as a sum of contributions with a particular separation of frequency and slowness dependence, which allows the inversion integrals to be performed, in part, analytically.

9.3.1 Reflectivity methods

The approach we have adopted in this book for the characterisation of the response of a stratified medium to excitation by a source has been to work in terms of the reflection and transmission properties of portions of the stratification. As a result, the techniques we have discussed for the generation of complete theoretical seismograms can be regarded as 'reflectivity' methods.

The name was, however, introduced by Fuchs & Müller (1971) to describe a technique in which all multiple reflections and conversions between wave types were retained in part of the structure. The stratification was split at a level z_J as in our treatment in Section 9.2, and attention was concentrated on just those waves reflected back from the region beneath z_J. The approximation employed by Fuchs & Müller may be derived from our expression for ${}^{de}\boldsymbol{w}_0$ (9.41) by retaining only the downward radiation from the source and working with just transmission terms in the region above z_J. This eliminates any reverberatory effects in the receiver and source operators $\mathbf{W}_U^{fJ}$, $\mathbf{C}_D^J$ and leads to the approximate form

$$ {}^{re}\boldsymbol{w}_0 = \mathbf{W}_F \mathbf{T}_U^{0J} \mathbf{R}_D^{JL} \mathbf{T}_D^{SJ} \boldsymbol{\Sigma}_D^S. \tag{9.57}$$

In the original treatment further approximations were made to allow only for direct interfacial transmission losses, and a single *P* wave component of $\mathbf{R}_D^{JL}$ was included. The PP reflection coefficient for the region $z \geq z_J$ was constructed by propagator matrix methods for a stack of uniform layers, which allowed for all multiples in (z_J, z_L). Later Fuchs (1975) allowed for *P* to *S* conversions on reflection from the region 'JL'.

The representation (9.57) is once again in the slowness-frequency domain and to generate theoretical seismograms at particular stations we must perform integrals over slowness and frequency. We have considered this problem for the full response in Section 7.3 and suitable techniques parallel those we have already discussed.

Since free-surface effects have been removed, there are no surface wave poles but there will be branch points at the *P* and *S* waveslownesses at the surface, source level, and the top and bottom of the reflection zone 'JL'. The absence of poles means that a *spectral* method with integration along the real p axis as in (7.75) or (7.77) with the approximation ${}^{re}\boldsymbol{w}_0$ instead of the the full vector will be suitable. The numerical integral has the form

$$ \bar{\mathbf{u}}(r, 0, \omega) = \tfrac{1}{2}\omega|\omega| M(\omega) \int_{-\infty}^{\infty} dp\, p [{}^{re}\boldsymbol{w}_0]^T \mathbf{T}_m^{(1)}(\omega p r), \tag{9.58}$$

where we have extracted a common source spectrum $M(\omega)$. Once the computational labour of generating the approximation to $\boldsymbol{w}_0$ has been completed, calculations may be made for many ranges for only the cost of performing the integrations. Once again, however, there is the problem of computing the integral of a highly oscillatory integrand which limits the achievable range for a given frequency band. The properties of the integrand will, however, be improved slightly if physical attenuation is included in the velocity model.

The integral (9.58) represents a convolution of the source-time function with the reflectivity response function for the stratification. In practice we will impose some upper limit on the frequency in (9.58) so that we get a filtered version of the response. In (9.45) we have represented the seismogram spectrum in terms of outgoing wave components and to isolate particular features in the response the integration is restricted to a band of slownesses

$$\bar{\mathbf{u}}(r, 0, \omega) = \tfrac{1}{2}\omega|\omega|M(\omega)\int_{p_1}^{p_2} dp\, p[{}^{re}\mathbf{w}_0]^T\mathbf{T}_m^{(1)}(\omega p r), \tag{9.59}$$

For example, if we consider the *P* wave response at large distances, we do not anticipate that there should be a large response from reverberations at near vertical incidence. We would therefore choose p_1 to be about 0.025 s/km and take p_2 to be larger than $1/\alpha_J$, where α_J is the *P* wavespeed just above the reflection zone. This allows for weakly evanescent waves in the upper zone which may have a significant amplitude (especially at low frequencies) but excludes strongly evanescent waves. When calculations are to be made at short ranges we would choose $p_1 = 0$, and adopt the 'standing wave' form of the integral (9.59) in terms of $\mathbf{T}_m(\omega p r)$, cf. (7.75).

Injudicious choice of the integration interval can give quite large numerical arrivals at the limiting slownesses, but these can be muted by applying a taper to ${}^{re}\mathbf{w}_0$ near these limits. For problems where there is a thick layer of very low wavespeed material overlying the reflection zone, the exclusion of evanescent waves in this overburden is satisfactory since all the arrivals of interest occur in a limited range of slowness. Such is, for example, the case for compressional arrivals returned from the structure beneath the sea bed in the deep ocean (Orcutt, Kennett & Dorman, 1976).

In order to work with a fixed time interval at varying ranges it is convenient to work in terms of reduced time $t - p_{red}r$ (Fuchs & Müller 1971) and follow the arrivals with distance. This may be achieved by multiplying the spectrum at each range by $\exp(-i\omega p_{red}r)$ before taking the inverse Fourier transform over frequency. As in our discussion of complete theoretical seismograms (Section 7.3.3) it is often useful to build up the response for a wide range of slownesses by combining the results for different slowness panels with appropriate time shifts.

The 'reflectivity' calculations can be recast in terms of the *slowness* approach of Section 7.3.2, in which the transform over frequency is performed first and is then followed by a slowness integral (7.86), (7.103). The integrand in space and time is reasonably well behaved (figures 6.4, 6.5) even though long time series may be needed to avoid aliasing, as a result the sampling in slowness can be reduced compared to what is required for the spectral integral (Fryer 1980). This approach also allows the calculation of a specified time window on the final theoretical seismograms. For moderate to large ranges the slowness integral can be simplified and various source-time functions introduced via an 'effective source' operator (7.115)-(7.117).

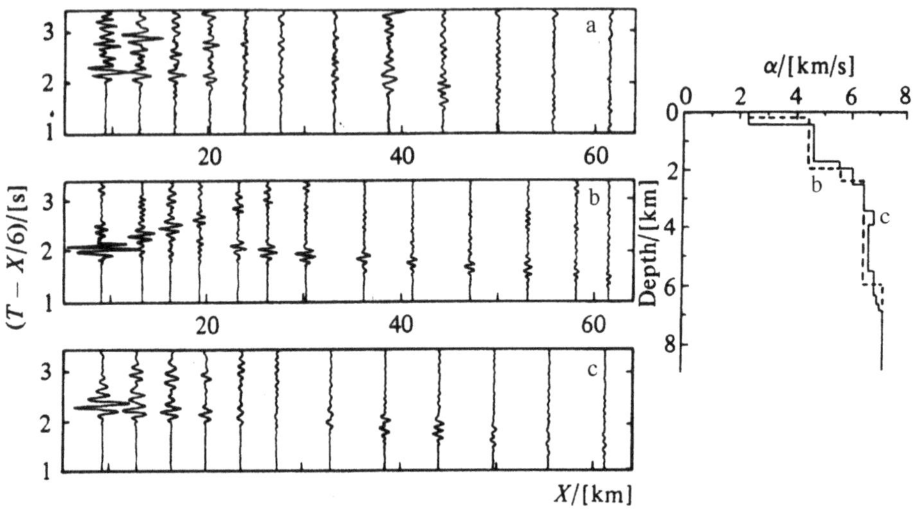

Figure 9.6. Illustration of use of reflectivity method in matching main features of observed seismograms: a) observed marine seismic records; b) calculations based on travel-time modelling alone; c) after model refinement.

The numerical reflectivity method was introduced as an aid to the interpretation of crustal refraction profiles, where it has enabled the incorporation of amplitude information and the travel-times of the earliest arriving phases. As an example of this application we show in figure 9.6 observed and theoretical seismograms for a marine seismic refraction survey on the Reykjanes Ridge, southwest of Iceland (Bunch & Kennett, 1980). The upper record section shows the beginning of the compressional wavetrains on the experimental records, up to the onset of the first multiple in the relatively shallow water (1 km deep) The middle section shows the theoretical seismograms calculated for the structure deduced from travel-time analysis alone, which is indicated by a dotted line on the velocity display. These seismograms do not reproduce the main features of the observations. But, after careful refinement (basically by working with gradient zones rather than major jumps in wavespeed), a velocity model can be found for which theoretical seismograms give a good match to the character of the observations; these are shown in the lower record section. The final velocity model is indicated by the solid line.

The success of the reflectivity method in such crustal applications led to its use for many other classes of problem, and stimulated work on calculating complete theoretical seismograms. With the top of the reflection zone set at the Moho, reflectivity calculations have been used to examine propagation in the lithosphere (Hirn et al., 1973), and with a slightly deeper separation level, P wave propagation in the upper mantle (Kennett, 1975; Burdick & Orcutt, 1979). Müller (1973) has

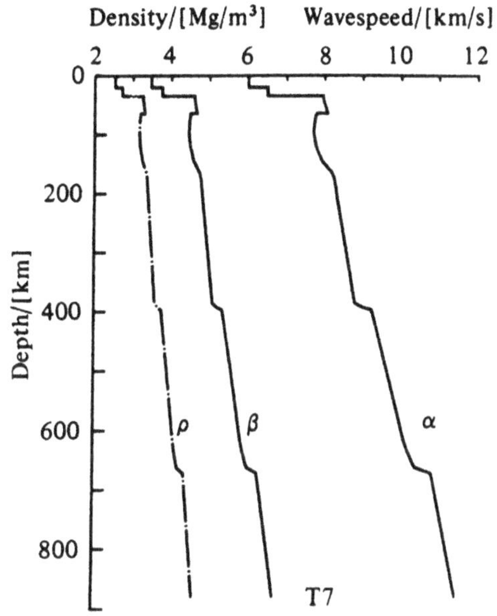

Figure 9.7. The upper mantle model *T7*

examined seismic wave propagation in the earth's core with a flattened wavespeed model and taken the top of the reflection zone a few hundred kilometres above the core-mantle interface.

Müller & Kind (1976) moved the reflection zone right up to the surface and restored, in part, surface reflection effects by taking a model with a very dense fluid having the acoustic wavespeed of air overlying a stratified elastic half space. This gives a reasonable approximation to the free-surface reflection coefficients, but does not lead to any singularities on the real slowness axis. Müller & Kind have used this model to generate long-period seismograms for the whole earth.

Although we have based our discussion on the 'reflectivity' approximation (9.44), we may make a comparable development with almost all of the approximate results we have generated in Sections 9.1 and 9.2. Thus, for example, if we wish to allow for surface reflected phases near the source (*pP* etc.) and a comparable reflection level near the receiver, we would use

$$^{re}w_0 = \mathbf{W}_F(\mathbf{I} + \mathbf{R}_D^{0J}\mathbf{R}_F)\mathbf{T}_U^{0J}\mathbf{R}_D^{JL}\mathbf{T}_D^{SJ}(\mathbf{\Sigma}_D^S + \mathbf{R}_U^{fS}\mathbf{\Sigma}_U^S). \tag{9.60}$$

Similar approximations can be generated to study other problems, e.g., Faber & Müller (1980) have constructed a form of displacement response which enables them to examine *Sp* conversions from upper mantle discontinuities.

Although most applications of reflectivity methods have been made using models composed of uniform layers, the essential ingredient is an accurate representation of the reflection matrix for the reflection zone $(z > z_J)$. For a piecewise smooth

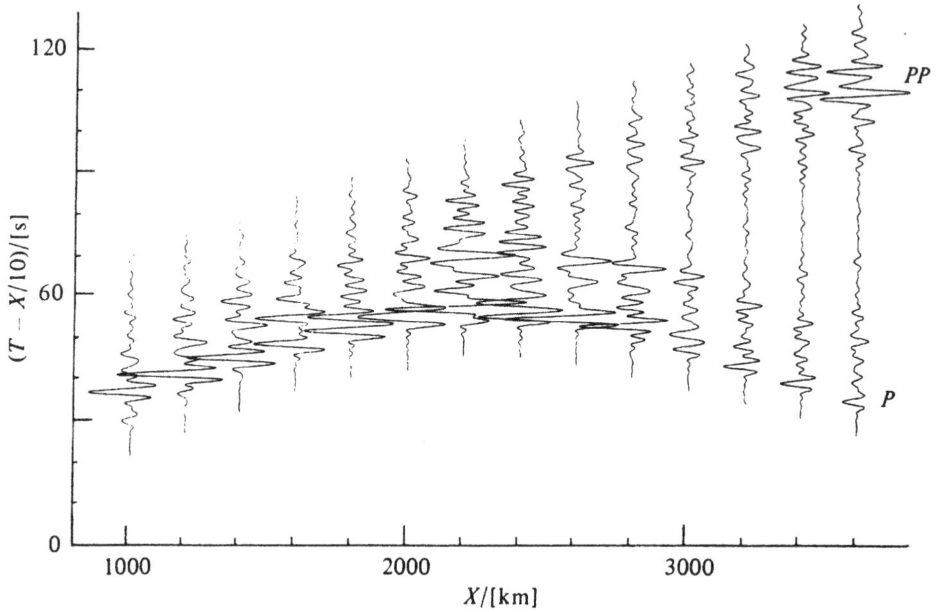

Figure 9.8. Theoretical seismograms for the model *T7* using a piecewise smooth velocity structure and a reflectivity slowness integral.

wavespeed profile we can therefore use the recursive method of Section 6.3 to construct $\mathbf{R}_D^{JL}$. An important case is provided by the wavespeed distribution in the upper mantle, and we illustrate the model *T7* (Burdick & Helmberger, 1978) in figure 9.7. We represent the upper mantle discontinuities in this model by closely spaced changes in wavespeed gradient and small jumps in wavespeed. In figure 9.8 we show a record section of theoretical seismograms calculated for a surface source with an allowance for a single surface reflection to describe crustal effects. The approximate response function we have used is

$$w_0 \approx \mathbf{W}_F[\mathbf{I} + \mathbf{R}_D^{0L}\mathbf{R}_F](\mathbf{\Sigma}_U^0 + \mathbf{R}_D^{0L}\mathbf{\Sigma}_D^0). \tag{9.61}$$

and the integration has been carried out to generate the vertical component for the range of *P* waveslowness in the upper mantle. As a result we get some *S* waves appearing in the vertical component records as a result of *P* to *S* conversion at the surface. At the larger ranges there is in addition a clear appearance of the *PP* phase. For the frequency band we have considered (0.03–0.6 Hz) the combination of the upper-mantle triplications with intracrustal reflections give a rather complex pattern of behaviour.

For a smoothly varying wavespeed profile, it is preferable to work directly with calculation schemes designed for such distributions. It is possible to use a staircase of uniform layers to approximate smooth variation but this is effective only at moderate frequencies (Choy et al., 1980). At high frequencies, thin layers must be used to avoid separation of internal multiples in time and so many layers are needed, for example about one hundred layers down to 1000 km for 1 Hz seismograms.

9.3.2 Approximations for specific seismic phases

In the reflectivity method the emphasis is on the representation of the seismic waves returned by an entire region, which frequently consists of a complex interference of many different reflection processes.

For deep propagation within the earth, the time separation between seismic pulses seen on distant seismographs is such that we can concentrate on an individual seismic phase or a related group of phases. Such a phase is described within the reflection treatment by a particular combination of reflection and transmission elements (Scholte, 1956).

A convenient starting point is the expression ${}^{de}w^0$ (9.41) for deeply propagating waves. The shallow propagation elements $\mathbf{W}_U^{fJ}$, $\mathbf{C}_D^J$ should be retained in full to allow for any crustal reverberation effects, but are often approximated by eliminating these terms to give

$$ {}^{de}w_0 = \mathbf{W}_F \mathbf{T}_U^{0J} \mathbf{R}_D^{JL} \mathbf{T}_D^{SJ} (\boldsymbol{\Sigma}_D^S + \mathbf{R}_U^{fS} \boldsymbol{\Sigma}_U^S). \tag{9.62} $$

This form allows for surface reflections near the source. We now seek to extract the appropriate reflection effects from the reflection matrix $\mathbf{R}_D^{JL}$. Below the separation level z_J we will suppose the model consists of smoothly varying gradient zones separated by discontinuities in the elastic parameters or their derivations. This model which we have already considered in Section 6.3, provides a very effective description of the wavespeed distribution in the mantle and core.

As an example we consider the wave system comprising the *P* wave returned from the deep mantle, the *PcP* reflection from the core-mantle boundary and their associated surface reflected phases *pP*, *sP* etc. For a *P* wave turning point beneath the interface at z_G we are interested in the contribution

$$ (\hat{\mathbf{T}}_U^{JG})_{PP} (\mathbf{R}_D^{GL})_{PP} (\hat{\mathbf{T}}_D^{JG})_{PP}. \tag{9.63} $$

The transmission term $(\hat{\mathbf{T}}_D^{JG})_{PP}$ describes direct transmission from z_J to z_G with no allowance for internal multiples

$$ (\hat{\mathbf{T}}_D^{JG})_{PP} = \prod_{j=J}^{G} \mathbf{t}_d^j(p,\omega) \mathbf{t}_d^{j-1,j}(p,\omega) \tag{9.64} $$

where $\mathbf{t}_d^j$ is the *generalized* transmission coefficient for *P* waves at the jth interface derived from (6.60). Such coefficients are frequency dependent since they allow for the effects of gradients bordering the interface through the generalized vertical slownesses $\eta_{\alpha u}$, $\eta_{\alpha d}$ (3.104). The terms $\mathbf{t}_d^{j-1,j}(p,\omega)$ allow for transmission loss and phase delay for propagation between the j-1th and jth interfaces and, far from turning points, can often be represented by asymptotic forms. We have a similar expression to (9.64) for the upward transmission operator $(\hat{\mathbf{T}}_{\mathrm{U}}^{\mathrm{JG}})$.

The portion of $(\mathbf{R}_{\mathrm{D}}^{\mathrm{SL}})_{\mathrm{PP}}$ which corresponds to P waves turned back by the gradients or reflected from the core-mantle boundary is given by

$$\mathbf{r}_{\mathrm{d}}^{\mathrm{GC}} + \mathbf{t}_{\mathrm{u}}^{\mathrm{GC}}\mathbf{r}_{\mathrm{d}}^{\mathrm{C}}\mathbf{t}_{\mathrm{d}}^{\mathrm{GC}}, \tag{9.65}$$

where $\mathbf{r}_{\mathrm{d}}^{\mathrm{C}}$ is the generalized PP reflection coefficient at the core mantle boundary. The terms $\mathbf{r}_{\mathrm{d}}^{\mathrm{GC}}$, $\mathbf{t}_{\mathrm{d}}^{\mathrm{GC}}$ represent the leading order contributions to reflection and transmission from the gradient zone in (z_G, z_C). From (6.51) if the P wave turning point lies above z_C

$$\begin{aligned} \mathbf{r}_{\mathrm{d}}^{\mathrm{GC}} &= \mathrm{Ej}_{\alpha}(z_G)e^{-i\pi/2}[\mathrm{Fj}_{\alpha}(z_G)]^{-1}, \\ \mathbf{t}_{\mathrm{d}}^{\mathrm{GC}}\mathbf{t}_{\mathrm{u}}^{\mathrm{GC}} &= 2\mathrm{Aj}_{\alpha}(z_C)\mathrm{Ej}_{\alpha}(z_G)[\mathrm{Fj}_{\alpha}(z_G)\mathrm{Bj}_{\alpha}(z_C)]^{-1}, \end{aligned} \tag{9.66}$$

which include the $\pi/2$ phase shift associated with the caustic generated on total reflection. The presence of $\mathrm{Aj}_{\alpha}(z_C)[\mathrm{Bj}_{\alpha}(z_C)]^{-1}$ in the transmission terms means that there is very rapid decay in the evanescent region and so at high frequencies the first term in (9.65) dominates. As p decreases towards α_{c-}^{-1}, where α_{c-} is the P wavespeed just above the core-mantle boundary, the leakage through the evanescent region to the boundary increases and so the second term in (9.65) becomes much more important (Richards, 1973). Once p is less than α_{c-}^{-1} we have the representation

$$\begin{aligned} \mathbf{r}_{\mathrm{d}}^{\mathrm{GC}} &\approx 0 \\ \mathbf{t}_{\mathrm{d}}^{\mathrm{GC}}\mathbf{t}_{\mathrm{u}}^{\mathrm{GC}} &= \mathrm{Ej}_{\alpha}(z_G)\mathrm{Fj}_{\alpha}(z_C)[\mathrm{Fj}_{\alpha}(z_G)\mathrm{Ej}_{\alpha}(z_C)]^{-1}, \end{aligned} \tag{9.67}$$

and reflection contributions from the gradient zone will only arise from the higher order terms (6.43)-(6.44). In this case we are left with just the interface reflection term in (9.65).

The final expression for the frequency-slowness domain response for the P and PcP phases is

$$^{f}w_0 = (\mathbf{W}_F)_P(\mathbf{T}_{\mathrm{U}}^{0\mathrm{J}})_{\mathrm{PP}}(\hat{\mathbf{T}}_{\mathrm{U}}^{\mathrm{JG}})_{\mathrm{PP}}(\mathbf{r}_{\mathrm{d}}^{\mathrm{GC}} + \mathbf{t}_{\mathrm{u}}^{\mathrm{GC}}\mathbf{r}_{\mathrm{d}}^{\mathrm{C}}\mathbf{t}_{\mathrm{d}}^{\mathrm{GC}})(\hat{\mathbf{T}}_{\mathrm{D}}^{\mathrm{JG}})_{\mathrm{PP}}(\mathbf{T}_{\mathrm{D}}^{\mathrm{SJ}}[\mathbf{\Sigma}_{\mathrm{D}}^{\mathrm{S}} + \mathbf{R}_{\mathrm{U}}^{\mathrm{fS}}\mathbf{\Sigma}_{\mathrm{U}}^{\mathrm{S}}])_{\mathrm{PP}}, \tag{9.68}$$

where $(\mathbf{W}_F)_P = [(W_F)_{UP}, (W_F)_{VP}]^T$. Viewed as a function of complex p, at fixed ω the generalized reflection and transmission coefficients at the jth interface have strings of poles associated with zeroes of the generalized Stoneley denominator. These pole strings depart from the real p-axis close to the slownesses $\alpha_{j-}^{-1}, \alpha_{j+}^{-1}$ and $\beta_{j-}^{-1}, \beta_{j+}^{-1}$ (see figure 9.9), and lie close to the Stokes' lines for the Airy functions in $\eta_{\alpha u,d}$ and $\eta_{\beta u,d}$. The strings will lead off into the upper half p plane from α_{j-}^{-1} and β_{j-}^{-1} since these lie close to the singularities of the upward generalized slownesses. From α_{j+}^{-1} and β_{j+}^{-1} the poles depart into the lower half p plane.

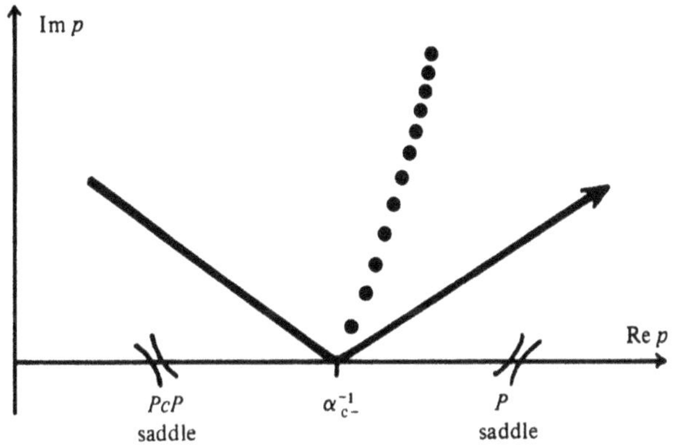

Figure 9.9. The saddle points for the *P* and *PcP* system as a function of slowness and a suitable numerical contour Γ.

In the spectral method, with an outgoing wave representation we have to evaluate integrals of the form (9.58) in terms of ${}^f\mathbf{w}_0$ i.e.

$$\bar{\mathbf{u}}(r, 0, \omega) = \tfrac{1}{2}\omega|\omega|M(\omega) \int_\Gamma dp\, p[{}^f\mathbf{w}_0]^T \mathbf{T}_M^{(1)}(\omega p r), \tag{9.69}$$

where the contour of integration Γ can be chosen to exploit the character of ${}^f\mathbf{w}_0$ (Richards, 1973).

If we take the asymptotic form (7.79) for the Hankel functions in $\mathbf{T}_m^{(1)}(\omega p r)$ and the asymptotic forms for the Airy function terms in (9.66)–(9.67), then we find that the integrand has two main saddle points. The first for $p < \alpha_{c-}^{1}$ arises from the term $\mathbf{t}_u^{GC}\mathbf{r}_d^{C}\mathbf{t}_d^{GC}$ in (9.65) and thus corresponds to *PcP* reflections. The second saddle at fixed range r occurs for $p > \alpha_{c-}^{-1}$ and corresponds to the *P* wave (figure 9.9). Slightly shifted saddles are associated with the surface reflected phases *pPcP*, *sPcP* and *pP*, *sP*.

If we take a contour of integration which crosses both main saddles and make a steepest descent approximation at each one, we recover the geometric ray theory results for two isolated rays *PcP* and *P* .

At finite frequencies we can exploit the rapid decay of the integrand away from the saddle points by choosing a numerical integration contour Γ that follows the general character of the steepest descent path whilst avoiding the singularities of the integrand (Richards, 1973,1976). This approach which has been termed 'full-wave' theory has the disadvantage that calculations are required for complex p, although numerical convergence is improved, and considerable care may be needed to find the singularities of the integrand. Since the full frequency dependence of ${}^f\mathbf{w}_0$ is retained in (9.69), attenuation may be included via complex wavespeed profiles. A comparable approach was used by Chapman & Phinney (1972) with a numerical solution of the differential equations (2.24) to find $\mathbf{w}_0$, but involved considerable computational expense. The use of the Langer approximations for the reflection elements as described here substantially reduces the effort needed. The approximations needed to use the Langer approach for a radially heterogeneous model are equivalent to an earth flattening transformation of the type described in Section 1.5.

An alternative to the complex contour of the full-wave approach is to adopt an approach similar to the reflectivity method (9.62) by integrating along a finite portion of the real p axis with the response specified by ${}^{f}w_0$. This approach has been used by Ward (1978) but has the disadvantage that numerical truncation phases with the slownesses p_1, and p_2 can arise from the ends of the integration interval.

In the problem we have discussed so far the propagation pattern was fairly simple but this approach may be extended to deal with groups of phases for which the time separation is insufficient to allow separate analysis (Choy, 1977; Cormier & Richards, 1977).

When a wave-propagates into a lower velocity medium beneath an interface, e.g. an upper mantle velocity inversion or *P* waves entering the core, we can expand the reflection matrix as seen at z_C (cf. 9.28)

$$\mathbf{R}_D(z_C-) = \mathbf{R}_D^C + \mathbf{T}_U^C\mathbf{R}_D^{CL}\mathbf{T}_D^C + \mathbf{T}_U^C\mathbf{R}_D^{CL}\mathbf{R}_U^C\mathbf{R}_D^{CL}\mathbf{T}_D^C + \dots. \qquad (9.70)$$

where $\mathbf{R}_D^{CL}$ is the reflection matrix for the region below z_C and $\mathbf{R}_U^C$ etc. are the interface matrices. Each term in (9.70) may now be treated separately and, for example, only *P* waves retained. Such an expansion justifies Richard's (1973) treatment of the *P4KP* phase. Richards showed that for *P* turning points above the core-mantle boundary ($p > \alpha_{c-}^{-1}$) tunnelling in evanescence represented above through $\mathbf{t}_u^{GC}\mathbf{t}_d^{GC}$ (9.66) is sufficient to excite multiple reflections within the core.

When, however, there is an increase in velocity across an interface with a gradient zone below we have an interference head wave situation. In addition to the sub-Moho case discussed in Section 9.2.1, a similar situation arises in the inner core for *PKIKP* and in the multiple *SmKS* system generated by conversion at the core-mantle interface. In this case we wish to retain the full reverberation operator describing the interference, but since the first reflection tends to separate from the rest in time we make a partial expansion

$$\mathbf{R}_D(z_{C-}) = \mathbf{R}_D^C + \mathbf{T}_U^C\mathbf{R}_D^{CL}\mathbf{T}_D^C + \mathbf{T}_U^C\mathbf{R}_D^{CL}\mathbf{R}_U^C\mathbf{R}_D^{CL}[\mathbf{I} - \mathbf{R}_U^C\mathbf{R}_D^{CL}]^{-1}\mathbf{T}_D^C. \qquad (9.71)$$

The main interference head wave is then described by the final composite term. With an integrand ${}^{f}w_0$ containing such a term the integration path would approach the real axis near the saddle associated with $\mathbf{T}_U^C\mathbf{R}_D^{CL}\mathbf{T}_D^C$, follow the real axis past the limit point for the multiple reflections (the slowness in the upper medium) and then proceed into the complex plane (Choy, 1977). A similar approach is appropriate for triplications where once again multiple saddles occur in a small region of the slowness axis.

Although we have based our discussion on a representation based on a shallow source the extension of the treatment to deep sources presents no difficulty. We would then start with (9.48) and extract the required reflection terms from $\mathbf{R}_D^{SL}$.

If sufficient contributions are used, with a separate integration step (9.69) in each case, this phase by phase method may be used to study quite complex systems. Cormier & Choy (1981) have used full-wave calculations to generate theoretical seismograms for upper mantle models by constructing *P* wave reflections from the gradient zones and interfaces and then superposing the results. If low velocity zone effects or crustal reverberations are included it is probably simpler to use a piecewise smooth model with a reflectivity type approach in which internal multiples are automatically included, as in figure 9.8.

9.3.3 Teleseismic P and S phases

The principal information used in many studies of seismic sources is the polarity and shape of the onset of the *P* and *S* wavetrains. For long-period records the interference of the direct wave (*P*) and the surface reflected phases (*pP*, *sP*) give a pattern which depends on source depth (figure 9.11) and so by comparison with observations we may hope to constrain the depth of the source. At high frequencies, if records from a number of stations are available, the relative amplitudes of *P*, *pP* and *sP* can help to constrain the focal mechanism of the source (Pearce, 1980). However, at these frequencies crustal effects complicate the records and full calculations for crustal reverberations at source and receiver are desirable. Such calculations have been used in studies of methods to discriminate between underground nuclear events and shallow earthquakes (see, e.g., Douglas, Hudson & Blamey, 1973). With broad-band records it is worthwhile to use realistic source models, e.g. a propagating fault, and these may be included in our treatment by modifying the slowness and frequency dependence of our source terms $\mathbf{\Sigma}_D^S, \mathbf{\Sigma}_U^S$.

For crustal sources we take a separation level just below the Moho (z_M) and then the approximation (9.41) has the explicit form

$$^{de}w_0 = \mathbf{W}_F[\mathbf{I} - \mathbf{R}_D^{0M}\mathbf{R}_F]^{-1}\mathbf{T}_U^{0M}\mathbf{R}_D^{ML}\mathbf{T}_D^{SM}[\mathbf{I} - \mathbf{R}_U^{fS}\mathbf{R}_D^{SM}]^{-1}(\mathbf{\Sigma}_D^S + \mathbf{R}_U^{fS}\mathbf{\Sigma}_U^S). \tag{9.72}$$

To recover the seismograms at a given range r we have now to evaluate an integral over slowness as in (9.69) with $^{de}w_0$ as the response term.

For receiver ranges between 3500 km and 9500 km it is common to represent the PP or SS element of $\mathbf{R}_D^{ML}$ via approximations of the form (6.52) e.g.

$$(\mathbf{R}_D^{ML})_{PP} \sim \exp\{i\omega\tau_m(p) - i\pi/2\}Q(\omega), \tag{9.73}$$

where $\tau_m(p)$ is the phase delay for the mantle (6.52). The effect of attenuation in the mantle $Q(\omega)$ is normally included via an empirical operator based on the assumption that for a particular wave type the product of travel-time and overall loss factor is a constant t^*. For *P* waves $t_\alpha^* = T_\alpha Q_\alpha^{-1}$ is frequently taken to be 1.0

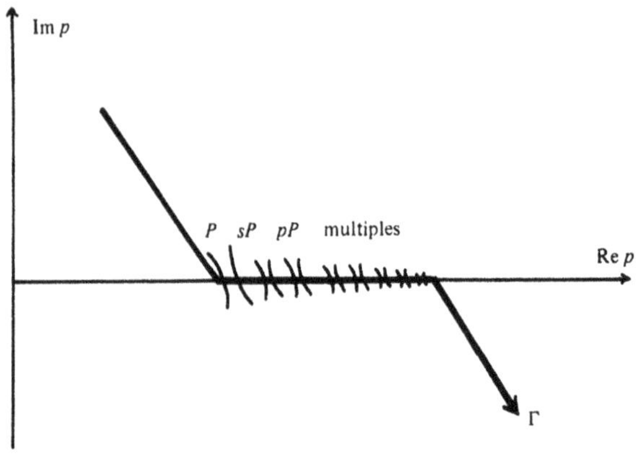

Figure 9.10. The multiple saddles associated with teleseismic propagation and a suitable integration path Γ.

s but for low loss paths can be about 0.3 s. For *S* waves t^*_β is rather higher e.g. Langston & Helmberger (1975) have assumed $t^*_\beta = 3.0$ s. In terms of frequency

$$Q(\omega) \approx \exp\{-i\omega t^* \pi^{-1} \ln(\omega/2\pi)\} \exp\{-\tfrac{1}{2}|\omega| t^*\}, \tag{9.74}$$

where the effects of velocity dispersion are included, cf. (1.25).

With the approximation (9.73) the slowness integral in (9.69) has normally been evaluated at high frequencies via a steepest descent approximation at the saddle point p_r for the direct wave given by

$$r + \partial_p\{\tau_m(p_r) + \tau_c(p_r)\} = 0, \tag{9.75}$$

where $\tau_c(p)$ is the phase delay in the crust corresponding to the transmission terms $\mathbf{T}^{OM}_U \mathbf{T}^{SM}_D$. The expression in braces is the geometrical ray theory expression for the range $-X(p_r)$ for slowness p_r. The resulting expression for the spectrum of the seismogram is

$$\bar{\mathbf{u}}(r, 0, \omega) = \frac{-i\omega M(\omega)}{[p_r r\, |\partial_p X(p_r)|]^{1/2}} e^{i\omega T(p_r)} Q(\omega) C_{RS}(p, \omega), \tag{9.76}$$

where $T(p_r)$ is the travel-time for the direct wave, and C_{RS} includes amplitude and phase terms associated with the crustal terms at source and receiver. The term $-i\omega M(\omega)$ is the far-field radiation from the source as seen through the appropriate instrument. This approximation has been justified by treating the radiation leaving the source crust as seen at large ranges as a plane wave, using ray theory in the mantle and a plane wave amplification factor $\mathbf{W}^{fM}_U$ at the receiver (see, e.g., Hudson 1969b). No allowance is made, however, for the slightly different paths for the surface reflected phases or the wavefront divergence associated with multiple crustal reflections.

If we make an expansion of the reverberation operators in (9.72) we get a

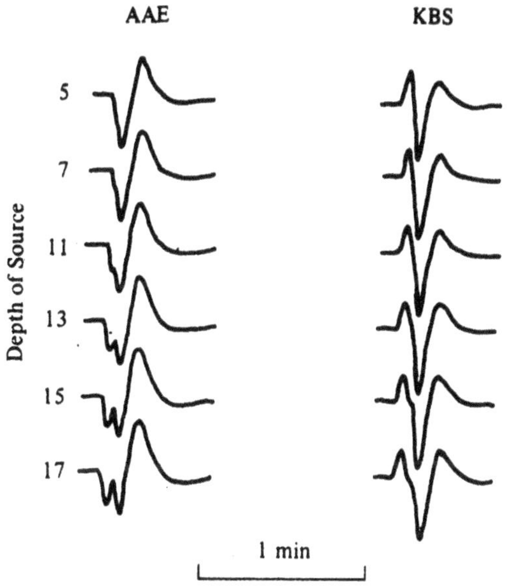

Figure 9.11. The effect of source depth on teleseismic *P* waveform.

sequence of terms corresponding to the surface reflections and crustal multiples each of which is associated with a subsidiary saddle point. The situation is represented schematically for *P* waves in figure 9.10. The range of slownesses occupied by the saddles is not large and the most accurate representation of the teleseismic wavefield will be obtained by a 'full-wave' treatment with a contour of integration Γ as indicated in figure 9.10. This corresponds to treating a bundle of slownesses clustered around p_T rather than the single slowness in (9.76). The result is that we get a better representation of amplitude effects due to conversion, and also of the decay of the crustal reverberations. For teleseismic *SV* waves conversion to *P* at the Moho is important and using a range of slownesses we can also model shear coupled *PL* waves. For shorter distances than 3500 km we need to make a more accurate representation of the reflection terms in $\mathbf{R}_D^{JL}$ to account for the detailed structure in the upper mantle. The presence of triplications in the travel time curves means that a band of slownesses is needed to represent the response. The 'full-wave' approach or a reflectivity treatment are required for accurate seismograms.

For comparison with long-period records the main interest is in the interference of the direct wave and the surface reflected phases. Langston & Helmberger (1975) have introduced a simple approximation based on a composite source term to model these effects. For *P* waves, for example, they construct the downward radiation term

$$\Sigma_D^C = (\Sigma_D^S)_P + (\mathbf{R}_F)_{PP} e^{i\omega\Delta\tau_1} (\Sigma_U^S)_P + (q_{\alpha 0}/q_{\beta 0})(\mathbf{R}_F)_{PS} e^{i\omega\Delta\tau_2} (\Sigma_U^S)_S, \quad (9.77)$$

where $\Delta\tau_1$ is the phase lag of *pP* relative to the direct wave and $\Delta\tau_2$ is the phase lag of *sP*. The ratio $(q_{\alpha 0}/q_{\beta 0})$ allows for the change in wavefront divergence on conversion at the surface. With this composite source a single slowness is used to calculate receiver effects and mantle propagation is included only through the attenuation operator. Thus the receiver displacement is approximated by

$$\bar{\mathbf{u}}(r, 0, \omega) = -(\mathbf{W}_F)_P \mathbf{\Sigma}_D^C(p_r, \omega) Q(\omega) i\omega M(\omega), \tag{9.78}$$

where p_r is the geometric slowness given by (9.75). Langston & Helmberger suggest the use of a far-field source time function consisting of a trapezoid of unit height described by three time parameters, which allows relative time scaling of rise time, fault duration and stopping time. When convolved with a long-period response, this may be used for $\partial_t M(t)$.

An illustration of the effect of depth of source with this procedure is shown in figure 9.11, showing the significant variation in *P* waveform.

All the expressions for the displacements which we have generated in this chapter are linear in the force or moment tensor components describing the point source which we have introduced to represent the physical source of seismic radiation. When we have a good model of the wavespeed distribution with depth we can calculate the contribution to a seismic phase for a number of ranges from each moment tensor component. With observed seismograms at the same ranges we can set up a linear inverse problem for the relative weighting of the moment tensor components (see e.g. Ward 1980), since we have only a few parameters describing the source. If the matching of theoretical and observed seismograms is performed in the frequency domain the estimates of $M_{ij}(\omega)$ give an indication of the time evolution of the source.

As we have noted in chapter 4, such a procedure will give us the moment tensor elements appropriate to our reference model, rather than the real Earth, so that there can be systematic bias. For large events the higher order moments of the source can be significant, but their effect can be reduced by working with a point source at the centroid of the disturbance and then allowing the position of the centroid as a function of frequency to be a free parameter in the inversion (Woodhouse 1981 - private communication).

Chapter 10

Generalized ray theory

In the previous chapter we have developed approximations to the seismic response in which, in essence, we retain frequency dependence in the amplitude of any reflection effects and so a numerical integration over slowness is needed. We now turn our attention to a further class of approximation in which the factorisation of the seismic response is carried even further.

The seismic displacement is represented as a sum of 'generalized ray' contributions for which the amplitude depends only on slowness and the phase has a slowness dependent term multiplied by frequency. For each of the generalized rays we are able to make use of the functional form of the integrand in the transform domain to reduce the space-time response to a slowness integral. For a medium composed of uniform layers an exact representation of each generalized ray may be made using the Cagniard-de Hoop method (Helmberger, 1968) with a complex slowness contour chosen to give a certain combination of phase variables the attributes of time. For uniform layers, or smoothly varying media, Chapman (1978) has proposed an alternative, approximate, method with a real slowness contour.

The success of these generalized ray techniques depends on an adroit choice of 'rays', from the infinite expansion of possible generalized rays to represent the portion of the seismogram of interest. If no conversions are included all rays with a given multiple level in any region can be generated by combinatorial techniques; and these methods can be extended with more difficulty to rays with a limited number of converted legs (Hron, 1972; Vered & BenMenahem, 1974).

10.1 Generation of generalized ray expansions

In Chapter 9 we have made use of partial expansions of the response to generate various classes of approximation for the seismic wavefield in a stratified half space. In order to obtain a generalized ray sum we carry this expansion process much further and now represent all reverberation operators appearing within the response by their infinite series representation (6.17).

As in our previous discussion of approximations a convenient starting point is provided by (7.66) which displays the surface reflection operator explicitly

$$w_0 = \mathbf{W}_F[\mathbf{I} - \mathbf{R}_D^{0L}\mathbf{R}_F]^{-1}\mathbf{T}_U^{0S}[\mathbf{I} - \mathbf{R}_D^{SL}\mathbf{R}_U^{0S}]^{-1}(\Sigma_U^S + \mathbf{R}_D^{SL}\Sigma_D^S), \tag{10.1}$$

We make an expansion to the full infinite sequence of surface reflection terms

$$[\mathbf{I} - \mathbf{R}_D^{0L}\mathbf{R}_F]^{-1} = \mathbf{I} + \sum_{k=1}^{\infty}(\mathbf{R}_D^{0L}\mathbf{R}_F)^k, \tag{10.2}$$

and also expand the stratification operator

$$[\mathbf{I} - \mathbf{R}_D^{SL}\mathbf{R}_U^{0S}]^{-1} = \mathbf{I} + \sum_{l=1}^{\infty}(\mathbf{R}_D^{SL}\mathbf{R}_U^{0S})^l. \tag{10.3}$$

This gives a representation of the surface displacements in terms of a doubly infinite sequence of reflection terms. The way in which we now extract a generalized ray expansion depends on the assumptions we make about the nature of the stratification.

10.1.1 Uniform layer models

In section 6.2.1 we have shown how the reflection and transmission matrix elements which appear in (10.1) can be constructed for a stack of uniform layers by a recursive application of the addition rules in two stages to allow for phase delays and interface effects (6.24), (6.26). In going from interface $k+1$ to interface k we have for example,

$$\mathbf{R}_D(z_k-) = \mathbf{R}_D^k + \mathbf{T}_U^k E_D^k \mathbf{R}_D(z_{kr+1}-)E_D^k[\mathbf{I} - \mathbf{R}_U^k E_D^k \mathbf{R}_D(z_{k+1}-)E_D^k]^{-1}\mathbf{T}_D^k, \tag{10.4}$$

where E_D^k is the phase income for downgoing waves in crossing the kth layer (3.46) and $\mathbf{R}_D^k$ etc are the interface matrices at z_k.

The contribution introduced in crossing this kth layer can now itself be expanded into an infinite sequence of terms representing internal multiples within the kth layer superimposed on the reflection behaviour beneath z_{k+1} by writing

$$[\mathbf{I} - \mathbf{R}_U^k E_D^k \mathbf{R}_D(z_{k+1}-)E_D^k]^{-1} = \mathbf{I} + \sum_{r=1}^{\infty}\{\mathbf{R}_U^k E_D^k \mathbf{R}_D(z_{k+1}-)E_D^k\}^r. \tag{10.5}$$

When such an expansion is made in each layer, $\mathbf{R}_D(z_{k+1}-)$ appearing in (10.5) will itself be an infinite sequence of reflection terms. The overall reflection matrix for a zone, e.g., $\mathbf{R}_D^{SL}$ will then consist of a nested sequence of infinite expansions. If all the expansions are carried out we get finally an infinite sequence of terms representing all possible classes of reflection processes within the multilayered stack.

The nature of the individual terms is conveniently examined by looking at the first two terms in the expansion of (10.4)

$$\mathbf{R}_D^k + \mathbf{T}_U^k \mathsf{E}_D^k \mathbf{R}_D(z_{k+1}-)\mathsf{E}_D^k \mathbf{T}_D^k. \tag{10.6}$$

We suppose the thickness of the kth layer is h_k and now introduce the explicit phase dependence of the second term in (10.6) for the *P-SV* case to obtain

$$\begin{bmatrix} T_U^{PP} & T_U^{PS} \\ T_U^{SP} & T_U^{SS} \end{bmatrix} \begin{bmatrix} R_D^{PP} e^{2i\omega q_{\alpha k} h_k} & R_D^{PS} e^{i\omega(q_{\alpha k}+q_{\beta k})h_k} \\ R_D^{SP} e^{i\omega(q_{\alpha k}+q_{\beta k})h_k} & R_D^{SS} e^{2i\omega q_{\beta k} h_k} \end{bmatrix} \begin{bmatrix} T_D^{PP} & T_D^{PS} \\ T_D^{SP} & T_D^{SS} \end{bmatrix}, \tag{10.7}$$

where, e.g., R_D^{PP} is the PP element of $\mathbf{R}_D(z_{k+1}-)$. As we have shown in section 6.1, the inclusion of higher terms in the expansion, i.e. $r > 0$, corresponds to the introduction of multiple reflections within the kth layer, and thus further phase delays in the additional terms.

We may assess the error introduced by truncation of the infinite sequences by using the partial expansion identity (9.1). If an overall accuracy level ϵ is derived for, say, $\mathbf{R}_D^{SL}$ a convenient working criterion for the number of terms (L) to be retained in the expansion is that, if there are N layers in 'SL',

$$[R_A R_B]^L \le \epsilon/N, \qquad \text{i.e. } L \ge \ln(\epsilon/N)/\ln(R_A R_B), \tag{10.8}$$

where R_A, R_B are the moduli of the largest reflection and transmission coefficients at the roof and floor of the layer (Kennett, 1974). If the layer has lower wavespeeds than its surroundings, R_A and R_B can be quite large and many terms are needed. For near-grazing incidence at an interface R_B will approach unity and for high accuracy L should be quite large.

For simplicity we will assume that all the elements of the source moment tensor M_{ij} have a common time dependence $M(t)$. Then we may write the vector $\boldsymbol{w}_0$ as the infinite sequence

$$\boldsymbol{w}_0(p, m, \omega) = M(\omega) \sum_I \mathbf{g}_I(p, m) \exp\{i\omega\tau_I(p)\}. \tag{10.9}$$

The individual 'ray' terms corresponding to a particular reflection process have

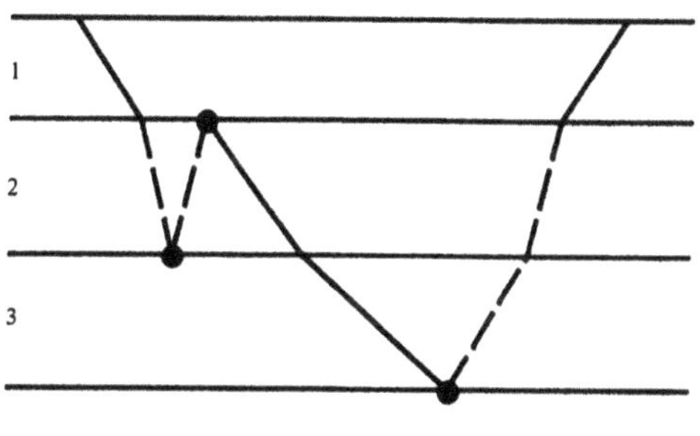

Figure 10.1. A generalized ray in a layered medium, *P* wave legs are indicated by solid lines, *S* wave legs by dashed lines. Reflection points are marked by circles.

been factored to show their amplitude and phase dependence. The phase delay term for the uniform layers is

$$\tau_I(p) = \prod_r n_r q_r h_r, \tag{10.10}$$

where n_r is the number of times the rth layer, with thickness h_r, is crossed in the same mode of propagation by the 'ray' path which specifies the Ith processes whereby energy can pass from source to receiver (see figure 10.1). The vertical slowness q_r is taken as $q_{\alpha r}$ for *P* waves and $q_{\beta r}$ for *S* waves. The expression $g_I(p, m)$ factors into two parts: the first $f_I(p, m)$ represents the way in which the source radiation and receiver displacement operators depend on slowness, and the second is the product of all reflection and transmission coefficients along the Ith path. Thus

$$g_I(p, m) = f_I(p, m) \prod_j T_j(p) \prod_k R_k(p), \tag{10.11}$$

where T_j is the plane wave transmission coefficient for an interface crossed by the Ith ray and R_k is the reflection coefficient for an interface at which the Ith ray changes direction. In each case conversion is taken into account, if appropriate, and we have exploited the frequency independence of the interface coefficients. The directivity function $f_I(p, m)$ depends on the azimuthal order m through the source terms $\mathbf{\Sigma}_D^S, \mathbf{\Sigma}_U^S$. If, for example, we consider a ray path which starts with a downgoing *P* wave and also ends at the surface as *P* we would have

$$f_I(p, m) = (\mathbf{W}_F)^P(p)\mathbf{\Sigma}_D^S, \tag{10.12}$$

where the vector $(\mathbf{W}_F)^P = [(W_F)^{UP}, (W_F)^{VP}]^T$ includes the free surface amplification factors for *P* waves, since we have extracted the frequency dependence of the source in $M(\omega)$.

With the expression (10.9) for $\boldsymbol{w}_0$ the surface displacement as a function of space and time (7.75) takes the form, for *P-SV* waves,

$$\mathbf{u}_P(r, \phi, 0, t) = \frac{1}{2\pi}\int_{-\infty}^{\infty} d\omega\, e^{-i\omega t}\omega^2 M(\omega) \sum_m \int_0^{\infty} dp\, p \sum_I g_I^T e^{i\omega\tau_I(p)} \mathbf{T}_m(\omega p r). \tag{10.13}$$

The separation of the frequency dependence of the Ith ray into the cumulated phase term will enable us in Sections 10.3–10.5 to use analytic techniques to construct expressions for $\mathbf{u}$.

10.1.2 Piecewise smooth media

For a medium consisting of a stack of uniform layers the representation (10.13) is exact when the full infinite ray expansion is present. Approximations are only introduced when we truncate the expansion to a finite number of terms so that it is possible to make computations.

When, however, we have a medium consisting of a sequence of smooth gradient zones separated by discontinuities in the elastic parameters or their derivatives, the contribution from the Ith ray path can only be represented as in (10.9) in a high frequency approximation. Nevertheless we will find that this will prove to be a useful form.

To see how the approximation arises consider a gradient zone in (z_A, z_B) bounded by uniform half spaces with continuity of properties at z_A, z_B. This is the model we have considered in section 6.3.1 and we have shown there that we may build up the reflection and transmission matrices $\mathbf{R}_D^{AB}$, $\mathbf{T}_D^{AB}$ from elements associated with entry and exit from the gradient zone and the nature of the wavefield within the zone.

At z_A there will normally be a discontinuity in wavespeed gradient and so partial reflection can occur. Within the gradient zone we can describe the wavefield via generalized vertical slownesses $\eta_{u,d}(p, \omega)$ (3.104) and phase terms which depend on Airy function entries. The partial reflection terms at z_A depend on the difference between η_u and η_d and the vertical slowness in a uniform medium q. The frequency dependence of $\eta_{u,d}$ arises from the character of the wavefield away from z_A, and in the high frequency limit, when turning points are far from z_A and z_B we have, e.g.,

$$\eta_{\alpha u}(p, \omega) \sim q_\alpha(p), \quad \eta_{\alpha d}(p, \omega) \sim q_\alpha(p), \tag{10.14}$$

with a similar relation at z_B. In this limit, parameter gradient discontinuities at z_A and z_B appear to be transparent and we may use the asymptotic forms for the Airy function terms to generate approximations for the reflection and transmission terms. To the leading order approximation there is no coupling between wave types in the gradient zone. Thus for *P* waves if there is no turning point in (z_A, z_B) from (9.66) we have

$$\mathbf{R}_D^{AB}|_{PP} \sim 0, \quad \mathbf{T}_D^{AB}\mathbf{T}_U^{AB}|_{PP} \sim \exp\left\{2i\omega \int_{z_A}^{z_B} d\zeta\, q_\alpha(\zeta)\right\}, \tag{10.15}$$

whereas if there is a turning level at $Z_\alpha(p)$, from (9.65) we find

$$\mathbf{R}_D^{AB}|_{PP} \sim \exp\left\{2i\omega \int_{z_A}^{Z_\alpha(p)} d\zeta\, q_\alpha(\zeta) - i\pi/2\right\}, \tag{10.16}$$

$$\mathbf{T}_U^{AB}\mathbf{T}_D^{AB}|_{PP} \sim \exp\left\{2i\omega \int_{z_A}^{Z_\alpha(p)} d\zeta\, q_\alpha(\zeta)\right\} \exp\left\{-2|\omega| \int_{Z_\alpha(p)}^{z_B} d\zeta\, |q_\alpha(\zeta)|\right\}. \tag{10.17}$$

Now we have already shown in section 6.3 that we can build the reflection matrix for a piecewise smooth medium from the reflection and transmission matrices for the gradient zones and the interface matrices between two uniform half spaces with the properties just at the two sides of the interface.

Thus, to the extent that (10.15) and (10.17) are valid we have the representation

$$w_0(p) \sim M(\omega) \sum_m g_I(p) \exp\{i\omega\tau_I(p)\}. \tag{10.18}$$

We must now account for turning points so that

$$\tau_I(p) = \sum_r \left\{ n_r \int_{z_r}^{z_{r+1}} d\zeta\, q_r(\zeta) + 2n_r^* \int_{z_r}^{Z^*} d\zeta\, q_r(\zeta) \right\}, \tag{10.19}$$

where n_r is once again the number of transmissions through the rth layer (now a gradient zone) in a particular propagating wave type and n_r^* is the number of legs in the rth zone in which total reflection occurs at the level Z^*. The term $g_I(p)$ includes all the factors in (10.11), but in addition includes a factor $\exp\{in_r^*\pi/2\}\text{sgn}(\omega)$ to allow for the phase shifts for the turning rays.

As we can see from the discussion above we would expect the right hand side of (10.18) to be a poor approximation to the full field at low frequencies and when a turning point lies close to one of the boundaries of a gradient zone. With the form (10.19) for the phase delays we cannot account for tunnelling phenomena into low wavespeed zones, as for example in *P4KP* (Richards, 1973).

For most high frequency problems we can, however, adopt (10.18) and then the surface displacements can asymptotically be represented as in (10.13).

10.2 Ray Selection and generation

We may describe an individual generalized ray path within a multilayered medium by a code indicating the nature of the ray, and for this purpose it makes no difference whether a layer is uniform or has smoothly varying properties. The layer number and wave type may be described by assigning an ordered pair of integers to each layer

$$\{C_j, i_j\} \tag{10.20}$$

where i_j is the layer number and C_j indicates the wave type in that layer ($C_j = 1, P$ waves; $C_j = 2, S$ waves). For n ray segments there will be n ordered pairs $\{C_j, i_j\}$; for example, the ray in figure 10.1 can be represented as

$$\{1, 1; 2, 2; 2, 2; 1, 2; 1, 2; 2, 3; 2, 2; 1, 1\} \tag{10.21}$$

Rays which do not include conversions of wave type are completely described by the layer indices $\{i_j\}$,

The properties of such ray codes have been extensively studied by eastern European seismologists and a convenient summary of results and algorithms is presented by Hron (1972).

Since we wish to achieve an economical means of generating generalized rays it is advantageous to consider groups of rays which have properties in common. If we consider a class of rays with permutations of the same ray codes we may divide these into:
kinematic groups, for which the phase term τ_I will be the same; and,
dynamic groups, which form subclasses of the kinematic groups and have the same products of interface coefficients $g_I(p)$.

For generalized rays for which all legs are in a single wave type it is possible to enumerate all the possible kinematic and dynamic groups. We will consider a surface source and receiver and then we will have an even number of ray segments. The extension to upgoing and downgoing rays from a buried source has been discussed by Vered & Ben Menahem (1974).

For a ray without conversions the time characteristics τ_I can be described by the set

$$\{n_1, n_2, ..., n_j\}, \quad J \geq 2, \tag{10.22}$$

where n_j is half the number of segments in the jth layer since each downgoing leg is matched by an upgoing. All ray numbers of a kinematic group will share the same set (10.22). If $J = 2$ the number N_k of different rays in the kinematic group built from $2n_1$, segments in the first layer and $2n_2$ segments in the second layer is equal to the number of ways of distributing n_2 objects into n_1 cells, where any number can occupy one cell with the result

$$N_k(n_1, n_2) = \frac{(n_1 + n_2 - 1)!}{n_2!(n_1 - 1)!} = \binom{n_1 + n_2 - 1}{n_2}, \tag{10.23}$$

in terms of the combinatorial coefficient $\binom{n}{r}$. If $J = 3$ we now have to intermesh the $2n_2$ segments in layer 2 with the $2n_3$ segments in the third layer whilst still having $N_k(n_1, n_2)$ possibilities from the top two layers, thus

$$N_k(n_1, n_2, n_3) = N_k(n_1, n_2) \binom{n_2 + n_3 - 1}{n_3}. \tag{10.24}$$

In general the number of rays in a kinematic group for $J \geq 2$ will be

$$N_k(n_1, n_2, ..., n_j) = \prod_{j=1}^{J-1} \binom{n_j + n_{j+1} - 1}{n_{j+1}}. \tag{10.25}$$

For the kinematic group $\{2, 2\}$ we illustrate the 3 possible rays in figure 10.2, and this set divides into two dynamic groups: one with two members and the other with one, characterised by the number of reflections at the first interface. Since we assume that rays are continuous we can describe the members of a dynamic group by the numbers of reflections from interfaces. We therefore define m_j to be the

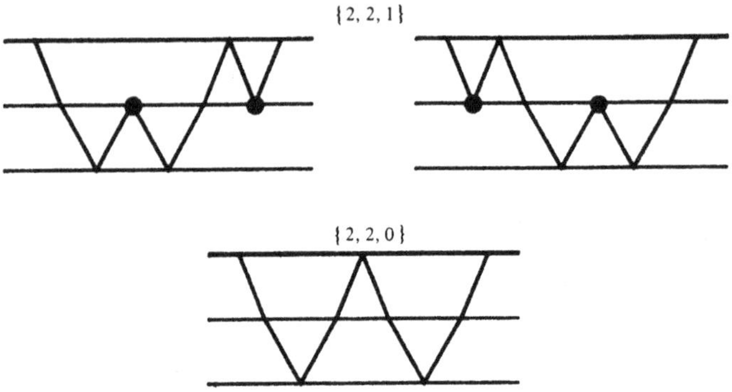

Figure 10.2. The three members of the kinematic group 2,2 separate into two dynamic groups: the first 2,2;1 has two members, the second 2,2;0 only one.

number of reflections from the jth interface when the ray is in the jth layer. The set of $2J - 1$ integers

$$\{n_1, n_2, \ldots.., n_j; m_1, m_2, \ldots.., m_{J-1}\} \tag{10.26}$$

completely describes the function $\mathbf{g}_I(p)$ for all the members of the same dynamic group, since $m_j \equiv n_j$. The number of members in each dynamic group is (Hron 1972)

$$N_{dk}(n_1, \ldots, n_j; m_1, \ldots, m_{J-1}) = \prod_{j=1}^{J-1} \binom{n_j}{m_j} \binom{n_{j+1} - 1}{n_j - m_j - 1}. \tag{10.27}$$

When we seek to generate rays we can effect considerable savings by only taking one ray from each dynamic group and then using the multiplicity factor N_{dk} to account for all the other rays in the group since they give equal contributions to (10.13).

We may organise the ray sum in (10.13) to exploit the benefits of the kinematic and dynamic groupings by writing the slowness integral as

$$\int dp\, p \sum_k \left\{ \sum_d N_{dk} \mathbf{g}_d^T(p) \right\} e^{i\omega\tau_k(p)} \mathbf{T}(\omega p r). \tag{10.28}$$

The frequency dependent portions are then the same for each kinematic group k, and the inner sum over dynamic groups accounts for different reflection processes with the same phase delays.

The concepts of dynamic and kinematic groups are just as useful for rays with converted legs, but the combinatorial mathematics becomes very difficult for more than the converted leg (Hron, 1972). To get over this problem Vered & Ben Menahem (1974) have specified the interfaces at which conversion can occur, they have then, in effect, worked out rays from the source to a receiver at the conversion point and then started the ray generation system again from the conversion point.

Since it is computationally very expensive to generate more than a limited number of rays, care has to be taken in their selection. In most problems the generalized rays which make the largest contribution to (10.13) are those with the least number of reflections from interfaces and thus the most transmissions. For piecewise smooth media, turning rays are particularly important. However, rays that have most of their path in the near-surface zone of low wavespeeds can have significant amplitude even though they have suffered many reflections (cf., Helmberger & Engen, 1980). A similar effect can occur with other waveguides.

Generalized rays including conversion from P to S usually make most contribution to (10.13) when conversion occurs at reflection. Conversion at transmission is typically small, unless the P wavespeed on one side of an interface is fairly close to the S wavespeed on the other side. This can occur, for example, with water and hard rock at the seafloor to give significant S wave-propagation in the sub-seafloor rocks.

If a stack of uniform layers are used to simulate a gradient zone, a turning ray is represented by the superposition of a system of multiple reflections at near grazing incidence within the uniform layers near the turning level. Commonly, only a very limited sequence of rays is employed, e.g. the expansion in (10.5) is truncated at the level (10.6) (see, e.g., Helmberger 1968), but Müller (1970) has forcibly demonstrated the need for retaining high order reflections for accurate results in even simple models.

10.3 Slowness results for generalized rays

The surface displacement contribution with azimuthal order m from a single generalized ray is given by

$$\mathbf{u}^{I}_{m}(r,t) = \frac{1}{2\pi}\int_{-\infty}^{\infty} d\omega\, e^{-i\omega t}\omega^2 M(\omega)\int_0^{\infty} dp\, \mathbf{w}_I(p,m,\omega)\mathbf{T}_m(\omega p r), \quad (10.29)$$

with $\mathbf{w}_I(p,m,\omega) = \mathbf{g}_I(p,m)e^{i\omega\tau_I(p)}$. We now follow the *slowness* treatment of Section 7.3.2 and perform the frequency integral first so that we express (10.29) as a sequence of convolutions over time with a residual integral over p:

$$\mathbf{u}^{I}_{m}(r,t) = \partial_{tt}M(t) * \int_0^{\infty} dp\, p\{\check{\mathbf{w}}_I(p,m,t) * (1/pr)\check{\mathbf{T}}_m(t/pr)\}. \quad (10.30)$$

We have already tabulated the time transforms of the vector harmonics in (7.98), (7.100) and so we are left to evaluate the Fourier inverse of $\mathbf{w}_I(p,m,\omega)$.

The final seismograms must be real time functions and so $\mathbf{u}(r,\omega) = \mathbf{u}^*(r,-\omega)$, so that with the choice of physical Riemann sheet we have made for the radicals appearing in $\mathbf{g}_I(p,m)$, $\tau_I(p)$

$$\begin{aligned}\mathbf{g}_I(p,m) &= \mathbf{g}'_I(p,m) + i\,\mathrm{sgn}(\omega)\mathbf{g}''_I(p,m),\\ \tau_I(p) &= \tau'_I(p) + i\,\mathrm{sgn}(\omega)\tau''_I(p).\end{aligned} \quad (10.31)$$

We may now perform the inverse time transform for $\mathbf{w}_I$ to obtain (Chapman, 1978)

$$\check{\mathbf{w}}_I(p, m, t) = \frac{1}{\pi} \lim_{\epsilon \to 0} \mathrm{Im} \left[\frac{\mathbf{g}_I(p, m)}{t - \tau(p) - i\epsilon} \right], \tag{10.32}$$

where we have used the convergence factor ϵ to ensure the existence of the transform. As the imaginary part of τ_I tends to zero and we just have a real phase delay, $\mathbf{w}_I$ tends to a delta function:

$$\lim_{\epsilon \to 0} \mathrm{Im}[t - \tau(p) - i\epsilon]^{-1} \to \delta(t - \tau'(p)) \quad \text{as} \quad \tau'' \to 0, \tag{10.33}$$

and this property will enable us to simplify some subsequent results.

We will now restrict attention to the vertical component of displacement and azimuthal symmetry since this will enable us to illustrate the nature of the solution. The radial dependence now arises from $J_0(\omega p r)$ and its time transform is given by

$$\pi(1/pr)\check{J}_0(t/pr) = B(t, pr)(p^2r^2 - t^2)^{-1/2}, \tag{10.34}$$

from (7.98). We now perform the convolution of the two slowness dependent terms (10.32), (10.34) to obtain the explicit form

$$u^I_{z0}(r, t) = \partial_{tt}M(t) * \int_0^\infty dp\, p \times \lim_{\epsilon \to 0} \left\{ \frac{1}{\pi^2} \int_{-pr}^{pr} ds\, \mathrm{Im} \left[\frac{G_I(p)}{t - s - \tau_I(p) - i\epsilon} \right] \frac{1}{(p^2r^2 - s^2)^{1/2}} \right\}, \tag{10.35}$$

where $G_I(p)$ is the vertical component of $\mathbf{g}_I(p, 0)$. Along the real p axis only $G_I(p)$ and $\tau_I(p)$ may be complex and so the imaginary part operator can be abstracted to the front of the slowness integral.

The time and slowness elements in (10.34) are common to all the expressions for $\check{\mathbf{T}}_m(t/pr)$ (7.98)-(7.100) and so the form of the integral in (10.35) is modified for other components or angular orders by the addition of well behaved functions. Various methods of calculating theoretical seismograms can now be generated by using different techniques to evaluate the slowness integral in (10.35).

10.4 The Cagniard method

For a generalized ray in a stack of perfectly elastic layers, the contribution to the displacement field which we have so far expressed as a slowness integral of a convolution in time can be recast as an integral over time. This result was first obtained by Cagniard (1939) although the basic ideas are present in the work of Lamb (1904). The technique has subsequently been developed by a number of authors, notably Pekeris (1955) and de Hoop (1960). The first application to the calculation of theoretical seismograms seems to have been made by Helmberger (1968) and subsequently the method has been extensively used in the analysis of seismic records over a very wide range of distances.

For each generalized ray we extract the imaginary part operator to the front of the integral in (10.35) and then have to evaluate

$$I(r,t) = \frac{1}{\pi^2}\mathrm{Im}\int_0^{\infty} dp\, pG_I(p) \int_{-pr}^{pr} ds \frac{1}{[t - s - \tau_I(p) - 0i](p^2r^2 - s^2)^{1/2}}, \tag{10.36}$$

where we have included the $0i$ term in the denominator to remind us of the limiting procedure in (10.35). We now split the integral over s into the differences of the ranges $(-\infty, pr)$ and $(-\infty, -pr)$ and then with a change of variable in the convolution integrals we can rewrite (10.36) as

$$I(r,t) = \frac{1}{\pi^2}\mathrm{Im}\int_{-\infty}^{\infty} dp \int_{-\infty}^{0} dy \frac{pG_I(p)}{iy^{1/2}(y + 2pr)^{1/2}[t - y - \theta_I(p,r) - 0i]}, \tag{10.37}$$

and the slowness integration follows a path above the branch points in $G_I(p), \tau_I(p)$ for $\mathrm{Re}p < 0$ and below the branch points for $\mathrm{Re}p > 0$; the path C is illustrated in figure 10.3. We have here introduced the important auxiliary quantity

$$\theta_I(p,r) = \tau_I(p) + pr \tag{10.38}$$

which if r was the geometrical range for slowness p would just be the associated travel time. We now change the order of integration to give

$$I(r,t) = \frac{1}{\pi^2}\int_{-\infty}^{0} \frac{dy}{y^{1/2}}\mathrm{Im}\int_C dp \frac{pG_I(p)}{i(y + 2pr)^{1/2}[t - y - \theta_I(p,r) - 0i]} \tag{10.39}$$

We recall that the directivity and reflection function $G_I(p)$ and the phase delay $\tau_I(p)$ both depend on the vertical wave slownesses q_α, q_β at the interfaces and in the layers. Our original choice of branch cuts (3.8) was, e.g., $\mathrm{Im}(\omega q_\alpha) \geq 0$. We can however, rotate the branch cuts to lie along the real axis as in figure 10.3 and still maintain this condition on the contour C.

Following Burridge (1968) we now represent the integral over the slowness contour C as the sum of two contributions. The first contour C_1 lies along the two sides of the branch cut in $\mathrm{Re}p < 0$. The integrand in (10.39) is imaginary for slowness $|p| < p_l$, where p_I is the closest branch point to the origin; and so the integral along C_1 is real. There is, therefore, no contribution to $I(r,t)$ from this path. The second contour C_2 lies along the underside of the real p axis and its contribution can be evaluated by using Cauchy's theorem. The sole singularity in the lower half p plane is a simple pole $p(r,t)$ where

$$t - \theta_I(p,r) - y = 0i. \tag{10.40}$$

Since both t and y are real variables, $\theta_I(p,r)$ at this pole is real and can therefore

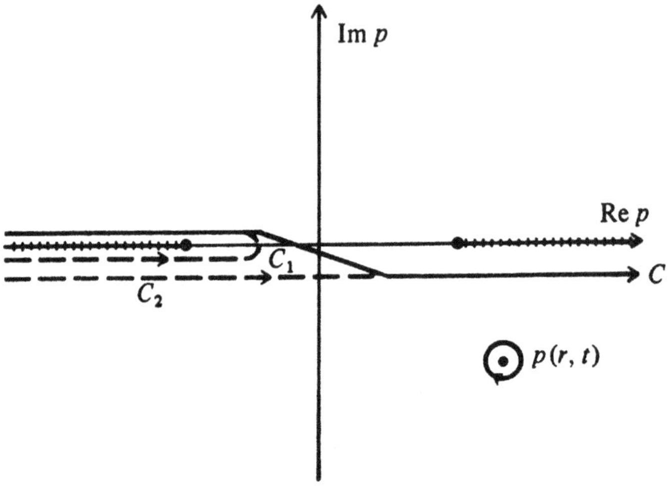

Figure 10.3. Branch cuts and contours in the complex p plane. The original contour C can be deformed into the sum of C_1 and C_2. There is a single pole in the lower half plane at $p(r,t)$.

take on the role of a real time. The integral (10.39) can be evaluated as just the residue contribution at the pole (10.40) and so

$$I(r,t) = \frac{2}{\pi}\mathrm{Im}\int_0^t d\theta_I \frac{pG_I(p)}{(t-\theta_I)^{1/2}(t-\theta_I+2pr)^{1/2}}\left[\frac{\partial p}{\partial \theta_I}\right], \tag{10.41}$$

where we have changed variables from y to θ_I. In (10.41) the slowness p is an implicit function of θ_I via the requirement that θ_I be real i.e.

$$\mathrm{Im}[\theta_I(p,r)] \equiv \mathrm{Im}[\tau_I(p) + pr] = 0. \tag{10.42}$$

The path of the pole specified by (10.42) will play an important role in our subsequent discussion and we will term this trajectory in the complex p plane the Cagniard path (H). There will be a different path for each generalized ray at each range.

When we reinstate the time dependence of the source we obtain the vertical displacement contribution from the Ith generalized ray as

$$u_{z0}^I(r,t) = \partial_{tt}M(t) * \frac{2}{\pi}\mathrm{Im}\int_0^t d\theta_I \frac{pG_I(p)[\partial_p\theta_I]^{-1}}{(t-\theta_I)^{1/2}(t-\theta_I+2pr)^{1/2}}. \tag{10.43}$$

This result is usually obtained by a rather different route in which the original transform integral (10.29) is manipulated into a form where the time dependence can be recognised directly (see, e.g., Gilbert & Helmberger, 1972).

For observations at large ranges so that pr is very much larger than the duration of the source it is an adequate approximation to replace $(t-\theta+2pr)^{1/2}$ by $(2pr)^{1/2}$ which leads to a considerable simplification in (10.43).

$$u^I_{z0}(r,t) \approx \partial_{tt}M(t) * \frac{1}{\pi}\mathrm{Im}\int_0^t d\theta_I \frac{2p^{1/2}G_I(p)[\partial_p\theta_I]^{-1}}{r^{1/2}(t-\theta_I)^{1/2}}. \tag{10.44}$$

The time integral is in the form of a convolution of $H(t)t^{-1/2}$ with a function of time along the Cagniard path. Thus in terms of the 'effective' source function $\mathcal{M}(t)$ introduced in (7.86)

$$\mathcal{M}(t) = \int_0^t dl\, \partial_l M(l)/(t-l)^{1/2} \tag{10.45}$$

we can express (10.43) as

$$u_{z0}(r,t) \approx \mathcal{M}(t) * \pi^{-1}\mathrm{Im}\, \partial_t\{G_I(p)(2p/r)^{1/2}[\partial_p\theta_I]^{-1}\}. \tag{10.46}$$

The effective source needs to be calculated only once for all generalized rays and the convolution in (10.46) can be carried out after the generalized ray sum has been formed.

The high-frequency result (10.44) can alternatively be derived directly by starting from the approximation of the time transform of the Bessel function by separated singularities (7.108) or from the asymptotic expansion of the Bessel function itself. For azimuthal orders $|m| > 0$ additional factors will appear in the inverse transforms (7.100) and the near-field terms need to be included. These aspects are discussed, with numerical comparisons, by Helmberger & Harkrider (1978).

Chapman (1974a, 1976) has shown how the Cagniard results can be extended to WKBJ solutions (3.52) in vertically varying media. Unfortunately the method cannot be applied directly to a turning ray because it is no longer possible to make the contour deformation into the lower half plane. It is however possible to make an iterative development via multiply reflected rays using (3.57) to approach the turning ray solution (Chapman, 1976) but the method becomes numerically unrewarding after the third order reflections.

10.4.1 The Cagniard path

The properties of the contribution made by a generalized ray are controlled by the character of the Cagniard path H in the complex p plane and the positions of the branch points appearing in the phase delay $\tau_I(p)$ and the directivity and reflection term $g_I(p)$.

The phase delay $\tau_I(p)$ will have branch points at the waveslownesses of the wave type in which each layer is traversed. The closest branch point to the origin in τ_I will be v_{max}^{-1}, where v_{max} is the largest wavespeed along the path. The term $G_I(p)$

has branch points at α_j^{-1}, β_j^{-1} for each side of an interface, for *P-SV* waves, and so the closest branch point to the origin will here be α_h^{-1}, where

$$\alpha_h^{-1} = \max[v_l,\ l = 1, ..., n+1] \tag{10.47}$$

if the deepest layer traversed is n. The point α_h^{-1} will be closer to the origin than v_{max}^{-1} for *S* waves, and also for *P* waves if the *P* wavespeed is greatest in layer $n+1$.

The Cagniard path H is defined by $\operatorname{Im}\theta_I(p, r) = 0$, (10.42) i.e.

$$\operatorname{Im}[pr + \tau_I(p)] = \operatorname{Im}[pr + \sum_r n_r q_r h_r] = 0. \tag{10.48}$$

For large p the slowness radicals $q_r \sim \pm ip$ and so since we choose p on H to be complex in such a way that its own imaginary part removes the $-i\Sigma n_r h_r p$ term, the asymptote to the Cagniard path for large p is

$$\arg p \sim \tan^{-1}(\sum_r n_r h_r / r). \tag{10.49}$$

At $p = 0$, $\theta_I(0, r)$ is independent of range r and is just the vertical travel time along the ray path. The Cagniard path starts off along the real axis but turns away from it at the saddle point corresponding to the geometrical slowness $p_{0I}(r)$ for which this combination of ray elements would arrive at the range r. This saddle will occur when $\partial_p \theta_I$ vanishes.

The second derivative,

$$\partial_{pp}\theta_I = -\sum_l n_l h_l / v_l^2 q_l^3, \tag{10.50}$$

and is real and negative in $0 < p < v_{max}^{-1}$. This root may be found efficiently numerically by, e.g., using Newton's method.

In the neighbourhood of the saddle point $p_{0I}(r)$ we make an expansion of $\theta_I(p, r)$ in a power series

$$\theta_I(p, r) = \theta_I(p_{0I}, r) + \tfrac{1}{2}(p - p_{0I})^2 \partial_{pp}\theta_I(p_{0I}, r) + \dots . \tag{10.51}$$

Along the Cagniard path we require $\operatorname{Im}\theta_I = 0$ and we want θ_I to increase away from the origin. For $p < p_0$ the path lies along the real axis. At the saddle point $\partial_{pp}\theta_I$ is negative and so, in order to maintain θ_I as an increasing function we have to choose $(p - p_{0I})^2 < 0$ at the saddle point. The Cagniard path therefore leaves the real axis at right angles at the saddle point p_{0I}.

The closest branch point to the origin for the ray will be α_h^{-1} which, as we have seen, can be closer than v_{max}^{-1} and so there may be a branch point closer to the origin than the saddle point p_{0I}. If $\alpha_h = v_{max}$, then $p_{0I} < \alpha_h^{-1}$ and for $t < \theta_I(p_{0I}, r)$ the path lies along the real axis where $G_I(p)$ and $\partial_p \theta_I$ are both real. In this case we see from (10.43) that there will be no contribution until the time $\theta_I(p_{0I}, r)$ associated with the geometric ray path, and the Cagniard path is as shown in figure 10.4a.

If, however, $\alpha_h > \theta_{max}$, we have two possibilities. When $p_{0I} < \alpha_h^{-1}$ the

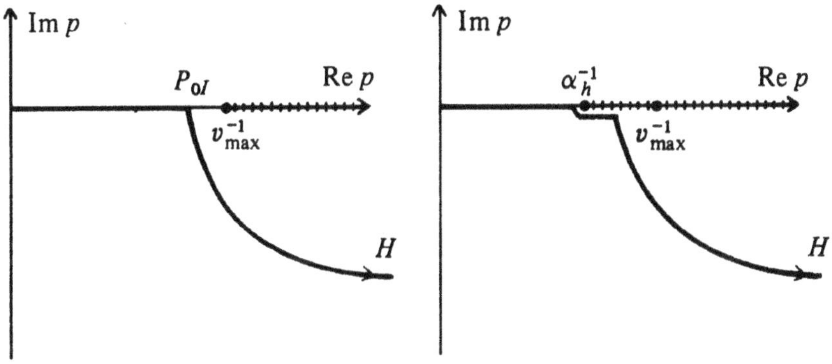

Figure 10.4. Cagniard paths in the complex p plane a) only reflected contributions; b) head wave segment in addition to reflections.

situation is as we have just described and there is no arrival before the geometric time. When $p_{0I} > \alpha_h^{-1}$ there is a head wave arrival before the geometric ray time. The Cagniard path is now as illustrated in figure 10.4b: for $t < \theta_I(\alpha_h^{-1}, r)$ both $G_I(p)$ and $\partial_p\theta_I$ are real and there is no contribution to (10.38), but for $\theta(\alpha_h^{-1}, r) < t < \theta_I(p_{0I}, r)$ although p and $\partial_p\theta_I$ are real, $G_I(p)$ is no longer real as some of the radicals will be complex. There is therefore a contribution to (10.43) from the segment H_h, and there is a separate interfacial head wave contribution for each branch point traversed in (α_h^{-1}, p_{0I}). Such head wave contributions appear moderately frequently for pure *P* wave paths and are very common for generalized rays with a significant portion of *S* wave legs.

Following Ben Menahem & Vered (1973) we can make an informative decomposition of the contribution (10.43) from a particular generalized ray path. The head wave contribution is

$$\mathrm{Im}\int_{t_h}^{t_r} d\theta_I \left\{ G_I(p)[\partial_p\theta_I]^{-1} U(p, \theta_I) \right\}, \tag{10.52}$$

where $t_h = \theta_I(\alpha_h^{-1}, r)$, $t_r = \min[t, \theta_I(p_{0I}, r)]$ and $U(p, t)$ represents the remainder of the integrand in (10.43).

After the geometrical arrival time $t_0 = \theta_I(p_{0I}, r)$ we can write the response as

$$\begin{aligned} &\mathrm{Im}\int_{t_0}^{t} d\theta_I \left\{ G_I(p_{0I})[\partial_p\theta_I]^{-1} U(p, \theta_I) \right\} \\ &+\mathrm{Im}\int_{t_0}^{t} d\theta_I \left\{ [G_I(p) - G_I(p_{0I})][\partial_p\theta_I]^{-1} U(p, \theta_I) \right\}. \end{aligned} \tag{10.53}$$

The first term represents the contribution from the geometric ray reflection, and since $\partial_p\theta_I$ vanishes at $p = p_{0I}$ the main contribution to the beginning of the reflected wave will come from the neighbourhood of p_{0I}. The second term represents the non-least-time arrivals, and since it vanishes at $\theta(p_{0I}, r)$ will have little contribution to the reflection. In Lamb's problem for, example, this term would include the Rayleigh wave contribution.

For numerical implementation of the Cagniard version of the generalized ray approach we must be able to find the Cagniard path H numerically and then achieve an adequate sampling in time. Different numerical schemes have been discussed by Wiggins & Helmberger (1974) and Vered & Ben Menahem (1974) and a general survey has been made by Pao & Gajewski (1977).

As we have seen in section 10.2, the effectiveness of generalized ray sums may be increased by making use of the ideas of kinematic and dynamic groups. In the present context all members of a kinematic group share the same Cagniard path H and members of a dynamic group have the same amplitude distribution along the path.

As an illustration of the way in which the contribution from a ray is determined by the Cagniard path, we consider in figure 10.5 the effect of inserting a thin higher speed layer into a model based on the work of Wiggins & Helmberger (1974). In figure 10.5a we show in solid lines the Cagniard contours for various ranges for a generalized ray corresponding to reflection from a small wavespeed jump in a uniform layer representation of an upper mantle model. The *P* wavespeed beneath the deepest interface is α_C and the corresponding critical range is r_C. The corresponding seismograms after passage through a low-pass filter are shown in figure 10.5b; a weak head wave separates at the largest ranges. If a thin (2 km) layer with *P* wavespeed α_L is introduced above the deepest interface the Cagniard paths are modified to those indicated in tone. All the paths leave the real p axis to the left of α_L^{-1} and then lie close to the real axis until the vicinity of α_C^{-1} when they bend away from the axis to follow the trend of the solid curves. The corresponding seismograms are shown in figure 10.5c with the same filtering as in b. The original interface is now in a shadow zone and the main contribution comes from the portions of the paths as they bend away from the real axis. In this case there is a much larger low frequency content and very little phase change occurs on reflection, as compared with figure 10.5b.

The Cagniard method has been employed to study a wide range of wave-propagation problems in models consisting of a stack of uniform layers. For example, Helmberger & Malone (1975) have looked at the effect of near-surface structure on local earthquake records and a number of studies have been made of upper mantle structure. Burdick & Orcutt (1979) have compared calculations made with the Cagniard technique and only primary reflections from each interface, with reflectivity calculations, including all multiples, in the same uniform layer model. Neglect of the multiples gives significant errors for strong transition zones, and where turning points occur near major discontinuities. For upper mantle structures these problems are not severe and good agreement is obtained at moderate frequencies (< 0.2 Hz).

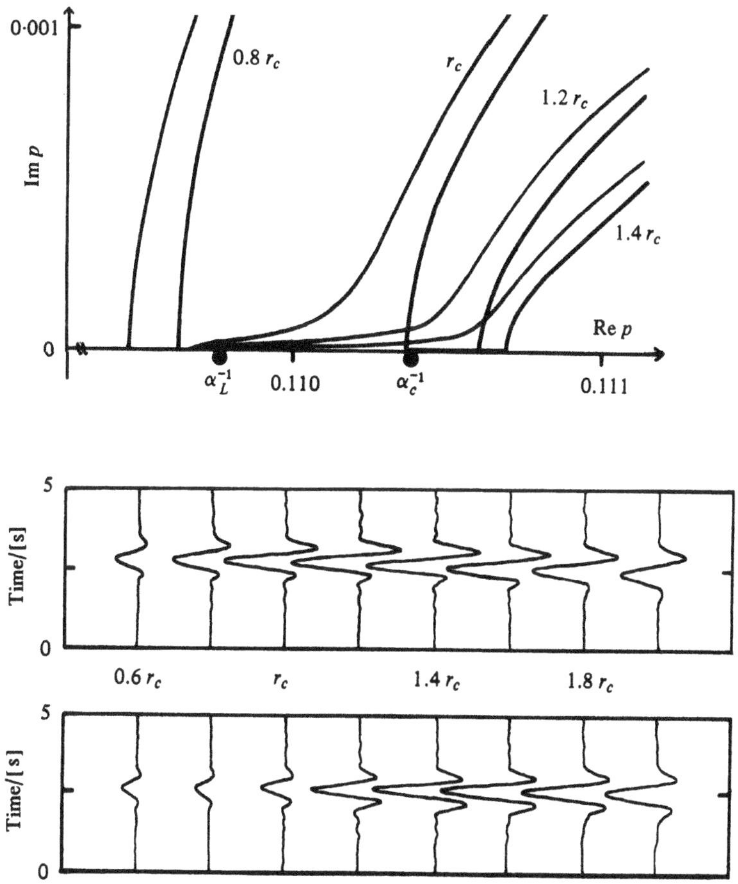

Figure 10.5. a) Cagniard paths for a generalized ray in an upper mantle model (Wiggins & Helmberger, 1974). The solid lines show the paths followed for various ranges (r_c is the critical range). The lines in tone show how the paths are modified by inserting a small high-velocity layer to give a shadow zone at the deepest interface. b) low passed seismograms for the solid paths; c) low passed seismograms for the paths in tone.

10.4.2 First-motion approximations

Consider the high frequency approximation (10.46) for a generalized ray corresponding to P wave reflection from the nth interface with transmission to and from the surface, for which the displacement contribution can be written

$$u_{z0} = \partial_t M(t) * H(t) t^{-1/2} * \\ \pi^{-1} \partial_t \mathrm{Im}\{f(p) R_n(p) T(p) (2p/r)^{1/2} [\partial_p t]^{-1}\}. \qquad (10.54)$$

The last term is to be evaluated along the Cagniard path H for the ray. Time increases as we move away from the origin along H. In the subsequent development we will often represent t as a function of slowness p and will be referring to the parameterisation of the Cagniard path. We have factored $G(p)$ to show the explicit dependence on $R_n(p)$ (the PP reflection coefficient at the nth interface); $T(p)$ is the product of transmission terms and $f(p)$ the source and receiver directivity.

We will assume that the *P* wavespeed increases with depth so that the closest branch point to the origin is $\alpha_{n+1}^{-1} (\equiv \alpha_h^{-1})$ so that for many ranges we have a head wave contribution, in addition to reflection terms.

We recall that $\partial_p t$ vanishes the saddle point p_0 at which the Cagniard path for range r turns away from the real p axis and so we seek an approximation to the reflection behaviour by examining the neighbourhood of p_0. For a wavespeed distribution that increases monotonically with depth, $T(p)$ and $f(p)$ will be real at p_0 and only $R_n(p)$ will be complex. Near $t(p_0)$

$$t - t_0 \approx -\tfrac{1}{2}A(p - p_0)^2, \tag{10.55}$$

where $A = |\partial_{pp} t(p_0)|$ and $t_0 = t(p_0)$. If we now differentiate (10.55) with respect to p we obtain

$$\partial_p t \approx [2A(t_0 - t)]^{1/2}. \tag{10.56}$$

Thus $\partial_p t$ is real for $t < t_0$ and imaginary for $t > t_0$ with the result that we may write the final term in (10.55) as

$$\Psi^R(r, t) = \pi^{-1}(p/Ar)^{1/2} T(p) f(p) \times \left[\operatorname{Im} R_n(p) \frac{H(t_0 - t)}{(t_0 - t)^{1/2}} + \operatorname{Re} R_n(p) \frac{H(t - t_0)}{(t - t_0)^{1/2}} \right], \tag{10.57}$$

and the displacement may be recovered by differentiation followed by convolution with the far-field time function for the source $\partial_t M(t)$ and $H(t)t^{-1/2}$,

$$u_{z0}(r, t) \approx \partial_t M(t) * H(t)t^{-1/2} * \partial_t \Psi(t). \tag{10.58}$$

We have previously noted, in connection with (7.108), that $H(-t)(-t)^{-1/2}$ is the Hilbert transform of $H(t)t^{-1/2}$ and so, on carrying out the convolutions in (10.58), we find

$$u_{z0}^R(r, t) = \pi^{-1}(p/Ar)^{1/2} \{\operatorname{Re} R_n(p) \partial_t M(t - t_0) + \operatorname{Im} R_n(p) \partial_t \hat{M}(t - t_0)\}, \tag{10.59}$$

where $\partial_t \hat{M}$ is the Hilbert transform of the far-field source function. The term in braces represents the scaling and phase distortion associated with reflection beyond the critical angle which we have already considered in figure 5.3. At precritical reflection, for ranges such that $p_0 < \alpha_h^{-1}$, we have only the contribution.

$$u_{z0}(r, t) = \pi^{-1}(p/Ar)^{1/2} T(p) f(p) \operatorname{Re}\{R_n(p) \partial_t M(t - t_0)\}, \tag{10.60}$$

and the reflected pulse shape is the same as the far-field source function. At the critical range $p_0 = \alpha_h^{-1}$ and we have a coincident saddle point and branch point and so a special treatment is necessary to obtain a 'first-motion' approximation similar to (10.60) (Zvolinskii, 1958) near this point.

Once p_0 separates from α_h^{-1} we can represent the post-critical reflections as in (10.59) but need now to take account of the head wave contribution. In the neighbourhood of the branch point α_h^{-1} we are unable to use a Taylor's series but we can approximate $\Psi(t)$ as

$$\Psi_h(r, t_h + \Delta t) \approx \pi^{-1}(p/r)^{1/2}T(p)f(p) \times \text{Im}\{[R_n(\alpha_h^{-1}) + \gamma\partial_\gamma R_n(\alpha_h^{-1})][\partial_p t(\alpha_h^{-1})]^{-1}\}, \tag{10.61}$$

where t_h is the time of arrival of the head wave, since only $R_n(p)$ for $p > \alpha_h^{-1}$. Along H_h, $i\gamma = (\alpha_h^{-2} - p^2)^{1/2}$ and so

$$p \approx \alpha_h^{-1} + \tfrac{1}{2}\gamma^2\alpha_h. \tag{10.62}$$

The small increment in slowness Δp associated with Δt is therefore

$$\Delta p = p - \alpha_h^{-1} \approx \tfrac{1}{2}\gamma^2\alpha_h. \tag{10.63}$$

This relation enables us to determine γ in terms of Δp and thus $\partial_p t$,

$$\gamma \approx (2\alpha_h^{-1}\Delta p)^{1/2} \approx (2\alpha_h^{-1}\Delta t[\partial_p t(\alpha_h^{-1})]^{-1})^{1/2}. \tag{10.64}$$

With this substitution, we find

$$\psi^h(r, t) \approx \pi^{-1}[\partial_p t(\alpha_h^{-1})]^{-3/2}[2\alpha_h^{-1}(t - t_h)]^{1/2} \times \text{Im}[\partial_\gamma R_n(\alpha_h^{-1})](p/r)^{1/2}T(p)f(p)H(t - t_h); \tag{10.65}$$

and also the distance travelled along the refractor, L, is $\partial_p t(\alpha_h^{-1})$ since

$$\partial_p t(\alpha_h^{-1}) = r - \alpha^{-1}\sum_r n_r h_r/q_r(\alpha_h^{-1}) = L. \tag{10.66}$$

The sum allows for the horizontal distance travelled in transmission. When we perform the convolution and differentiation in (10.58) to produce the displacement term we make use of the result

$$\partial_t M(t) * H(t)t^{-1/2} * \partial_t\{H(t - t_h)(t - t_h)^{-1/2}\} = \pi M(t - t_h), \tag{10.67}$$

to give

$$u_{z0}^h(r, t) \approx (2p\alpha_h^{-1}/rL^3)^{1/2}T(p)f(p)\text{Im}[\partial_\gamma R_n(\alpha_h^{-1})]M(t - t_h). \tag{10.68}$$

The pulse shape of the head wave is thus the integral of the far-field source time function and the rate of decay of the head wave is $r^{-1/2}L^{-3/2}$.

These first motion approximations have been used by a number of authors (e.g., Werth, 1967) to model the first few swings of *P* waveforms. The approximations are most effective when the significant phases are well separated in time but can be superimposed to allow for a number of arrivals.

The convenience of the first-motion approximations has lead Mellman & Helmberger (1979) to suggest modifications designed to link together the pre- and post-critical approximations as well as the head wave. They have used an approximate contour in the neighborhood of the saddle point and checked their results against full Cagniard calculations. For a wide variety of models it was possible to get good agreement and the modified first-motion approximations were much less expensive than the full calculations.

10.5 The Chapman method

The essence of the Cagniard method for calculating the contribution of a generalized ray is that the slowness integral is taken along such a path in the complex p plane that the time dependence can be easily recognised. In contrast Chapman (1978) has advocated that the slowness integral should be carried out along the real p axis as in (10.30) and (10.35).

In this section we will show how Chapman's idea can be used to generate a simple approximation for the displacement contribution from a generalized ray which can be used even when turning points and caustics are involved. We consider the vertical component of displacement and azimuthal symmetry as in (10.35) so that

$$u^{I}_{z0}(r,t) = \partial_{tt}M(t) * \pi^{-1} \int_0^{\infty} dp\, p \int_{-pr}^{pr} ds\, U_I(p, t-s)(p^2r^2 - s^2)^{-1/2}, \quad (10.69)$$

and

$$U_I(p,t) = \pi^{-1}\mathrm{Im}[G_I(p)/(t - \tau_I(p) - 0i)]. \quad (10.70)$$

We aim to produce an approximation which is valid for large ranges r and so we follow the procedure discussed in (7.108)-(7.114). We approximate $B(t, pr)(p^2 - t^2)^{-1/2}$ by two isolated singularities along the lines $t = pr$ and $t = -pr$ and then for larger ranges will retain only the outgoing term associated with the singularity along $t = pr$. The displacement is then represented by

$$u^{I}_{z0}(r,t) \approx \partial_{tt}M(t) * \frac{1}{\pi(2r)^{1/2}} \int_0^{\infty} dp\, p^{1/2} \times \int_{-\infty}^{\infty} ds\, \hat{U}_I(p, s - pr)H(t-s)/(t-s)^{1/2}, \quad (10.71)$$

and here $\hat{U}_I$ is the Hilbert transform of $\check{U}_I$ with the form

$$\hat{U}_I(p,t) = \pi^{-1}\mathrm{Re}[G_I(p)/(t - \tau_I(p) - 0i)]. \quad (10.72)$$

The convolution in (10.71) can be rearranged by shifting the Hilbert transform from the generalized ray term to give

$$\int_{-\infty}^{\infty} ds\, \check{U}_I(p, s - pr)H(s-t)(s-t)^{-1/2} \quad (10.73)$$

with the same slowness integral as before. Since we have forced the phase factors associated with turning points into $G_I(p)$, the choice of (10.71) or (10.73) is dictated by the properties of $G_I(p)$. If G_I were real then (10.71) would be the best choice, whereas if G_I were imaginary we would prefer (10.73). In practice $G_I(p)$ is real for part of the range and imaginary in others and we may exploit the linearity of the problem to produce a combination of (10.71) and (10.73) which is best suited to a real p contour. In terms of the explicit forms of the generalized ray term this is

$$u_{z0}^{I}(r,t) \approx \partial_{tt}M(t) * \frac{1}{\pi(2r)^{1/2}} \int_0^{\infty} dp\, p^{1/2}$$
$$\times \mathrm{Im}\left[\int_{-\infty}^{\infty} ds\, L(t-s) G_I(p) \mathrm{Im}(s - \tau_I(p) - pr - 0i)^{-1}\right], \quad (10.74)$$

where we have introduced the analytic time function

$$L(t) = H(t)t^{-1/2} + iH(-t)(-t)^{-1/2}, \quad (10.75)$$

which combines the inverse square root operator and its Hilbert transform. In (10.74) we see within the slowness integral the term $\theta_I(p,r) = \tau_I(p) + pr$ which played such an important role in the Cagniard method.

We now restrict attention to perfectly elastic media, so that $\tau(p)$ is real for the range of slowness for which we have propagating waves all along the ray path. We will denote the slowness at which some portion of the ray becomes evanescent by p_I and then

$$\int_0^{p_I} dp\, p^{1/2} G_I(p) \mathrm{Im}[t - \theta_I(p,r) - 0i] = \int_0^{\infty} dp\, p^{1/2} G_I(p) \delta(t - \theta_I(p,r)). \quad (10.76)$$

This integral may be evaluated by splitting the slowness range intervals containing just one root of $t = \theta(p,r)$ and then changing integration variable to θ to give

$$\int_0^{\infty} dp\, p^{1/2} G_I(p) \delta(t - \theta_I(p,r)) = \sum_j G_I(\pi_j) \pi_j^{1/2} [\partial_p \theta_I(\pi_j, r)]^{-1}, \quad (10.77)$$

where the sum is taken over the slowness roots $\pi_j(t)$ of $t = \theta_I(\pi_j, r)$. There will be one root for a simple turning ray and multiple roots in the neighbourhood of triplications. The full integral (10.74) can now be written as

$$u_{z0}^{I} \approx \partial_{tt}M(t) * \frac{1}{\pi(2r)^{1/2}} \mathrm{Im}\{L(t) * \sum_j G_I(\pi_j)\pi_j^{1/2}[\partial_p\theta_I(\pi_j, r)]^{-1}$$
$$+ L(t) \int_{p_I}^{\infty} dp\, p^{1/2} G_I(p) \mathrm{Im}[t - \theta_I(p,r) - 0i]^{-1}\}. \quad (10.78)$$

The integral includes those values for which $\tau(p)$ becomes complex and may be regarded as a correction to the main approximation represented by the sum.

Chapman (1978) has termed the displacement contribution

$$u^I_{z0}(r,t) = \partial_{tt}M(t) * \frac{1}{\pi(2r)^{1/2}} \mathrm{Im}\{L(t) * \sum_j G_I(\pi_j(t))\pi_j(t)^{1/2}[\partial_p\theta_I(\pi_j(t),r)]^{-1}\} \quad (10.79)$$

the *WKBJ seismogram*. He has shown that with a locally quadratic approximation to $\theta(p,r)$ we recover the results of geometrical ray theory for a single turning ray. The expression (10.79) is, however, still usable at caustics and at shadow boundaries where geometrical ray theory fails. For interface problems (10.79) gives the same results as the first-motion approximation for the head wave (10.68). For reflected waves a complete representation requires the inclusion of the correction terms to allow for evanescent waves, but near the geometric arrival time (10.79) is a good approximation.

For numerical evaluation it is convenient to use a smoothed version of (10.79), to generate a discrete time series. For a digitisation interval Δt we employ an operator $F(t,\Delta t)$ which is zero outside $(t-\Delta t, t+\Delta t)$ to smooth each time point. This smoothing eliminates the effects of singularities in (10.79) associated with apparent details in the model. The interpolation scheme used to define the wavespeed profile can often give discontinuities in parameter gradients which will lead to singularities in (10.79); there is also a chance that small triplications may be introduced. These features will have no true physical significance and will be unobservable with realistic source terms.

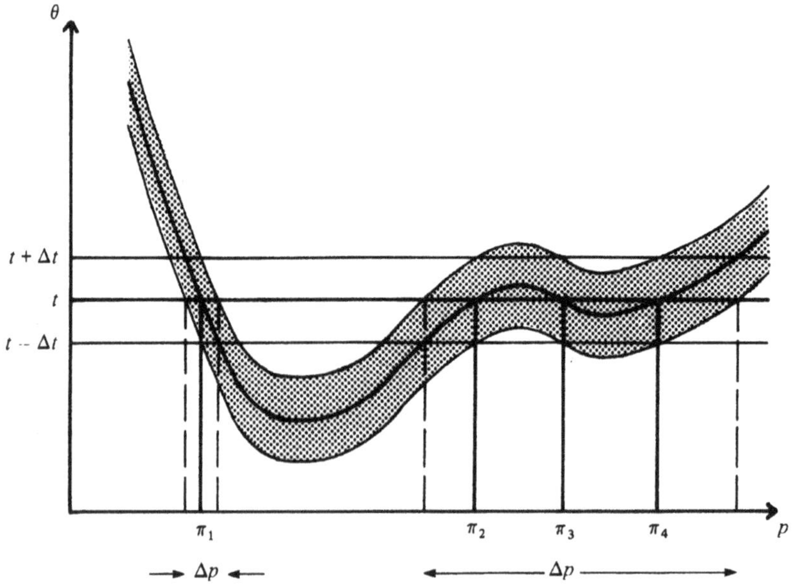

Figure 10.6. The construction of the WKBJ seismogram by smoothing over an interval $t-\Delta t, t+\Delta t)$ about the desired time.

We now convolve the displacement with the unit area smoothing operator $F(t, \Delta t)$, and restrict attention to the range of slownesses for which $\tau_I(p)$ is real as in (10.79),

$$u_{z0}^I(r,t) * F(t,\Delta t) = \partial_{tt} M(t) * \frac{1}{\pi(2r)^{1/2}} \mathrm{Im}\{L(t) * \int_0^\infty dp\, p^{1/2} G_I(p) F(t - \theta_I(p,r), \Delta t)\}. \tag{10.80}$$

The slowness integral will reduce to a sum of contributions from bands such that

$$t - \Delta t < \theta_I(p,r) < t + \Delta t, \tag{10.81}$$

in the neighbourhood of $\pi_j(t)$ (see figure 10.6). Thus

$$u_{z0}^I(r,t) * F(t,\Delta t) = \partial_{tt} M(t) * \frac{1}{\pi(2r)^{1/2}} \mathrm{Im}\{L(t) * \sum_j \int dp\, p^{1/2} G_I(p) F(t - \theta_I, \Delta t)\}, \tag{10.82}$$

where the integral is to be taken over the span of slowness values about π_j for which (10.81) is satisfied, with width Δp_j (figure 10.6).

Chapman (1978) has suggested using a boxcar filter over the interval $(t-\Delta t, t+\Delta t)$ in which case, when $G_I(p)$ is slowly varying, the sum in (10.82) can be approximated by

$$\sum_j \pi_j G_I(\pi_j) \Delta p_j / \Delta t, \tag{10.83}$$

and this is the form which Dey-Sarkar & Chapman (1978) have used for computations. Figure 10.6 illustrates the way the sum is formed, for an isolated arrival near π_1, Δp_1 is small. But for the triplication π_2, π_3, π_4 which is unresolvable at the discretisation level Δt, there is a long effective Δp.

In the approximations which enabled us to generate the frequency-slowness response of a turning ray we have taken $\mathbf{g}_I(p)$ to be determined by products of frequency independent plane-wave reflection and transmission coefficients. All frequency dependent propagation effects associated with changes in parameter gradients and with turning points close to interfaces are ignored. Although the WKBJ seismogram will give a good representation of the major features of the seismic phases via the behaviour of $\theta_I(p, r)$, secondary features can be in error or missing. This can be well illustrated by examining the slowness-time map for a full frequency dependent calculation. In Figure 10.7 we show such a projective display of the slowness-time response for the *SS* reflection from the upper mantle model *T7* (figure 9.7). The continuous refraction, the main feature for large times is approximated well in the WKBJ scheme (10.82)–(10.83). The reflection from the Moho will also be represented quite well. However, the reflections from the discontinuity in wavespeed gradient at 170 km, and the complex transition

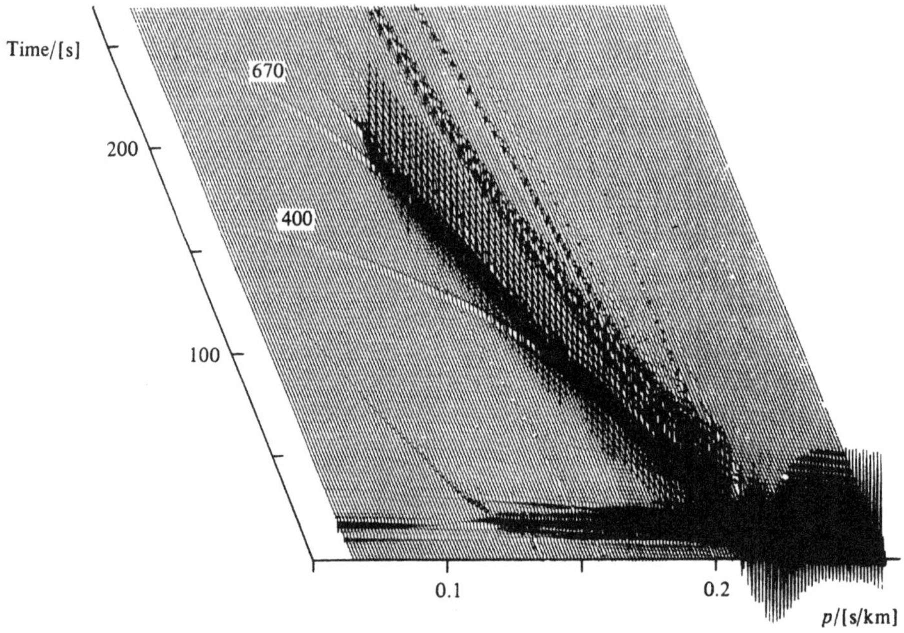

Figure 10.7. Projective display of the *SS* reflection from the upper model *T7* as a function of slowness and time.

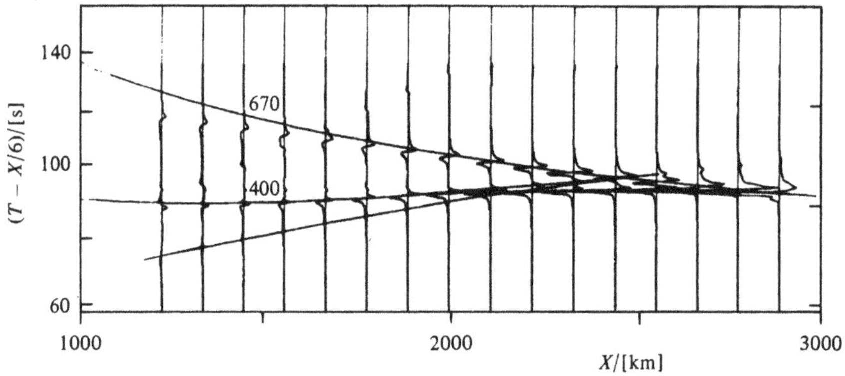

Figure 10.8. Record section of WKBJ theoretical seismograms for *S* waves in the mantle model *T7*.

zones at 400 and 700 km, in the flattened model, will not be well approximated. These features depend strongly on frequency dependent gradient effects and have a significant effect on the seismograms for waves reflected from the upper mantle.

Although the WKBJ seismograms do not have the full accuracy which can be attained with more sophisticated techniques, they are inexpensive to compute. With a knowledge of the travel-time and reflection characteristics of a model it is possible to get a quick idea of the character of the seismic wavefield. In figure 10.8 we show a record section of WKBJ theoretical seismograms (10.82)–(10.83) calculated for a simple oceanic crustal model. These seismograms display the main features of the crustal response represented by the refracted branch of the travel time curve, but do not include reflections arising from changes in wavespeed gradient.

10.6 Attenuation and generalized rays

In both Cagniard's and Chapman's methods for determining the response of a generalized ray we have had to assume that the medium is perfectly elastic.

As discussed in section 1.3 we can model the effects of attenuation on seismic propagation by letting the elastic wavespeeds become complex. In general they will also be frequency dependent because of the frequency dispersion associated with causal attenuation. In an attenuative medium to first order, for *P* wave

$$\tau(p) \to 2\int_0^{Z_\alpha} dz\, q_\alpha(p,z,\omega) + i\,\mathrm{sgn}\,\omega \int_0^{Z_\alpha} dz\,[\alpha^2 Q_\alpha q_\alpha]^{-1} \tag{10.84}$$

and so the separation of frequency and slowness effects we have employed above is no longer possible.

For Cagniard's method attenuation can be introduced into the final ray sum by applying an attenuation operator, such as (9.74), to each ray contribution allowing for the nature of the path. Burdick & Helmberger (1978) have compared this approach with applying a single attenuation operator to the full ray sum, and suggest that often the simpler approximation is adequate.

In Chapman's method, a good far-field approximation may be obtained using (10.71) but now evaluating the inverse transform $U_I(p,t)$ numerically. For a delta function source the result will be a broadened pulse following a trajectory similar to the τ curve in slowness. If dispersion can be neglected

$$\tau(p) = \tau_0(p) + i\,\mathrm{sgn}\,\omega\lambda(p), \tag{10.85}$$

where τ_0 is the perfectly elastic value, and the attenuative term $\lambda(p)$ is small. For the Ith generalized ray

$$\boldsymbol{w}_I(p,t) = \mathrm{Im}[G_I(p)/(t-\tau_I(p))], \tag{10.86}$$

but this is non-causal. Along the real p axis $\boldsymbol{w}_I(p,t)$ will still have its maximum value near $\tau_{0I}(p)$ and a comparable approximation to the smoothed WKBJ seismogram can be made including a convolution with the broadened pulse form.

Chapter 11

Modal Summation

The various expressions which we have derived for the receiver response for general point source excitation have singularities associated with the properties of the reflection and transmission matrices for portions of the stratification and the corresponding reverberation operators. In particular we have a set of poles associated with the vanishing of the secular function for the half space $\det\{\mathsf{T}_{DL}(0)\}$ (7.12). This secular function is independent of the depth of the source and depends on the elastic properties in the half space.

For the combinations of frequency and slowness for which $\det\{\mathsf{T}_{DL}(0)\}$ vanishes we have non-trivial solutions of the equations of motion satisfying both the boundary conditions: the vanishing of traction at the surface, and decaying displacement at depth ($|\boldsymbol{w}| \to 0$ as $z \to \infty$). For our choice of structure this latter property arises from the presence of only exponentially decaying waves in the lower half space $z > z_L$ (7.3). The detailed character of these eigenfunctions for displacement will depend on the actual wavespeed distribution within the half space. Above the *S* wave turning level z_s for the slowness p (i.e., the depth at which $\beta^{-1}(z_s) = p$), the character of the eigensolutions will be oscillatory. Below this level we will have evanescent decaying behaviour. For slowness p less than the inverse of the *S* wavespeed in the lower half space (β_L^{-1}) we will have travelling waves at depth; in this case we can no longer match both sets of boundary conditions. The surface wave poles are thus restricted to $p > \beta_L^{-1}$ and the maximum slowness depends on the wave type.

Since we have a semi-infinite domain, the roots of the dispersion equation $\det\{\mathsf{T}_{DL}(0)\} = 0$ are ordered into continuous strings in frequency-slowness (ω–p) space, ordered by overtone number (see figures 11.1, 11.3, 11.4). For the *SH* case, the overtone number for Love waves represents the number of zero crossings in the eigenfunctions. In the coupled *P-SV* system the overtone number for higher Rayleigh modes has a more abstract significance, but in many cases is one less than the number of zero crossings for the horizontal displacement.

For perfectly elastic media the poles reside on the real slowness axis at fixed frequency and the work of Sezawa (1935) and Lapwood (1948) shows that at large distances the contribution from the residues at the poles response includes the surface wavetrain. The stress-displacement fields associated with these residues show no discontinuity across the source level (Harkrider 1964), so that the excitation of the modal contributions should not be thought of as occurring directly, but by the interaction of the entire wavefield with the stratification and the surface.

11.1 The location of the poles

The location of the surface wave poles in frequency slowness space is given by the vanishing of the secular function. From the surface source expressions for the seismic field (7.12) and the representation of the stress component $\mathsf{T}_{DL}(0)$ in terms of the reflection and transmission properties of the stratification we require

$$\det\{\mathsf{T}_{DL}(0,\omega,p)\} = \det(\mathsf{n}_{D0} + \mathsf{n}_{U0}\mathbf{R}_D^{0L})/\det\mathbf{T}_D^{0L} = 0. \tag{11.1}$$

If we have a uniform half space we would require $\det\mathsf{n}_{D0}(p) = 0$. No root is possible for *SH* waves but for *P-SV* waves we require

$$(2p^2 - \beta_0^{-2})^2 + 4p^2 q_{\alpha 0} q_{\beta 0} = 0; \tag{11.2}$$

and this is the usual equation for the Rayleigh waveslowness p_{R0} on a uniform half space with the properties α_0, β_0 (Rayleigh, 1887). For any increase in velocity within the half space we have the possibility of dispersive wave-propagation with the slowness depending on frequency for both *SH* and *P-SV* waves.

We may rewrite (11.1) as

$$\Delta = \det\mathsf{n}_{D0}\det(\mathbf{I} - \mathbf{R}_F\mathbf{R}_D^{0L}) = 0, \tag{11.3}$$

although the impression that there is always a root when $\det\mathsf{n}_{D0} = 0$ is illusory since it is matched by a singularity in $\mathbf{R}_F$ at the same slowness. However, for slowness $p > \beta_0^{-1}$, such that all wave types are evanescent at the surface, with increasing frequency $\mathbf{R}_D^{0L} \to 0$ as the penetration into the stratification decreases, and so

$$\Delta \to \det\mathsf{n}_{D0}, \quad \text{as} \quad \omega \to \infty, \quad p > \beta_0^{-1}. \tag{11.4}$$

Thus for *P-SV* wave-propagation the limiting slowness for dispersive surface waves is the Rayleigh slowness p_{R0} for a half space with the surface properties throughout. Only one root of Δ exists in $p > \beta_0^{-1}$.

For slownesses $p < \beta_0^{-1}$, so that *S* waves at least are propagating, we may use the simpler dispersion relation (Kennett, 1974)

$$\det(\mathbf{I} - \mathbf{R}_F\mathbf{R}_D^{0L}) = 0. \tag{11.5}$$

This relation constitutes a constructive interference condition between waves reflected back from the stratification and those reflected from the surface.

When we start from the expressions (7.32) and (7.35) for the displacement from a buried source the secular function may be expressed as

$$\det(\mathbf{I} - \mathbf{R}_U^{fS}\mathbf{R}_D^{SL}) = 0, \tag{11.6}$$

where the source depth z_S is arbitrary. Equation (11.6) represents a further constructive interference condition between waves reflected above and below z_S. This form is inadequate and we must use (11.5) if the receiver contribution becomes singular i.e. $\det(\mathbf{I} - \mathbf{R}_U^{fR}\mathbf{R}_D^{RS}) = 0$ which corresponds to the existence of trapped waves on just the truncated structure above the level z_S.

11.2 SH wave dispersion: Love waves

For *SH* waves the free-surface reflection coefficient is unity so that (11.5) becomes

$$\mathbf{R}_D^{0L}|_{HH} = 1, \tag{11.7}$$

i.e. the secular relation requires us to seek those combinations of frequency and slowness for which an incident downgoing wave is reflected from the stratification without change of amplitude or phase. This result holds even for attenuative media and so at fixed frequency ω the poles will lie in the first and third quadrants of the slowness plane.

11.2.1 A layer over a half space

It is interesting to see how more familiar dispersion relations can be obtained from (11.7). Consider the simple example of a layer with density ρ_0, shear wavespeed β_0 and thickness h_0, overlying a uniform half space with density ρ_1 and shear wavespeed β_1. The reflection coefficient for the interface at h_0 is given by (5.12) and allowing for phase delay in the upper layer between the surface and the interface we require

$$\exp(2i\omega q_{\beta 0}h_0)\{\mu_0 q_{\beta 0} - \mu_1 q_{\beta 1}\}/\{\mu_0 q_{\beta 0} + \mu_1 q_{\beta 1}\} = 0, \tag{11.8}$$

and with a slight rearrangement we have

$$\tan \omega q_{\beta 0}h_0 = -i\mu_1 q_{\beta 1}/\mu_0 q_{\beta 0}, \tag{11.9}$$

which is the conventional Love wave dispersion relation (Ewing, Jardetsky & Press, 1957).

We can see directly from (11.8) that roots will only be possible when the interface coefficient has unit modulus and waves propagate in the upper layer i.e. the Love wave slowness is restricted to

$$\beta_1^{-1} < p < \beta_0^{-1}. \tag{11.10}$$

Also, since the reflection coefficient is independent of frequency, for fixed slowness the roots of the dispersion equation in frequency are

$$\omega_n = [n\pi - \tfrac{1}{2}\chi_{01}(p)]/q_{\beta 0}h_0, \tag{11.11}$$

where $\chi_{01}(p)$ is the phase of the interface reflection coefficient. It is therefore easy to generate multimode dispersion curves and the first twelve branches are illustrated

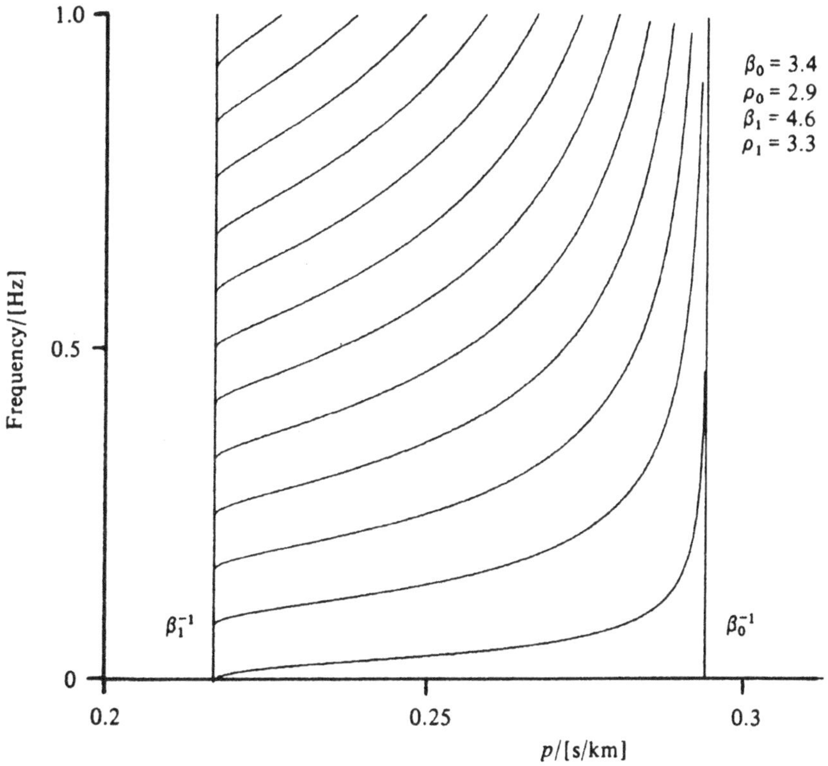

Figure 11.1. Love wave dispersion curves for the first twelve mode branches as a function of frequency and slowness.

in figure 11.1. $\chi_{01}(p)$ decreases steadily with increasing p (figure 5.2) approaching $-\pi$ when $p \to \beta_0^{-1}$ and is zero when $p = \beta_1^{-1}$. The successive modes have therefore a lower cutoff in frequency at

$$\omega_{n0} = n\pi/(\beta_1^{-2} - \beta_0^{-2})^{1/2}h_0 \tag{11.12}$$

and at high frequencies have slownesses which asymptote to β_0^{-1} (figure 11.1). For frequencies less than the cutoff for any mode, roots cannot be found on our chosen Riemann sheet $\text{Im}\, \omega q_{\beta 1} \geq 0$, but move off to complex p values on the lower sheet (Gilbert, 1964).

11.2.2 Love waves in a stratified medium

We may extend the approach we have used for the single layer to deal with a stratified half space. The reflection coefficient at $z = 0$ will have unit modulus for slownesses in the range

$$\beta_L^{-1} < p < \beta_0^{-1}, \tag{11.13}$$

for which we have evanescent waves in the underlying half space and travelling waves at the surface. In this interval we can recast the dispersion equation (11.7) as

$$\chi_0(\omega, p) = 2n\pi, \tag{11.14}$$

where χ_0 is the phase of the unimodular reflection coefficient $\mathbf{R}_D^{0L}|_{HH}$ at the level $z = 0$. We may therefore build up the reflection coefficient from the base of the stratification or varying frequencies at fixed slowness; and then interpolate to find the frequencies at which (11.14) is satisfied.

At high frequencies we can get approximate results by making use of asymptotic representations of the reflection behaviour. As discussed in Section 10.1, for a smooth monotonically increasing wavespeed distribution and slowness p such that the turning point $Z_\beta(p)$, for which $p = \beta^{-1}(Z_\beta)$, is not close to 0 or z_L, we have

$$\mathbf{R}_D^{0L}|_{HH} \sim \exp\{i\omega\tau_\beta(p) - i\pi/2\}, \tag{11.15}$$

where the intercept time

$$\tau_\beta(p) = 2\int_0^{Z_\beta(p)} dz\, q_\beta(z), \tag{11.16}$$

is the integrated vertical slowness down to the turning level. The Love wave dispersion relation corresponding to (11.15) is thus

$$\omega_n\tau_\beta(p) \sim (2n + \tfrac{1}{2})\pi. \tag{11.17}$$

This result is in the same form as (11.11) when we take account of the phase change of $\pi/2$ on internal reflection and turning level at the interface h_0 in the single layer case.

Although in general we need to take account of the full behaviour in each region at high frequencies an approximate iterative procedure for piecewise continuous media may be developed based only on the phase behaviour. Gradient effects at interfaces are ignored. Starting at the turning point the phase delay to the next higher interface (z_J say) is formed

$$\chi_T(\omega, p) = 2\omega\int_{z_J}^{Z_\beta(p)} dz\, q_\beta(z) - \tfrac{1}{2}\pi, \tag{11.18}$$

and then the effect of the interface is included to form an approximate reflection coefficient phase χ_J just above z_J; from

$$\tan(\tfrac{1}{2}\chi_J) = |Q_J|\tan(\tfrac{1}{2}\chi_T) \tag{11.19}$$

cf. (11.9). Q_J is the ratio of *SH* wave impedances at z_J, e.g., $Q_1 = \mu_1 q_{\beta 1}/\mu_0 q_{\beta 0}$. The phase delay between z_{J-1} and z_J must be included before constructing χ_{J-1}, as

$$\tan(\tfrac{1}{2}\chi_{J-1}) = |Q_{J-1}|\tan(\omega\int_{z_{J-1}}^{z_J} dz\, q_\beta(z) + \tfrac{1}{2}\chi_J). \tag{11.20}$$

This process is then carried out right up to the surface to form $\chi_0(\omega, p)$ then the dispersion equation (11.14) may be rewritten as

$$\omega_n \tau_\beta(p) \sim (2n + \tfrac{1}{2})\pi + \chi_0'(\omega_n, p), \tag{11.21}$$

where χ_0' arises from the cumulative effect of interfaces. Starting with a trial estimate of ω_n, the right hand side of (11.21) is calculated and a revised estimate for ω_n obtained by equating $\omega_n \tau_\beta$ to this quantity. Iteration is continued until convergence of the frequency estimates occurs and is usually quite fast.

This approach which has been used by Kennett & Nolet (1979) in studies of the high frequency normal modes of the Earth is a generalisation of a procedure suggested by Tolstoy (1955) for uniform layers. For a multilayered medium $\frac{1}{2}\pi$ would be replaced by the phase of the reflection coefficient at the deepest interface at which an *S* wave is propagating. Numerical comparisons between the iterative results and direct calculations for the dispersion show that for the lower mantle good agreement is obtained for frequencies higher than 0.03 Hz. For stronger gradient zones higher frequencies are needed before (11.21) gives an adequate approximation.

For this dispersive wave system, pulse distortion will occur as an individual mode propagates across the stratified half space (see Section 11.5), and a wave packet will travel with the group slowness

$$g = \frac{\partial}{\partial \omega}(\omega p) = p + \omega \frac{\partial p}{\partial \omega}. \tag{11.22}$$

For a secular function $Y(p, \omega) = 0$ we may calculate g by implicit differentiation

$$g = \{p \partial_p Y - \omega \partial_\omega Y\}/\partial_p Y. \tag{11.23}$$

It is interesting to see the form taken by the group slowness for the high frequency approximation (11.17) for a smooth monotonic increase in wavespeed. We recall the ray theory results that the range $X_\beta(p)$ for slowness p is $-\partial\tau_\beta/\partial p$ and the travel time $T_\beta(p) = \tau_\beta(p) + pX_\beta(p)$. For (11.17),

$$\omega \partial_\omega Y \sim i\omega\tau_\beta e^{i\omega\tau_\beta(p) - i\pi/2}, \tag{11.24}$$

$$\partial_p Y \sim i\omega(\partial\tau_\beta/\partial p) e^{i\omega\tau_\beta(p) - i\pi/2}, \tag{11.25}$$

and so the group slowness g is asymptotically

$$g \sim \{i\omega\tau_\beta(p) + i\omega p X_\beta(p)\}/i\omega X_\beta(p) = T_\beta(p)/X_\beta(p), \tag{11.26}$$

which is just the apparent slowness for an *S* body wave. We will explore this relation between modes and body waves further in Section 11.7.

11.3 P-SV wave dispersion: Rayleigh waves

For the *P-SV* wave system the analysis of dispersion is more complicated since we have to take account of the coupling between the wave types. From (11.1) the roots of the secular function are given by

$$\det(\mathbf{n}_{D0} + \mathbf{n}_{U0}\mathbf{R}_D^{0L}) = 0 \tag{11.27}$$

In terms of the components R_D^{PP}, R_D^{PS} etc. of $\mathbf{R}_D^{0L}(\omega, p)$ we have

$$\begin{aligned}&[\upsilon\epsilon_{\alpha 0}(1 + R_D^{PP}) - 2ipq_{\beta 0}\epsilon_{\beta 0}R_D^{SP}][\upsilon\epsilon_{\beta 0}(1 + R_D^{SS}) - 2ipq_{\alpha 0}\epsilon_{\alpha 0}R_D^{PS}]\\ &-[2ipq_{\alpha 0}\epsilon_{\alpha 0}(1 - R_D^{PP}) + \upsilon\epsilon_{\beta 0}R_D^{SP}][2ipq_{\beta 0}\epsilon_{\beta 0}(1 - R_D^{SS}) + \upsilon\epsilon_{\alpha 0}R_D^{PS}] = 0.\end{aligned} \tag{11.28}$$

where, as in (5.82), we have set

$$\upsilon = (2p^2 - \beta_0^{-2}) \tag{11.29}$$

When we have a uniform half space beneath the surface, $\mathbf{R}_D^{0L}$ vanishes and the dispersion relation (11.28) reduces to the Rayleigh function (11.2). For a stratified medium $\mathbf{R}_D^{0L}$ will no longer be zero and (11.28) shows the way in which the departures from uniformity affect the dispersion behaviour.

In general we will have to find the roots of (11.28) numerically by searching for the combinations of frequency ω and slowness p for which the left hand side of (11.28) is less than some preassigned threshold.

The nature of the dispersion behaviour varies with slowness as the character of the seismic wavefield changes. We will suppose that the minimum *P* and *S* wavespeeds occur at the surface and then the different regimes are controlled by the slownesses α_0^{-1}, β_0^{-1}. For a perfectly elastic medium we can illuminate the physical nature of the dispersion relations by making use of the unitarity properties derived in the appendix to Chapter 5.

11.3.1 P and S waves evanescent

When the slowness p is greater than β_0^{-1} both *P* and *S* waves are evanescent throughout the stratification. For perfectly elastic media all the reflection coefficients are real and less than unity. At moderate frequencies the *P* and *S* wave components will decay with depth but the rate of decay is much more for *P* waves; with the result that only R_D^{SS} has significant size and R_D^{PP}, R_D^{PS} are negligible in comparison.

To this approximation (11.28) reduces to

$$(2p^2 - \beta_0^{-2})^2(1 + R_D^{SS}) - 4p^2|q_{\alpha 0}||q_{\beta 0}|(1 - R_D^{SS}) = 0, \tag{11.30}$$

where R_D^{SS} is small. There is only one root for this dispersion relation and as the frequency increases R_D^{SS} tends to zero and so (11.30) tends to the Rayleigh function (11.2) for a uniform half space with the surface properties α_0, β_0. At high frequencies the slowness will be close to p_{R0} and so we can recognise (11.25) as the dispersion relation for *fundamental* mode Rayleigh waves.

At low frequencies R_D^{PP}, R_D^{PS} will no longer be negligible and R_D^{SS} will be close to unity and so the fundamental Rayleigh wave slowness will depart significantly from p_{R0} (figure 11.3). The modal eigenfunction at low frequency penetrates deeply into the half space and so the dispersive waves are influenced by the greater wavespeeds at depth and therefore the Rayleigh waveslowness decreases with frequency.

11.3.2 S propagating, P evanescent

For slownesses such that $\alpha_0^{-1} < p < \beta_0^{-1}$, *P* waves are still evanescent throughout the half space but *S* waves have travelling wave character at the surface and a turning point above z_L. Once again at high frequencies the reflection coefficients R_D^{PP} and R_D^{PS} will be very small compared with R_D^{SS}.

Since we are now some way from the surface Rayleigh slowness p_{R0}, we use the secular function (11.5) in terms of the reflection matrices which may be written as

$$1 - R_F^{SS}(R_D^{PP} + R_D^{SS}) - 2R_F^{PS}R_D^{PS} + (R_D^{PP}R_D^{SS} - (R_D^{PS})^2) = 0. \tag{11.31}$$

We have used the symmetry of the reflection matrices (5.75), (5.82) and $R_F^{PP} = R_F^{SS}$, $\det \mathbf{R}_F = 1$ in constructing this form. At high frequencies we can neglect the R_D^{PS}, R_D^{PP} terms and (11.31) reduces to

$$R_D^{SS} = (R_F^{SS})^{-1}, \tag{11.32}$$

which is very similar to the Love wave equation (11.7).

However, at low frequencies the *P* wave still influences the character of the dispersion and we may investigate this behaviour for a perfectly elastic half space. At $z = 0$ we have propagating *S* waves and evanescent *P*, with both wave types evanescent at z_L; from appendix 5(d) the unitarity relations require

$$\begin{aligned} &|R_D^{SS}| = 1, \quad \det \mathbf{R}_D^{0L} = R_D^{SS}(R_D^{PP})^*, \\ &|R_D^{PS}|^2 = 2\,\mathrm{Im}\,(R_D^{PP}), \quad \arg R_D^{PS} = \tfrac{1}{4}\pi + \tfrac{1}{2}\arg R_D^{SS}. \end{aligned} \tag{11.33}$$

For the surface reflections

$$|R_F^{SS}| = 1, \quad \arg R_D^{PS} = \tfrac{1}{4}\pi + \tfrac{1}{2}\arg R_F^{SS} \tag{11.34}$$

With these expressions for the reflection coefficients, we may rewrite (11.31) as

$$R_D^{SS} = (1 - 2R_F^{PS}R_D^{PS} + R_F^{SS}R_D^{PP})/[R_F^{SS} + (R_D^{PP})^*]. \tag{11.35}$$

As the frequency increases R_D^{PP} and R_D^{PS} diminish rapidly. This can be seen from figure 11.2 which displays the modulus of the coefficient $R_D^{PP}(p, \omega)$ for the mantle model *T7*. The onset of *P* wave evanescence at a slowness 0.76 s/km is marked by a very rapid decrease in the size of the PP coefficient. Only at the lowest frequencies displayed (~0.02 Hz) is there any significant reflection in the evanescent region and from (11.33) R_D^{PS} will also be small.

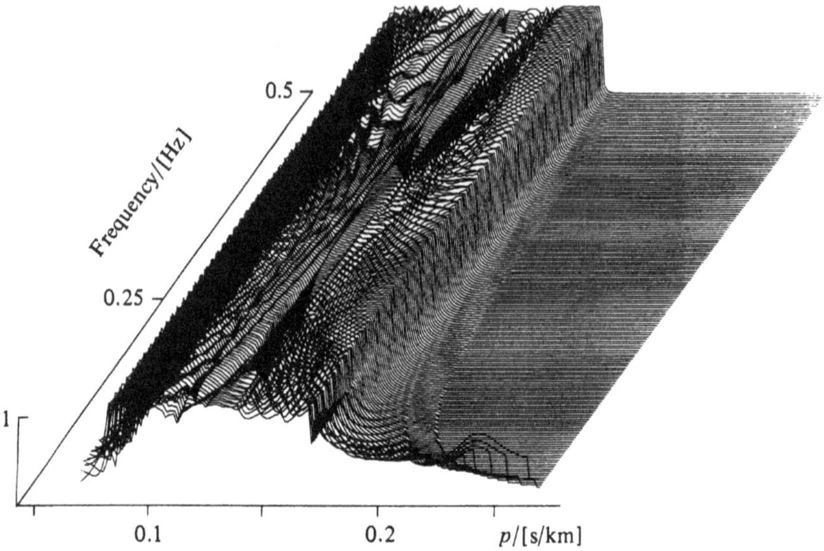

Figure 11.2. The amplitude of the PP reflection coefficient for the upper mantle model *T7* as a function of frequency and slowness.

Since both R_D^{SS} and R_F^{SS} are unimodular, we set

$$R_D^{SS} = \exp i\psi, \quad R_F^{SS} = \exp i\psi_0, \tag{11.36}$$

and also extract the phase dependence of R_D^{PP} as

$$R_D^{PP} = |R_D^{PP}| \exp i\phi. \tag{11.37}$$

The phase of the R_D^{PS}, R_F^{PS} coefficients are determined by the unitarity properties and the dispersion relation (11.31), for a perfectly elastic medium, is equivalent to

$$\sin \tfrac{1}{2}(\psi + \psi_0) + |R_D^{PP}| \sin \tfrac{1}{2}(\phi + \psi_0 - \psi) = |R_D^{PS}||R_F^{PS}|, \tag{11.38}$$

where from (11.33)

$$|R_D^{PS}|^2 = 2|R_D^{PP}| \sin \phi. \tag{11.39}$$

As the frequency increases, $|R_D^{PP}|$ tends to zero and so we are left with a relation between the phase of R_D^{SS} and that of the free surface,

$$\psi(\omega, p) \sim 2n\pi + \psi_0(p). \tag{11.40}$$

Comparison with (11.14) shows that there is just a slowness dependent phase shift of $\psi_0(p)$ from the Love wave result. The dispersion curves for Love and Rayleigh waves are very similar at high frequencies (cf. figures 11.3, 11.4) and the character of the higher Rayleigh modes, in this slowness range dominated by *S* wave behaviour, is akin to *SV* Love modes. The higher modes have a high frequency asymptote of β_0^{-1} for phase slowness and only the fundamental Rayleigh mode continues on into the evanescent *S* wave region.

For a smoothly increasing wavespeed profile the asymptotic form of the *SV* reflection has the same form as for *SH* waves

$$R_D^{SS} \sim \exp\{i(\omega\tau_\beta(p) - \pi/2)\}, \tag{11.41}$$

so that to this approximation the frequency of the nth higher Rayleigh mode is to be found from

$$\omega_n\tau_\beta(p) \sim (2n + \tfrac{1}{2})\pi + \tfrac{1}{2}\arg R_F^{SS}(p). \tag{11.42}$$

For a piecewise smooth medium an iterative technique similar to that discussed for Love waves may be employed and a further frequency dependent phase contribution $\Psi(\omega, p)$ will be added to the right hand side of (11.42). We can see how this term arises by examining the temporal-slowness display $R_D^{SS}(p, t)$ for the *SS* reflection from the mantle model *T7* in figure 10.7. The approximation (11.41) follows the main refracted arrival in the upper mantle and ignores any reflection from the Moho or the upper mantle discontinuities which are prominent features of the behaviour. When we consider the phase response the most significant effect will arise from the relatively large Moho reflection as in the *SH* wave study of Kennett & Nolet (1979).

11.3.3 Propagating P and S waves

For slowness in the range $\beta_L^{-1} < p < \alpha_0^{-1}$, both *P* and *S* waves have travelling wave character at the surface and turning points within the stratification above the level z_L. This means that *P* waves play a significant role in determining the dispersion and *P* and *S* effects are coupled through the relation

$$1 - R_F^{SS}(R_D^{PP} + R_D^{SS}) - 2R_F^{PS}R_D^{PS} + (R_D^{PP}R_D^{SS} - (R_D^{PS})^2) = 0. \tag{11.43}$$

When we have a perfectly elastic medium we may once again make use of the unitarity relations in Appendix 5(c). The reflection matrix is now unitary and

$$\begin{aligned} &|R_D^{PP}|^2 + |R_D^{SP}|^2 = |R_D^{SS}|^2 + |R_D^{PS}|^2 = 1, \\ &\arg\det \mathbf{R}_D^{0L} = \phi + \psi, \quad \arg R_D^{PS} = \tfrac{1}{2}\pi + \tfrac{1}{2}(\phi + \psi), \end{aligned} \tag{11.44}$$

where

$$\phi = \arg R_D^{PP}, \quad \psi = \arg R_D^{SS}. \tag{11.45}$$

The symmetry of $\mathbf{R}_D^{0L}$ (5.60) requires $R_D^{PS} = R_D^{SP}$ and so

$$|R_D^{PP}(p, \omega)| = |R_D^{SS}(p, \omega)|. \tag{11.46}$$

This is a very interesting result since the turning points of *P* and *S* waves lie at very different levels. For example a slowness of 0.16 s/km corresponds to a crustal turning point for *P* waves and a turning level at about 600 km in the upper mantle for S waves. Although the amplitudes are the same for the PP and SS elements, the phase behaviour will be very different, and it will be very much greater than ϕ.

In this slowness interval the free-surface reflection coefficient R_F^{SS} is real and negative and R_F^{PS} has phase $3\pi/2$. In terms of the phases ϕ, ψ of the reflection coefficients we can recast the dispersion equation into the form

$$\cos\tfrac{1}{2}(\phi+\psi) + |R_F^{SS}||R_D^{SS}|\cos\tfrac{1}{2}(\phi-\psi) = |R_F^{PS}||R_D^{PS}|, \tag{11.47}$$

which displays explicitly the coupling of *P* and *S* waves at the surface and in the half space. Equation (11.47) reduces in the case of a single uniform layer over a half space to the equation derived by Tolstoy & Usdin (1953) using a constructive interference argument.

For slowness p a little less than α_0^{-1}, R_F^{SS} can be quite small and then (11.47) takes on the character of a dispersion equation for PS propagation. One *P* leg and one *S* leg coupled by a free-surface reflection is represented by $R_D^{PP}\, R_F^{PS}\, R_D^{SS}\, R_F^{PS}$ with phase $-(\phi+\psi)$. R_D^{PS} will in general be small and so the dominant term in (11.47) will be $\cos\frac{1}{2}(\phi+\psi)$ and the other terms may be regarded as a frequency dependent perturbation. For these slowness values, ϕ will be very much smaller than ψ and so the PS behaviour is mostly controlled by R_D^{SS} and we get a smooth continuation of the dispersion curves from the previous region ($p > \alpha_0^{-1}$).

However, the presence of the propagating *P* waves superimposes a modulation of the spacing of the dispersion waves. Since ϕ is small compared to ψ we have in (11.47) a sinusoidal term with superimposed a sinusoid of slightly shorter period and smaller amplitude. The average spacing of the dispersion curves in frequency will depend on $\phi+\psi$ but periodically the curves for different overtone branches will pinch together as we have two closely spaced roots of (11.47). This periodicity is controlled by $\phi(\arg R_D^{PP})$ and so we appear to get 'ghost' dispersion branches controlled by the near-surface *P* wave distribution, visible only because of a change in the spacing of the *S* wave dominated true dispersion curves. Such effects are seen at small slownesses in figure 11.3.

For a smoothly varying medium we have asymptotic representations of the reflection coefficients at high frequency

$$R_D^{PP} \sim e^{i\omega\tau_\alpha(p)-i\pi/2}, \quad R_D^{SS} \sim e^{i\omega\tau_\beta(p)-i\pi/2}, \tag{11.48}$$

where

$$\tau_\alpha(p) = \int_0^{Z_\alpha(p)} dz\, q_\alpha(z), \quad \tau_\beta(p) = \int_0^{Z_\beta(p)} dz\, q_\beta(z), \tag{11.49}$$

and Z_α, Z_β are the turning points for *P* and *S* waves. To this approximation $R_D^{PS} \sim 0$ and coupling between *P* and *S* only occurs at the free surface. On using (11.48) the dispersion relation (11.47) becomes

$$\sin\tfrac{1}{2}\omega[\tau_\alpha(p)+\tau_\beta(p)] - R_F^{SS}\cos\tfrac{1}{2}\omega[\tau_\alpha(p)-\tau_\beta(p)] = 0, \tag{11.50}$$

which was derived by a rather different approach by Kennett & Woodhouse (1978). When there is no *P* and *S* coupling in the half space, Kennett & Woodhouse show that the dispersion relation may be separated into *P* and *S* parts coupled through the surface eccentricity of the mode. In terms of

$$\begin{aligned} T_\alpha(p,\omega) &= i(1 - R_D^{PP})/(1 + R_D^{PP}), \\ T_\beta(p,\omega) &= i(1 - R_D^{SS})/(1 + R_D^{SS}), \end{aligned} \tag{11.51}$$

the dispersion relation may be expressed as

$$T_\alpha(p,\omega)T_\beta(p,\omega) = (2p^2 - \beta_0^{-2})^2/4p^2 q_{\alpha 0} q_{\beta 0}, \tag{11.52}$$

when $R_D^{PS} = 0$. If we introduce the displacement elements for a mode, vertical U and horizontal V, then (11.52) separates to

$$\begin{aligned} T_\alpha(p,\omega) &= -(2p^2 - \beta_0^{-2})U(p,\omega)/2pq_\alpha V(p,\omega), \\ T_\beta(p,\omega) &= -(2p^2 - \beta_0^{-2})V(p,\omega)/2pq_\beta U(p,\omega). \end{aligned} \tag{11.53}$$

Asymptotically T_α, T_β reduce to a rather simple form

$$\begin{aligned} T_\alpha &\sim \tan(\omega\tau_\alpha(p) - \pi/4), \\ T_\beta &\sim \tan(\omega\tau_\beta(p) - \pi/4), \end{aligned} \tag{11.54}$$

and then (11.53) provides a means of extracting both *P* and *S* wave information if the ratio U/V can be estimated.

For models with discontinuities in properties, R_D^{PS} will depart from zero, but at moderate frequencies (0.02 Hz) the wavelengths are sufficiently long that the details of crustal structure are unimportant and then (11.53) may be used as a reasonable approximation (Kennett & Woodhouse, 1978). At higher frequencies the simplicity of the preceding results will be lost, and in general the dispersion will have to be determined numerically.

11.4 Dispersion curves

In the previous sections we have attempted to illuminate the physical character of the modal dispersion. We now turn our attention to the behaviour of the different branches of the dispersion curves which is one of the major factors influencing seismograms created by modal summation.

11.4.1 Dispersion behaviour of an upper mantle model

The behaviour of the Rayleigh and Love mode dispersion may be well illustrated by the dispersion curves for the flattened upper mantle model *T7* shown in figures 11.3 and 11.4 for frequencies from 0.01 to 1.0 Hz. The model itself is presented on the same slowness scale in figure 11.5.

The boundaries between the various dispersion domains have been superimposed on figure 11.3 and the slowness values corresponding to significant discontinuities are indicated on both figures 11.3 and 11.4. For the largest slownesses we have, as expected, only the fundamental Rayleigh mode; and for frequencies greater than 0.5 Hz the slowness is barely distinguishable from p_{R0}. The limiting slowness for all higher modes is β_0^{-1} and the first Rayleigh mode and fundamental Love mode are seen to be approaching this asymptote. For slownesses greater than 0.222 s/km *S* waves are reflected in the crust or at the Moho and the $\tau_\beta(p)$ values are small. The frequency spacing of these crustal modes is therefore large (cf. 11.36, 11.12). For smaller slownesses τ_β increases rapidly and the frequency spacing is much tighter.

The presence of a velocity inversion at about 70 km depth in model *T7* leads to complications in the dispersion curves. For the band of slownesses marked '*LVZ*' in figures 11.3, 11.4 *S* waves have travelling character both in the crust and in the velocity inversion. This gives rise to two classes of dispersion behaviour: (i) significant slowness dispersion with frequency associated with crustal propagation and (ii) almost constant slowness with frequency for waves channelled in the velocity inversion. Any individual Rayleigh or Love mode alternately partakes of these two characters. At osculation points the dispersion curves for two modes almost touch and the ensemble of modes build up a pattern of intersecting crustal and channel dispersion branches. At low frequencies the influence of the *P* waves on the Rayleigh mode dispersion gives a less close approach between the first few mode branches.

At lower slownesses where *S* waves penetrate more deeply, and have turning points near the upper mantle discontinuities, crustal reverberation still leads to a set of distinct kinks in the dispersion curves which line up to give 'ghosts' of the crustal modes. These regular perturbations in the pattern of eigenfrequencies are often referred to as *solotone* effects (see, e.g., Kennett & Nolet, 1979). At high frequencies the presence of the 400 km discontinuity further disturbs the regularity of the spacing in the dispersion curves.

Once *P* waves become propagating, the dispersion for Rayleigh modes becomes even more complex and at high frequencies 'ghosts' of *P* wave reverberation effects asymptoting to α_0^{-1} appear, superimposed on the rest of the pattern.

The calculations in figures 11.3 and 11.4 were made using a flattened earth model. The most significant departures from the results for a spherically stratified model will be for the fundamental mode at the lowest frequencies, where the density distribution has a definite influence on the dispersion. As we have seen in Sections 11.2 and 11.3 the high frequency behaviour is controlled by the delay times τ in the model and these are preserved under the earth-flattening transformation (1.27).

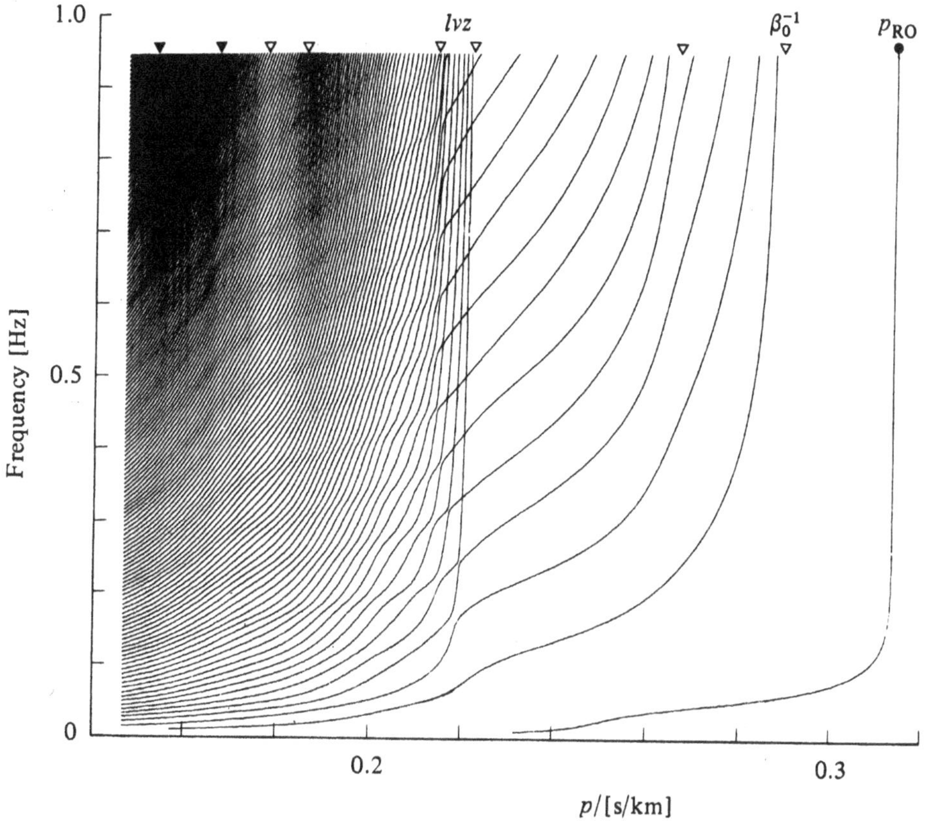

Figure 11.3. Rayleigh wave dispersion curves for the upper mantle model *T7*. Locations of discontinuities in the model are indicated.

11.4.2 Surface waves and channel waves

In both figures 11.3 and 11.4 we have seen how the presence of a velocity inversion leads to complex dispersion curves with interaction of waves propagating mainly in the crust or mainly in the low velocity channel. This behaviour is best understood by reference to the modal structure for a crust or velocity inversion alone.

If we split the stratification at a level z_S we see from (7.41), (7.46) that the singularities are determined by

$$\Delta = \det[\mathbf{I} - \mathbf{R}_D^{0S}\mathbf{R}_F]\det[\mathbf{I} - \mathbf{R}_D^{SL}\mathbf{R}_U^{fS}] = 0, \tag{11.55}$$

and as noted in Section 11.1 the second term is normally most important. When, however, the wavespeed profile for P and S waves is monotonically increasing with depth we can choose z_S to lie deep in the evanescent region. In this case $|\mathbf{R}_D^{SL}|$ will be very small and so (11.55) is approximately

$$\det[\mathbf{I} - \mathbf{R}_D^{0S}\mathbf{R}_F] \approx 0, \tag{11.56}$$

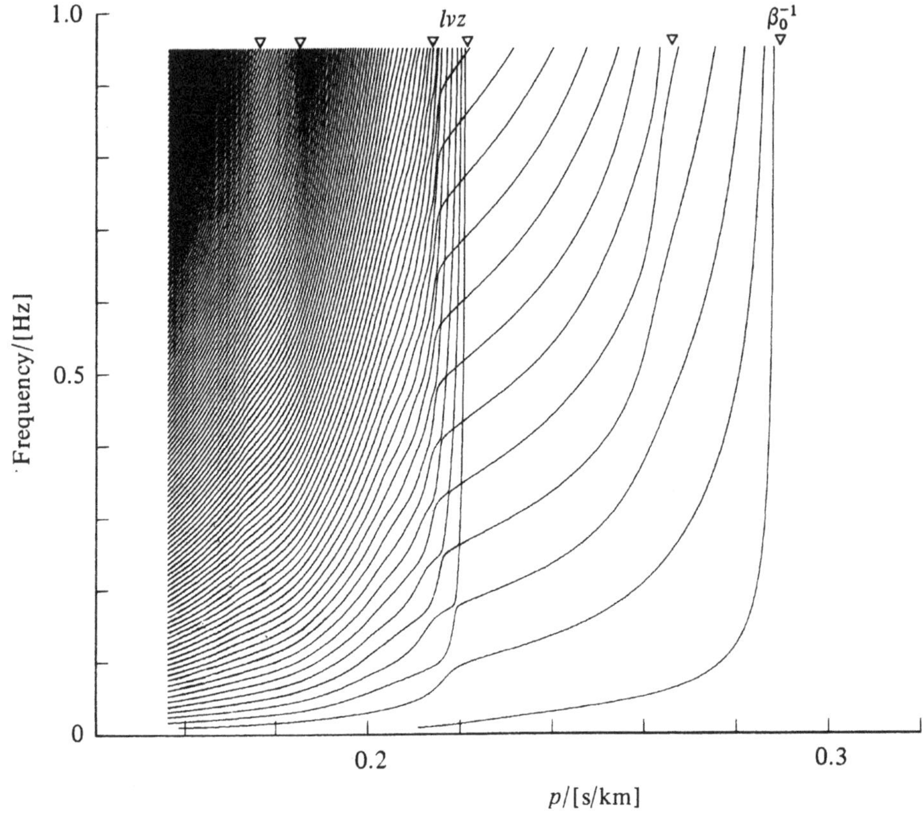

Figure 11.4. Love wave dispersion curves for the upper mantle model *T7*.

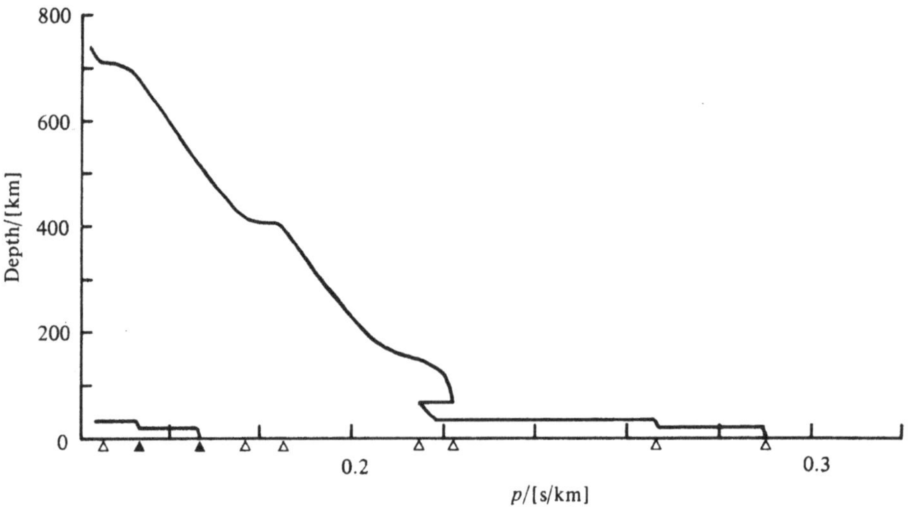

Figure 11.5. Wave slownesses for model *T7* as a function of depth.

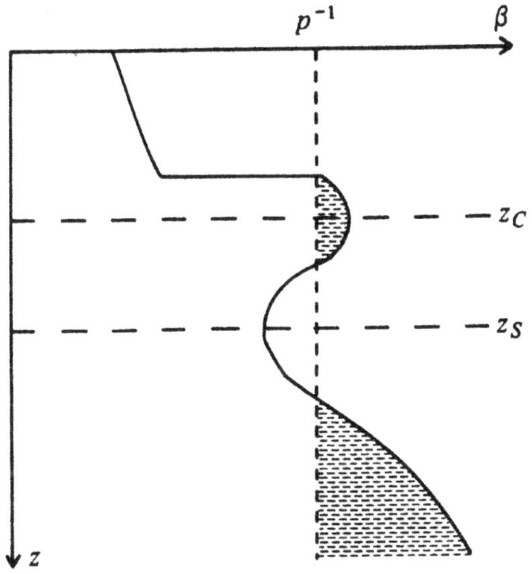

Figure 11.6. Decomposition of the surface wave secular function in the presence of a velocity inversion. For a slowness corresponding to the short dashed line *S* waves are evanescent in the shaded regions, and so the crustal waveguide and the deeper velocity inversion are partially decoupled.

which is the secular function for the truncated structure down to z_S terminated by a uniform half space in $z > z_S$ with continuity of properties at z_S. This result enables one to neglect the reflectivity of the deeply evanescent region and so as frequency increases allows a progressive simplification of the wavespeed model used for the calculation of dispersion curves. Such 'structure reduction' has been used by Schwab & Knopoff (1972), Kerry (1981).

The dispersion relation for the crust alone is given by, e.g., $\det[\mathbf{I} - \mathbf{R}_F\mathbf{R}_D^{0C}] = 0$ where z_C is a level just below the crust-mantle interface. For a wavespeed structure which is bounded by uniform half spaces in $z < z_A$ and $z > z_B$ with a wavespeed inversion in $z_A < z < z_B$ we have the possibility of localized *channel* waves for slownesses such that both *P* and *S* waves are evanescent in $z < z_A, z < z_B$. From the discussion in Section 7.2.1 we can recognise the secular function for this case as

$$\det(\mathbf{I} - \mathbf{R}_U^{KA}\mathbf{R}_D^{KB}) = 0, \tag{11.57}$$

for some level z_K in (z_A, z_B). When this condition is satisfied the reflection and transmission coefficients across the region (z_A, z_B) are singular.

For a stratified half space with low wavespeed crust and a velocity inversion at greater depth we have a rather complex behaviour. For slownesses such that the turning level for *S* waves lies well below the low velocity channel, a choice of z_S in the evanescent regime near the base of the stratification will give (7.79) once again. When, however, the slowness p is such that there are propagating *S* waves within the inversion but evanescent waves in regions above and below (figure 11.6) we have to be more careful.

We take a level z_C lying in the evanescent *S* wave region above the low velocity zone and then the two reflection contributions to (11.55) from above z_S can be expressed as

$$\begin{aligned} \mathbf{R}_D^{0S} &= \mathbf{R}_D^{0C} + \mathbf{T}_U^{CS}\mathbf{R}_D^{CS}[\mathbf{I} - \mathbf{R}_U^{0C}\mathbf{R}_D^{CS}]^{-1}\mathbf{T}_D^{CS}, \\ \mathbf{R}_U^{fS} &= \mathbf{R}_U^{CS} + \mathbf{T}_D^{CS}\mathbf{R}_U^{fC}[\mathbf{I} - \mathbf{R}_D^{CS}\mathbf{R}_U^{fC}]^{-1}\mathbf{T}_U^{CS}. \end{aligned} \tag{11.58}$$

In each of these expressions we have both upward and downward transmission through the evanescent region contained in the matrices $\mathbf{T}_D^{CS}$, $\mathbf{T}_U^{CS}$. At moderate frequencies these transmission terms will be quite small and diminish rapidly as the frequency increases. We thus have high frequency approximations

$$\mathbf{R}_D^{0S} \approx \mathbf{R}_D^{0C}, \quad \mathbf{R}_U^{fS} \approx \mathbf{R}_U^{CS}, \tag{11.59}$$

and so in this limit (11.55) becomes

$$\Delta \approx \det(\mathbf{I} - \mathbf{R}_D^{0C}\mathbf{R}_F)\det(\mathbf{I} - \mathbf{R}_U^{CS}\mathbf{R}_D^{SL}), \tag{11.60}$$

which is the product of the secular determinant for *surface* waves on the crustal structure above z_C,

$$\det(\mathbf{I} - \mathbf{R}_D^{0C}\mathbf{R}_F) \tag{11.61}$$

and the secular determinant for *channel* waves on the structure below z_C

$$\det(\mathbf{I} - \mathbf{R}_U^{CS}\mathbf{R}_D^{SL}). \tag{11.62}$$

At intermediate frequencies there will be coupling between the channel and the near surface through the transmission matrices $\mathbf{T}_D^{CS}$, $\mathbf{T}_U^{CS}$. A given surface wave mode will in certain frequency ranges be mainly confined to the near-surface region and then $\det(\mathbf{I} - \mathbf{R}_D^{0C}\mathbf{R}_F) \approx 0$; in other intervals a mode will be mostly a channel wave when $\det(\mathbf{I} - \mathbf{R}_U^{CS}\mathbf{R}_D^{SL}) \approx 0$ (Frantsuzova, Levshin & Shkadinskaya, 1972; Panza, Schwab & Knopoff, 1972).

Propagation in the channel is most efficient for slownesses close to β_m^{-1} where β_m is the minimum wavespeed in the inversion. As a result there is only very slight variation in the slowness of a channel wave with frequency.

A very good example of this effect is shown in figure 11.7 which shows the horizontal and vertical displacement elements (V, U) of the displacement eigenfunction for the sixth higher Rayleigh mode in the frequency range 0.35 to 0.50 Hz. At the lower frequency, propagation which is principally in the crust still shows some amplitude in the low velocity channel, but at the highest frequency virtually all the energy is confined to the crust.

A detailed study of the switch between crustal and channel behaviour has been made by Kerry (1981) who shows that in the neighbourhood of an osculation point between two dispersion branches the dispersion curves have hyperbolic form. The

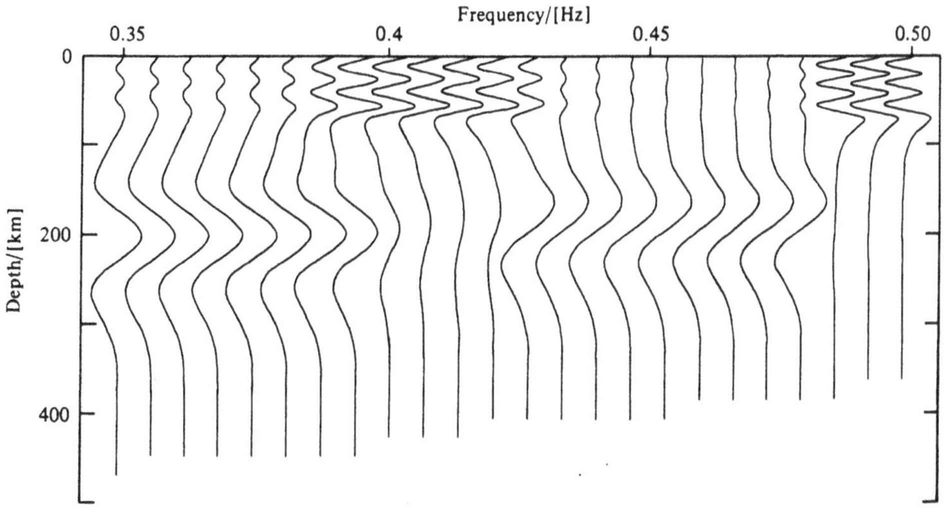

Figure 11.7. Rayleigh wave eigenfunctions for the sixth higher mode at equal frequency intervals from 0.35 to 0.50 Hz.

change from one class of propagation to the other takes place over a frequency interval which is of the same order of magnitude as the decay in two-way passage across the evanescent region above the inversion and so is very sharp at *high frequencies*.

11.4.3 Computation of surface wave dispersion

The reflection matrix representations of the modal secular equation which we have discussed in this chapter can be used as the basis of an efficient scheme for calculating modal dispersion (Kerry, 1981).

Although most calculations have previously been made with fixed frequency and variable slowness for a limited number of modes (see, e.g., Schwab & Knopoff, 1972), there are considerable advantages in fixing slowness p and varying frequency ω when seeking for roots of the secular equation. The zeroes of the secular function are approximately evenly spaced in frequency ω whilst the spacing in slowness p is rather irregular. Also for a piecewise smooth model or a model composed of uniform layers we can make use of the frequency independence of the interfacial reflection and transmission coefficients to reduce the computational effort of calculating the reflection matrices required for the secular function. This approach is well suited to finding the dispersion curves for a large number of higher modes.

The downward reflection matrix $\mathbf{R}_D^{0L}$ and $\det \mathbf{T}_D^{0L}$ are calculated by recursion from the base of the stratification. We then evaluate the secular function and use quadratic interpolation to iteratively refine an estimate of the frequency of the root. A suitable termination procedure for the iteration (Kerry, 1981) is when two estimates of the secular function with opposite sign are less than some threshold and the smallest eigenvector of $(\mathbf{n}_{D0} + \mathbf{n}_{U0}\mathbf{R}_D^{0L})$ evaluated at the estimated root is also less than a preassigned threshold.

This procedure works very well most of the time but, when we are close to a root of the channel wave function, $(\mathbf{I} - \mathbf{R}_U^{CS}\mathbf{R}_D^{SL})$ becomes nearly singular and it is very difficult to get accurate numerical evaluation of $\mathbf{R}_D^{0L}$ with finite accuracy arithmetic. To test for this possibility a level z_S in the low velocity zone is chosen and in a single pass through the structure $\mathbf{R}_D^{0L}$, $\mathbf{R}_D^{SL}$, $\det \mathbf{T}_D^{0L}$, $\det \mathbf{T}_D^{SL}$, $\det \mathbf{T}_D^{0S}$ are calculated. We construct the quantity

$$\eta = \det(\mathbf{I} - \mathbf{R}_U^{0S}\mathbf{R}_D^{SL}) = \det \mathbf{T}_D^{0S} \det \mathbf{T}_D^{SL} / \det \mathbf{T}_D^{0L} \qquad (11.63)$$

If η is large then we may use the secular function (7.1). If, however, η is small we may have a channel mode and then we calculate downwards from the surface to find $\mathbf{R}_U^{fS}$ and then terminate the iteration by requiring the smallest eigenvalue of $(\mathbf{I} - \mathbf{R}_U^{fS}\mathbf{R}_D^{SL})$ to be less than our threshold.

Once all the roots in a given frequency band are found the slowness is incremented by a small amount, and the calculation repeated.

The dispersion curves presented in figures 11.3 and 11.4 were calculated by the method described above for the upper mantle model *T7* (Burdick & Helmberger 1978). For this calculation the structure down to 950 km was represented in terms of 92 uniform layers. The reflection matrix approach is not restricted to representations in terms of uniform layers and can be extended to piecewise smooth models by using the treatment discussed in Chapter 6.

Most calculations for *surface wave* dispersion have been made for the fundamental mode and to a lesser extent for the first five higher modes. For these widely spaced modes it is probably more efficient to fix the frequency and then determine the slowness mode by mode, since as can be seen from figures 11.3 and 11.4 at low frequencies a small change in frequency gives a large change in slowness. The present scheme becomes most effective when many modes are to be found, particularly when one seeks all modes in a given frequency and slowness window. Working at fixed slowness the costs of calculating many frequency values are much reduced.

11.4.4 Variational results

The dispersion results we have presented so far depend on the differential properties of the stress-displacement field, however we may supplement these with some important integral results from a variational principle, for each modal solution.

We consider two displacement fields w_1, w_2 which satisfy the boundary condition at z_L for $p > \beta_L^{-1}$ and so have exponential decay as $z \to \infty$. For *SH* waves we consider the vertical variation of the product $W_1\omega T_2$ of displacement and traction:

$$\partial_z(W_1\omega T_2) = \omega^2\{(\rho\beta^2)^{-1}T_1T_2 + \rho(\beta^2p^2 - 1)W_1W_2\}, \tag{11.64}$$

and then integrate over the entire depth interval to give

$$\begin{aligned}\int_0^\infty dz\, \partial_z(W_1\omega T_2) &= \omega W_1 T_2|_{z=0}, \qquad (11.65)\\ &= \omega^2\int_0^\infty dz\{(\rho\beta^2)^{-1}T_1T_2 + \rho(\beta^2p^2 - 1)W_1W_2\}.\end{aligned}$$

A displacement eigenfunction W_e has vanishing traction at $z = 0$ and also decays in $z > z_L$, and so if we choose W_e for both fields in (11.66) we have

$$\begin{aligned}\omega^2\int_0^\infty dz\, \rho W_e^2 &= \omega^2\int_0^\infty dz\{(\rho\beta^2)^{-1}T_e^2 + \rho\beta^2p^2W_e^2\},\\ &= \int_0^\infty dz\, \rho\beta^2\{(\partial_z W_e)^2 + \omega^2p^2W_e^2\}.\end{aligned} \tag{11.66}$$

This identity is valid for a dissipative medium, and may be viewed as a variational result for the *SH* eigenfunction. If we require (11.66) to be a stationary functional of W_e, we obtain the Euler-Lagrange equation (cf. 2.30)

$$\partial_z(\rho\beta^2\partial_z W_e) - \rho\omega^2(\beta^2p^2 - 1)W_e = 0, \tag{11.67}$$

with the boundary conditions

$$T_e(0) = 0, \quad W_e(z) \to 0 \quad \text{as} \quad z \to \infty; \tag{11.68}$$

which are just the equations to be satisfied by a modal eigenfunction. For a perfectly elastic medium we can choose $W_1 = W_e$ and $W_2 = W_e^*$ and then from (11.66)

$$\int_0^\infty dz\, \rho\omega^2|W_e|^2 = \omega^2\int_0^\infty dz\{(\rho\beta^2)^{-1}|T_e|^2 + \rho\beta^2p^2|W_e|^2\}, \tag{11.69}$$

and this result shows the equality of kinetic and potential energy in the mode. This is just Rayleigh's principle and (11.66) represents the extension of this result to attenuative media.

For the *P-SV* wave case we can follow a similar development to the above by considering $\partial_z(\boldsymbol{w}_1\omega\mathbf{t}_2)$ and then for a displacement eigenvector $\boldsymbol{w}_e$ we have the variational result

$$\omega^2\int_0^\infty dz\, \rho[U_e^2 + V_e^2] = \omega^2\int_0^\infty dz\{\rho\nu p^2V_e^2 + (\rho\alpha^2)^{-1}P_e^2 + (\rho\beta^2)^{-1}S_e^2\}, \tag{11.70}$$

where, as in (2.24),

$$\nu = 4\beta^2(1 - \beta^2/\alpha^2). \tag{11.71}$$

The Euler-Lagrange equation here recovers (2.31) for the evolution of U_e, V_e. Kennett (1974) has shown that the first order equations (2.25), (2.24) which are equivalent to (2.30), (2.31) are the sets of Hamilton's equations associated with the Lagrangians (11.66), (11.70) in which displacement and stress quantities are conjugate variables.

The stationary property of the integrals (11.66), (11.70) for small perturbations in the eigenfunction $\mathbf{w}_e$ means that small changes may be made in ω and p or in the elastic parameters and the results obtained from (11.66), (11.70) will be correct to second order using the original eigenfunctions (Jeffreys, 1961). In particular we can avoid numerical differentiation in calculating the group slowness $g(= p + \omega\partial_\omega p)$, which describes the evolution of the dispersion curves. From the variational results we can represent the group slowness as a ratio of two integrals over the eigenfunctions

$$g = I/pJ, \tag{11.72}$$

where, for *SH* waves

$$I = \int_0^\infty dz\, \rho W_e^2, \quad J = \int_0^\infty dz\, \rho\beta^2 W_e^2; \tag{11.73}$$

and for *P-SV* waves

$$\begin{aligned} I &= \int_0^\infty dz\, \rho[U_e^2 + V_e^2] = \int_0^\infty dz\, \rho \mathbf{w}_e^T \mathbf{w}_e, \\ J &= \int_0^\infty dz\{\rho\nu V_e^2 + (\omega p)^{-1}[U_e S_e - (1 - 2\beta^2/\alpha^2)V_e P_e]\}. \end{aligned} \tag{11.74}$$

If we are working at fixed slowness and are already close to a root of the secular equation ω_k we may use (11.66), (11.70) with the eigenfunction estimate corresponding to our current value of ω to generate a closer approximation to the root. A similar procedure holds at fixed frequency to improve estimates of the phase slowness p.

As we have mentioned we may also obtain good estimates of the perturbation in the dispersion introduced by small changes in the elastic parameters although particular care must be taken if discontinuities in the parameters are moved (Woodhouse, 1976). At fixed frequency we may represent the change in phase slowness p as, e.g., for Rayleigh waves

$$\delta p = \int_0^\infty dz \left(\left[\frac{\partial p}{\partial \alpha}\right] \delta\alpha + \left[\frac{\partial p}{\partial \beta}\right] \delta\beta + \left[\frac{\partial p}{\partial \rho}\right] \delta\rho \right). \tag{11.75}$$

The quantities $[\partial p/\partial\alpha]$ etc. provide a measure of the sensitivity of the mode to the elastic parameters. These 'partial derivatives' are used in the linearised inversion of dispersion data to obtain a structural model by iterative updating of a trial model until the observed and computed dispersion agree to within the error of measurement.

If we consider small perturbations to the elastic wavespeeds introduced by including small loss factors Q_α^{-1}, Q_β^{-1} in a perfectly elastic model we are able to estimate the change in phase slowness at fixed frequency as in (11.60). The perturbation δp will have both real and imaginary parts. The imaginary part will specify the overall loss factor for the mode and the real part will contain the net effect of the wavespeed dispersion for *P* and *S* waves associated with attenuation (cf. 1.13).

11.5 Theoretical seismograms by modal summation

In Chapter 7 we have shown how the poles in the half space response contribute to the displacement field in slightly different ways depending on whether we perform the frequency or slowness integrals first when performing the inversion of the transforms. With an inner slowness integral the modal contribution to the surface displacement at a range r for *P-SV* waves may be represented as

$$\mathbf{u}_P(r,\phi,0,t) = \tfrac{1}{2}\int_{-\infty}^{\infty} d\omega\, e^{-i\omega t} i\omega^2 \sum_m \times \left(\sum_{j=0}^{N(\omega)} p_j \operatorname*{Res}_{\omega,p=p_j} [\boldsymbol{w}_0^T(p,m,\omega)]\mathbf{T}_m^{(1)}(\omega p_j r) \right), \quad (11.76)$$

where we have used the spectral response in (7.63), and $N(\omega)$ is the number of modes of frequency ω. The azimuthal dependence of $\boldsymbol{w}_0$ arises from the source which we have specified via a discontinuity in displacement and traction across the source plane $z = z_S$

$$\mathbf{S}(p,m,\omega,z_S) = [\mathbf{S}_W^m, \mathbf{S}_T^m]^T \quad (11.77)$$

From (7.32) we can represent the half space response in the transform domain as

$$\boldsymbol{w}_0(p,m,\omega) = -i\boldsymbol{W}_{1S}(0)[\mathbf{I}-\mathbf{R}_D^{SL}\mathbf{R}_U^{fS}]^{-1}\{\mathbf{T}_{2S}^T(z_S)\mathbf{S}_W^m - \boldsymbol{W}_{2S}^T(z_S)\mathbf{S}_T^m\}, \quad (11.78)$$

where the displacement matrix $\boldsymbol{W}_{1S}$ is chosen to give vanishing traction at the free-surface and $\boldsymbol{W}_{2S}$ satisfies the radiation condition of only downward propagating or decaying evanescent waves in $z > z_L$. The half space reverberation operator arises from linking the solutions above and below the source via the source jump **S**, and is related to the invariant generated from the matrices $\boldsymbol{W}_{1S}, \boldsymbol{W}_{2S}$ by

$$<\boldsymbol{W}_{1S}, \boldsymbol{W}_{2S}> = i[\mathbf{I} - \mathbf{R}_D^{SL}\mathbf{R}_U^{fS}] \quad (11.79)$$

We calculate the residue of $\boldsymbol{w}_0^T$ at the pole $p = p_j$, at fixed frequency ω, in terms of the displacement eigenfunction $\boldsymbol{w}_e$ following the treatment discussed in the Appendix to this chapter. The residue is

$$\operatorname*{Res}_{\omega,p=p_j} [\boldsymbol{w}_0^T(p,m,\omega)] = \frac{g_j}{2\omega I_j}\{\mathbf{t}_{ej}^T(z_S)\mathbf{S}_{Wj}^m - \boldsymbol{w}_{ej}^T(z_S)\mathbf{S}_{Tj}^m\}\boldsymbol{w}_{ej}^T(0), \quad (11.80)$$

in terms of the group slowness g_j for the mode and the 'kinetic energy' integral I_j introduced in the previous section.

The residue contribution to the surface displacement (11.76) is therefore

$$\mathbf{u}_P(r,\phi,0,t) = \int_{-\infty}^{\infty} d\omega\, e^{-i\omega t} i\omega \sum_{j=0}^{N(\omega)} \tfrac{1}{4} p_j g_j I_j^{-1} \sum_m \{\mathbf{t}_{ej}^T(z_S)\mathbf{S}_{Wj}^m - \mathbf{w}_{ej}^T(z_S)\mathbf{S}_{Tj}^m\}\mathbf{w}_{ej}^T(0)\mathbf{T}_m^{(1)}(\omega p_j r), \tag{11.81}$$

which is equivalent to the form derived by Takeuchi & Saito (1972) by a rather different approach. The *SH* wave residue contribution has a comparable form

$$\mathbf{u}_H(r,\phi,0,t) = \int_{-\infty}^{\infty} d\omega\, e^{-i\omega t} i\omega \sum_{l=0}^{M(\omega)} \tfrac{1}{4} p_l g_l I_l^{-1} \sum_m \{T_{el}^T(z_S)S_{Wl}^m - W_{el}^T(z_S)S_{Tl}^m\}W_e l^T(0)\mathbf{T}_m^{(1)}(\omega p_l r), \tag{11.82}$$

in terms of the scalar eigenfunction W_{el}. Frequently we wish to evaluate the modal contribution of large ranges and then it is usually adequate to use the asymptotic approximations (7.79) for $\mathbf{T}_m^{(1)}(\omega p r)$, $\mathbf{T}_m^{(1)}(\omega p r)$. To this approximation the *SH* wave contribution is purely transverse, and the *P-SV* waves appear on the vertical and radial components.

The expressions (11.82), (11.83) are particularly useful since we are able to relate the excitation of the various angular orders to force or moment tensor elements describing the source through $\mathbf{S}_{Wj}^m$ (4.63) and $\mathbf{S}_{Tj}^m$ (4.64). We also have a clear separation of the receiver contribution from $\mathbf{w}_{ej}(0)$ and the source effects

$$\{\mathbf{t}_{ej}^T(z_S)\mathbf{S}_{Wj}^m - \mathbf{w}_{ej}^T(z_S)\mathbf{S}_{Tj}^m\}. \tag{11.83}$$

Since the pole position $p_j(\omega)$ is independent of the source depth, the excitation of a particular mode as a function of the depth of the source is controlled by the size of the term in braces. In figure 11.8 we show the displacement eigenfunctions for the first five Rayleigh modes at a frequency of 0.06 Hz as a function of depth. As the mode number increases we see the increasing penetration of the eigenfunctions into the half space. Since the source excitation depends on these eigenfunction shapes, as the source depth increases higher order modes will be preferentially excited. The traction eigenfunction $\mathbf{t}_{ej}$ vanishes at the surface and especially for low frequencies ($<$ 0.02 Hz) increases only slowly away from the surface, as a result the contribution of $\mathbf{t}_{ej}(z_S)\mathbf{S}_{Wj}^m$ is very much reduced for shallow sources. This means that for near-surface sources the moment tensor components M_{xz}, M_{yz} play only a minor role in the excitation and so are difficult to recover if one attempts to invert for the source mechanism from distant observations. For a purely strike-slip fault, the normals to both the fault plane and the auxiliary plane line in a horizontal plane and so the only non-zero moment tensor elements are M_{xx}, M_{xy}, M_{yy} with the result that $\mathbf{S}_{Wj}^m$ vanishes, and so we have the simpler excitation term $\mathbf{w}_{ej}^T(z_S)\mathbf{S}_{Tj}^m$. Since the eigenfunctions are oscillatory with depth down to the turning level of an *S* wave with slowness p_j the actual mode excitation will vary

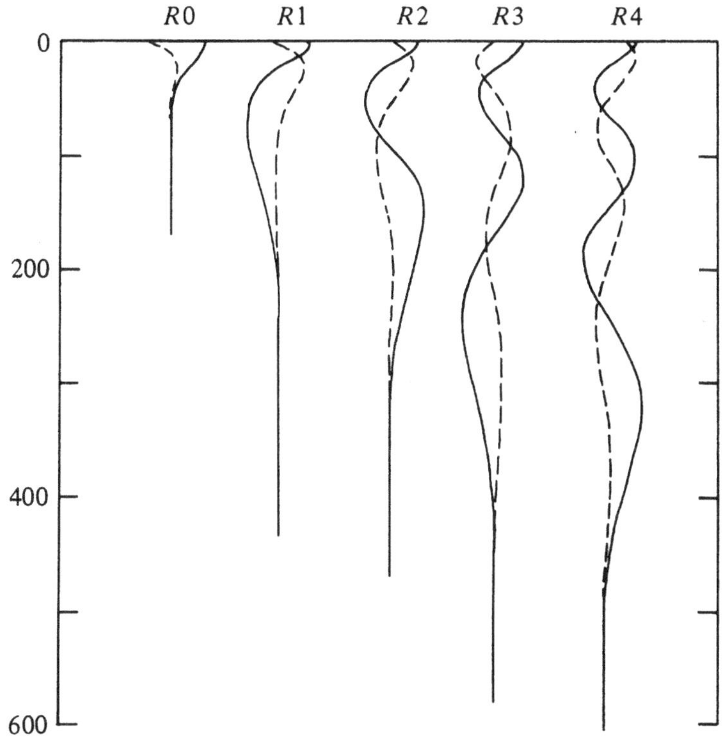

Figure 11.8. Eigenfunctions as a function of depth for the fundamental and first four higher mode Rayleigh waves for frequency 0.06 Hz. The vertical component is shown by solid lines and the horizontal component by dashed lines.

significantly with depth for a constant source mechanism and there can be near nulls in the excitation at certain depths.

The relative excitation of the modes depends on the surface expression of the modes through the receiver term $w_{ej}(0)I_j^{-1}$; the modes are effectively normalized by their energy content. Modes which are confined to the crustal region or which are sufficiently low frequency to penetrate through any wavespeed inversion will have significant amplitude at the surface. At low frequencies even modes which are mostly trapped in a wavespeed inversion still have some surface amplitude, but as the frequency increases the modes are almost entirely confined to the channel (figure 11.7).

For any given range r and azimuthal order m, we are now faced with the evaluation of the integrals (11.66), (11.70) over frequency. This is best performed numerically. If a very large number of modes are to be summed an effective approach is to use a fast Fourier transform with a large number of time points to encompass the length of the wave train. With a large number of time points the spacing in frequency is very fine and it is necessary to interpolate the behaviour of the dispersion curve branches to generate values at the required frequencies.

This fine sampling in frequency ensures an adequate representation of the rapidly oscillating Hankel functions $H^{(1)}_m(\omega p_j r)$ at large ranges. For each frequency point the contribution from all the modes are summed and then the transform is inverted.

If the frequency band and the ranges under consideration are such that $\omega_{min}\beta_L^{-1}r_{min}$ is large (> 6) we can use the asymptotic expressions for $\mathbf{T}^{(1)}_m, \mathbf{T}^{(1)}_m$ and then the range dependence of the integral is the same for all angular orders. If all components of the moment tensor have the same time dependence we have a single frequency integral modulated by an azimuthally varying factor.

The modal contributions to the seismograms are linear in the components of the source moment tensor. Thus, as for body waves, it is possible to make a linear inversion for the source mechanism using distant observations, if a good reference model of the wavespeed distribution with depth can be found for the paths in question (Mendiguren, 1977).

11.5.1 Mode branch contributions

The character of the contribution to the total seismogram from an individual mode branch will depend on the nature of the particular dispersion. The contribution to the vertical component in the asymptotic regime for large $\omega p r$ will for example be of the form

$$u_{zj}(r,\phi,0,t) = \int_{-\infty}^{\infty} d\omega\, |F(p_j(\omega),\omega)| e^{-i\omega t + i\omega p_j(\omega) r + i\psi_j(\omega)}, \tag{11.84}$$

where p_j depends on frequency. The 'initial phase' $\psi_j(\omega)$ arises from the effect of depth of source etc. in the model and also from the instrument response on recording.

We may get a qualitative picture of the modal behaviour by examining various approximations to the integral in (11.84). Away from an extremum in group slowness g_j we may use the stationary phase technique. The saddle point in frequency ω_s depends on range r and time t through

$$\partial_\omega\{\psi_j(\omega_s) + \omega_s p_j(\omega_s) r - \omega_s t\} = 0, \tag{11.85}$$

and the stationary phase approximation gives

$$\begin{aligned} u_{zj}(r,\phi,0,t) = {} & \frac{(2\pi)^{1/2}|F(p_j,\omega_s)|}{|\partial_{\omega\omega}\psi_j + r\partial_\omega g_j|^{1/2}} \\ & \times \exp\{i\omega_s(p_j r - t) + i\psi_j(\omega_s) \pm i\pi/4\}, \end{aligned} \tag{11.86}$$

with $\pm$ corresponding to

$$\partial_{\omega\omega}\psi_j + r\partial_\omega g_j \lesseqgtr 0. \tag{11.87}$$

The time of arrival of this mode branch contribution will vary with frequency and from (11.85)

$$t = (p_j + \omega_s \partial_\omega p_j) r + \partial_\omega \psi_j(\omega_s) \tag{11.88}$$

so that propagation is very nearly at the group slowness

$$g_j = p_j + \omega_s \partial_\omega p_j \tag{11.89}$$

for the frequency ω_s. Near extrema in the group slowness the above treatment is inadequate and an improved approximation may be found in terms of Airy functions, following analysis due to Pekeris (1948):

$$\begin{aligned} u_{zj}(r, \phi, 0, t) = &\frac{|F(p_j, \omega_a)|}{Q^{1/3}} \mathrm{Ai}\left[\frac{P\,\mathrm{sgn}(Q)}{|Q|^{1/3}}\right] \\ &\times \exp\{i\omega_a(p_j r - t) + i\psi_j(\omega_a)\}, \end{aligned} \tag{11.90}$$

where the frequency ω_a is determined by the condition

$$\partial_{\omega\omega}\psi_j(\omega_a) + r g_j(\omega_a) - t = 0 \tag{11.91}$$

and

$$\begin{aligned} P &= \partial_\omega \psi_j(\omega_a) + r g_j(\omega_a) - t \\ Q &= \tfrac{1}{2}\{\partial_{\omega\omega\omega}\psi_j(\omega_a) + r\partial_{\omega\omega} g_j(\omega_a)\}. \end{aligned} \tag{11.92}$$

In the case of a maximum in group slowness we have both higher and lower frequency arrivals appearing before the energy with the maximum slowness (the 'Airy' phase). A good example is provided by the fundamental Rayleigh mode and the Airy phase is often designated *Rg*, figure 11.9.

The relatively small dispersion in frequencies in the neighbourhood of an extremum in group slowness means that an Airy phase will often carry significant amplitude. The actual amplitude seen on a seismogram will depend on the instrument response as a function of frequency and also on the excitation of the mode. For the fundamental Rayleigh mode the maximum group slowness occurs at a frequency of about 0.06 Hz and this is well excited by relatively shallow earthquakes, e.g., at the base of the crust. Also conventional long-period seismometers have their peak response near to 0.05 Hz and so the *Rg* phase is often an important component of the surface wavetrain (figure 11.9).

Airy phases associated with shallow propagation can also be significant and a good example for the fundamental Rayleigh mode is seen on the short range seismograms in figure 7.5a for a crustal source at 2.5 km depth.

When we wish to synthesise the long-period contributions to the surface wavetrain we need only five or six mode branches to give a good representation. In this case it is most effective to calculate the contribution a mode branch at a time and so we are faced with the numerical evaluation of integrals such as (11.67). Aki (1960) suggested using a variable spacing in frequency and his work has been extended by Calcagnile et al. (1976). Between frequency points ω_l and ω_{l+1} the amplitude spectrum $|F(p_j, \omega)|$ is represented as a quadratic in ω

$$|F(p_j, \omega)| \approx f_0 + f_1(\omega - \bar{\omega}_l) + f_2(\omega - \bar{\omega}_l)^2, \tag{11.93}$$

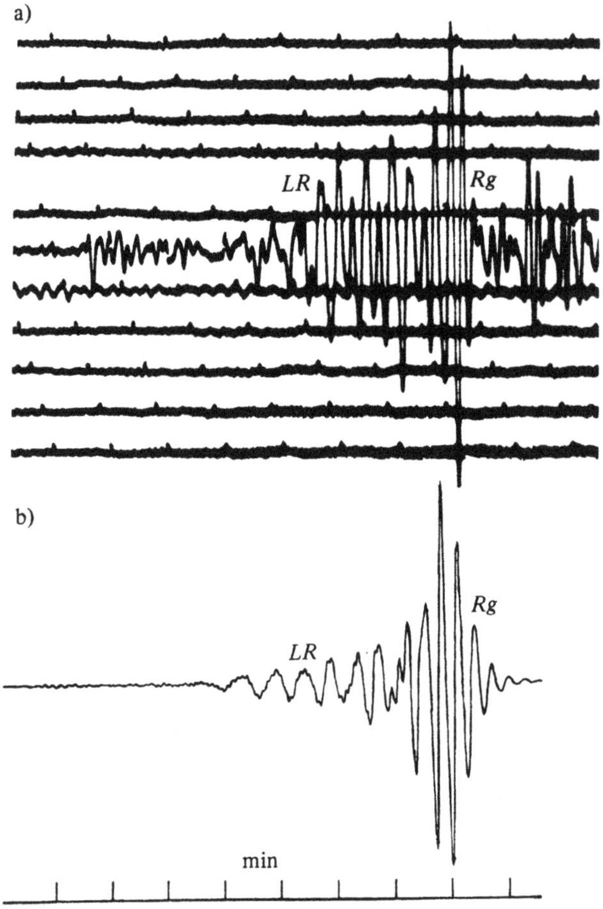

Figure 11.9. Observed and theoretical seismograms for Airy phases: a) *Rg* and long-period Rayleigh waves *LR* at 3000 km; b) modal sum to 0.25 Hz at 3000 km, higher modes ride on the long-period fundamental Rayleigh mode.

where $\bar{\omega}_l = \frac{1}{2}(\omega_l + \omega_{l+1})$; and the phase as a linear function of ω

$$\omega p_j(\omega) r + \psi_j(\omega) \approx \omega_l t_{pl} + (\omega - \bar{\omega}_l) t_{gl}, \tag{11.94}$$

where the mean phase delay

$$t_{pl} = \bar{\omega}_l p_j(\bar{\omega}_l) r + \psi_j(\bar{\omega}_l), \tag{11.95}$$

and the mean group delay

$$t_{gl} = g_j(\bar{\omega}_l) + \partial_\omega \psi_j(\bar{\omega}_l). \tag{11.96}$$

With the substitutions (11.90), (11.93) the integral (11.84) can be evaluated analytically over the panel (ω_l, ω_{l+1}) to give

$$\begin{aligned}\int_{\omega_l}^{\omega_{l+1}} & d\omega\, |F(p_j,\omega)| \exp\{-i\omega t + i\omega p_j(\omega) r + i\psi_j(\omega)\} \\ &= \Bigg[\left(\frac{|F(p_j,\omega)|}{i(t-t_{gl})} - \frac{\partial_\omega |F(p_j,\omega)|}{(t-t_{gL})^2} - \frac{2f_2}{i(t-t_{gL})^3}\right) \\ &\quad \times \exp\{-i\omega(t-t_{pl}) + i(\omega-\bar{\omega}_l)t_{gl}\}\Bigg]_{\omega_l}^{\omega_{l+1}} \end{aligned} \tag{11.97}$$

and the entire integral (11.84) may be obtained by summing the result from successive panels (a generalized Filon rule). The frequency points ω_l should be chosen to minimise the error which will be principally due to the phase representation (11.93). Calcagnile et al. (1976) suggest that the frequency points be chosen so that

$$r(g_{l+1} - g_l) < \tfrac{1}{2}\pi\bar{\omega}_l^{-1} \tag{11.98}$$

for reasonable accuracy ($\sim$ 0.5 per cent). As the period increases the accuracy of the asymptotic representation of the Hankel functions is reduced. A comparable development can be made with a quadratic approximation for the amplitude and a linear approximation for the initial phase and the explicit form for the Hankel function, but the simplicity of (11.98) is lost.

11.5.2 Examples of modal synthesis

As we can see from the results in the previous section the contribution of any particular mode to surface seismograms is heavily dependent on the variation of group slowness $g_j(\omega)$ as a function of frequency. When we have many modes present the character of the final seismogram depends on the relative excitation of the modes and the group slowness character shown by individual modes.

In figure 11.10 we illustrate the group slowness behaviour for all Rayleigh modes for model *T7* in figure 11.3 with frequency less than 0.33 Hz. This pattern derived from some 80 mode branches shows a number of coherent features which are associated with recognisable traits in the seismogram. With the exception of the behaviour for the fundamental branch the pattern for Love waves is very similar.

The prominent maximum in group slowness for the fundamental Rayleigh mode giving rise to the *Rg* phase has already been discussed. The group slowness extrema for the first few higher mode branches at frequencies above 0.10 Hz lie in the range 0.28–0.32 s/km and lead to a characteristic 'high frequency' train *Lg*. At low frequencies ($<$ 0.10 Hz) these same modes have a further set of group slowness extrema from 0.21-0.23 s/km; the corresponding Airy phases are well excited by intermediate depth earthquakes ($\sim$ 120 km deep) to give the *Sa* phase.

The existence of these two sets of group extrema can be understood if we refer

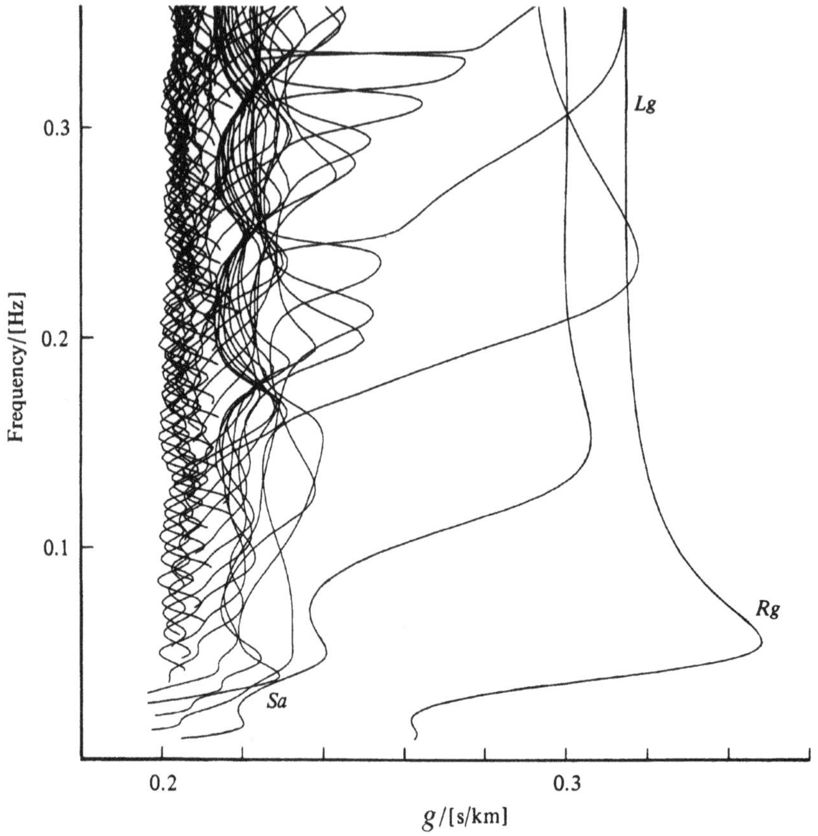

Figure 11.10. Group slowness dispersion for Rayleigh waves on model *T7* as a function of frequency. The group slowness extrema associated with phases illustrated in figures 11.9, 11.11-11.13 are indicated.

to the discussion of Section 9.2 in which we split the stratification at a level z_J. When we take z_J at 200 km, the contribution ${}^{sh}w_0$ (9.34) includes the crustal and channel structure alone, and the dispersion curves for this portion will reproduce the *Lg* maxima. The remaining part of the displacement response ${}^{d}w_0$ (9.38) contains the operator $[\mathbf{I} - \mathbf{R}_D^{JL}\mathbf{R}_U^{fJ}]^{-1}$ which will give the low frequency 'mantle branches' whose dispersion is dictated by the structure in the upper mantle. It is these latter branches which display the group extrema associated with *Sa*.

The tangled skein of group slowness curves near 0.20 s/km for all frequencies corresponds to *S* waves with turning points well into the upper mantle, which we would alternatively think of as 'body wave' phases. The complex behaviour at intermediate slownesses (0.22–0.26 s/km) with rapid changes in group slowness for an individual mode, arises from the presence of the wavespeed inversion in model *T7* and the switch in properties from channel to crustal guiding (see figure

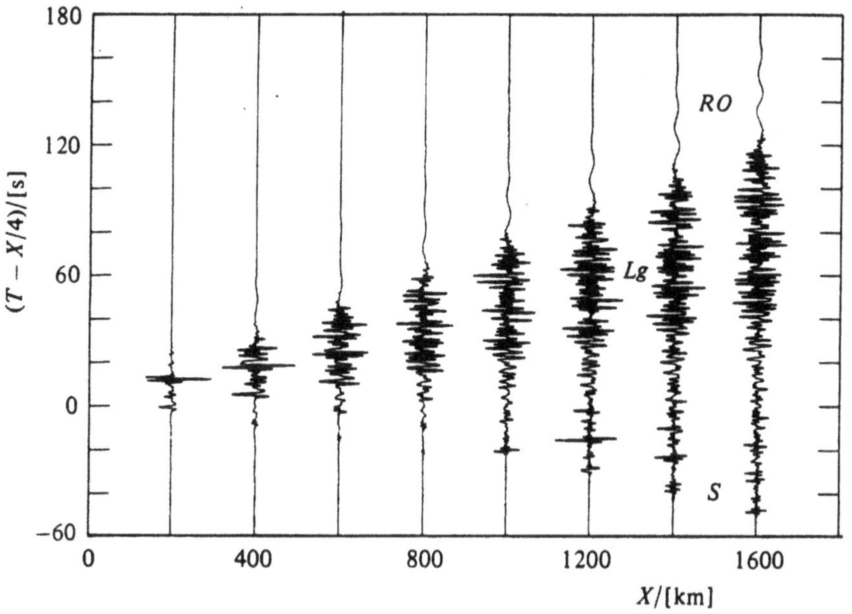

Figure 11.11. Theoretical seismograms for a 40 km deep vertical dip slip source. The vertical component is shown and is synthesised from all the modes illustrated in figure 11.3: there is a prominent *Lg* group and a clear separation of the fundamental Rayleigh mode *R0*, at short times *S* body waves are seen.

11.7). For a surface receiver, channel modes make very little contribution to the response except at low frequencies.

The group slowness behaviour in figure 11.10 shows very clearly the difficulties associated with trying to estimate group slowness for many mode branches from observed records. At moderate frequencies (> 0.15 Hz) we can have a number of branches with very similar group slownesses and it is very difficult to disentangle the behaviour near such cross-overs.

In figure 11.11 we illustrate the result of modal summation for all the mode branches shown in figure 11.3, with a vertical dip-slip event at a focal depth of 40 km. The wavetrains on these vertical component seismograms are quite complex but a clear low frequency *Rg* fundamental mode Rayleigh wave emerges from the rear of the disturbance. The display in figure 11.11 is plotted at a reduction slowness of 0.25 s/km, from figure 11.10 we see this group slowness separates out the 'surface waves' associated with the first few Rayleigh mode branches from the rest of the modes. The complex multimode interference in the high frequency *Lg* group is clearly seen in figure 11.11. The faster arrivals have a distinct pulse-like character and can be identified with *S* wave arrivals which mostly have turning points in the mantle. At short ranges we see upward radiated *S* which at larger ranges is superseded by *S* wave energy returned from beneath the source level,

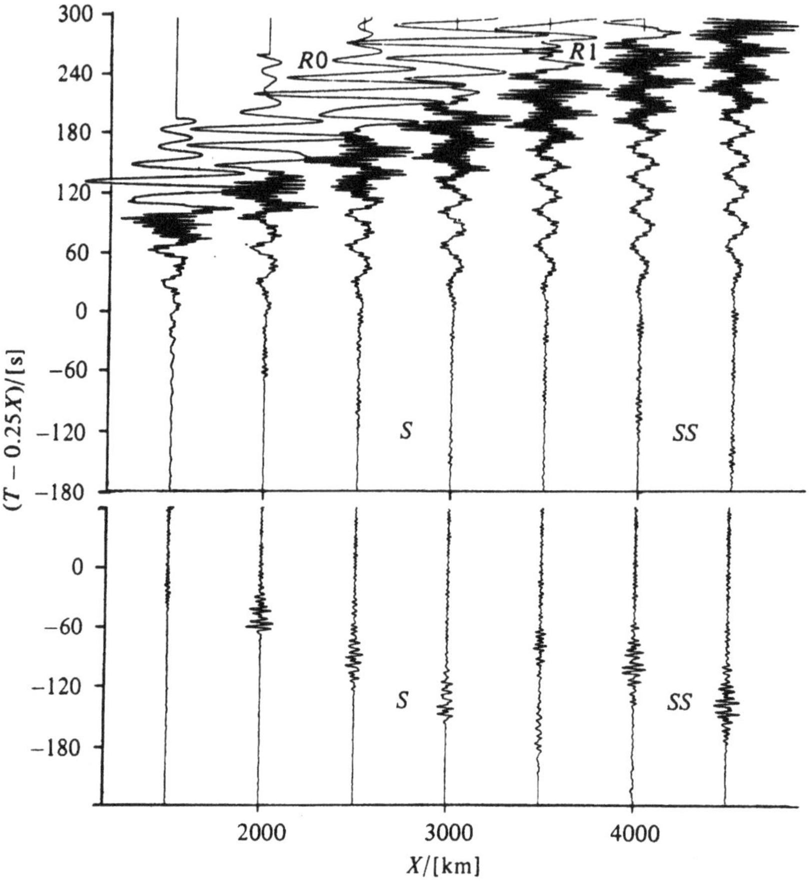

Figure 11.12. Theoretical seismograms, vertical component by summation of Rayleigh modes: a) all modes in figure 11.3 with a frequency less than 0.33 Hz; b) the difference between a) and a calculation with only 15 modes enlarged by a factor of 4.

both direct *S* and, with a small delay, a reflection from the 400 km discontinuity. For much greater source depths, e.g., 200 km the seismograms for the same source mechanism show almost no trace of surface wave character out to 1600 km. The modal summation for this deep source just synthesises the *S* body wavetrain.

For shallow sources the split into 'body wave' and 'surface wave' components persists as would be expected to greater ranges. In figure 11.12 we show theoretical seismograms for a focal depth of 10 km for ranges from 1500 to 4500 km. The upper frequency limit used for this calculation was 0.33 Hz so that we have used the mode set represented by figure 11.10. The lowest frequency fundamental mode Rayleigh waves have a group slowness close to 0.16 s/km and superimposed on this mode can be seen higher modes with shorter period. A distinct *Lg* packet

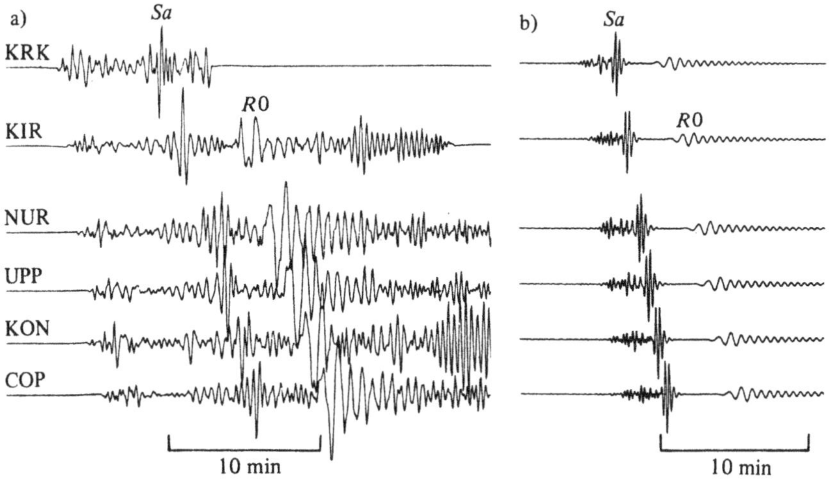

Figure 11.13. Observed and theoretical seismograms for the *Sa* phase and fundamental Rayleigh mode *R0* for the distance range 6500–9500 km: a) WWSSN-LP vertical component records for a Kuriles event recorded in Scandinavia; b) Theoretical seismograms with pass band 0.01–0.10 Hz.

of high frequency energy grades into an Airy phase for the first higher mode (*R1*) which is dwarfed by *Rg*. In the body wave field we see the emergence of the surface reflected phases *SS* at the larger ranges (3500–4500 km). These waves are enhanced in figure 11.12b where we have taken the difference of the seismograms calculated with the full mode set in figure 11.12a and a comparable set calculated with just the 15 lowest mode branches. From figure 11.3 we see that the 15 mode set gives a slowness dependent frequency window and misses nearly all high frequencies ($>$ 0.10 Hz) for slowness less than 0.2 s/km. The missing portion of the response corresponds to higher frequency *S* body waves, the long period response being tolerably well represented.

At even longer ranges, for intermediate depth events, the long-period ($<$ 0.10 Hz) character of the modal dispersion is most important and only a limited number of modes make a significant contribution. In figure 11.13 we show theoretical and observed vertical component seismograms for ranges from 6500 km to 9500 km calculated by the superposition of the first 16 mode branches. The theoretical seismograms have been calculated for a 160 km source in model *T7*, derived for western America. The observations are taken from vertical component WWSSN (LP) records in Scandinavia for an event in the Kuriles at 120 km depth. Although the relative excitation of the fundamental Rayleigh mode is different both sets of traces show the phase *Sa* arising from the interference of the first few higher modes with very similar group velocities. This phase, as noted by Brune (1965), shows varying waveform from station to station since the dominant mode branch varies with position. This variation can be exploited with an array of stations, as in figure 11.13, to use the spatial behaviour to separate mode branches. The traces are combined with appropriate delays to enhance an individual phase slowness (p), and then the group slowness behaviour as a function of frequency is determined by sweeping a narrow-band filter through the array sum. Individual mode contributions can be recognised in plots of the signal strength as a function of p and g at each frequency (Nolet, 1977) but with a limited array careful work is needed to avoid contamination from sidebands in the array response.

11.6 Separation of body wave phases

We have seen in figures 11.10–11.13 how *S* body wave phases may be generated by modal synthesis, but at high frequencies several hundred mode branches may need to be summed to give reliable mode shapes. Most of the *surface wavetrain*, with high group slowness, can be described by relatively few mode branches over the entire frequency band of interest. We would therefore like to find some means of splitting up the seismic response which conforms to the conventional seismic terminology and which enables us to relocate the modal summation results to our discussions of body waves in Chapters 9 and 10.

A convenient starting is provided by the representation (7.66) for the surface displacement w_0 which displays explicitly the effect of free-surface reflections

$$w_0(p,\omega) = \mathbf{W}_F[\mathbf{I} - \mathbf{R}_D^{0L}\mathbf{R}_F]^{-1}\boldsymbol{\sigma} \tag{11.99}$$

with

$$\boldsymbol{\sigma} = \mathbf{T}_U^{0S}[\mathbf{I} - \mathbf{R}_D^{SL}\mathbf{R}_U^{0S}]^{-1}(\mathbf{R}_D^{SL}\boldsymbol{\Sigma}_D^S + \boldsymbol{\Sigma}_U^S). \tag{11.100}$$

We may now make a partial expansion of the surface reflection operator (11.99) to give

$$w_0(p,\omega) = \mathbf{W}_F\boldsymbol{\sigma} + \mathbf{W}_F\mathbf{R}_D^{0L}\mathbf{R}_F\boldsymbol{\sigma} + \mathbf{W}_F\mathbf{R}_D^{0L}\mathbf{R}_F\mathbf{R}_D^{0L}\mathbf{R}_F[\mathbf{I} - \mathbf{R}_D^{0L}\mathbf{R}_F]^{-1}\boldsymbol{\sigma}. \tag{11.101}$$

The displacements specified by $\mathbf{W}_F\boldsymbol{\sigma}$ contain no interactions with the free-surface elements we would refer to as *S* (and *P* if appropriate). $\mathbf{W}_F\boldsymbol{\sigma}$ includes propagation directly from the source to the surface and reflection from beneath the source. The second term in (11.101) $\mathbf{W}_F\mathbf{R}_D^{0L}\mathbf{R}_F\boldsymbol{\sigma}$ allows for a single surface reflection and so the *S* wave portion of the response includes the phases *sS*, *pS* and *SS*. The remaining term will include all propagation effects which involve two or more reflections from the surface and thus the 'surface waves'.

We can now separate out the first two terms from (11.101) and recover the surface displacement in space and time using (7.75). The displacement spectrum, for *P*, *S* and their first surface reflections for angular order m, is given by

$$\bar{\mathbf{u}}_1(r,m,0,\omega) = \omega^2\int_0^\infty dp\, p\boldsymbol{\sigma}^T(\mathbf{I} + \mathbf{R}_F\mathbf{R}_D^{0L})\mathbf{W}_F^T\mathbf{T}_m(\omega pr). \tag{11.102}$$

Since the reverberation operator $[\mathbf{I} - \mathbf{R}_D^{0L}\mathbf{R}_F]^{-1}$ does not appear in (11.76), the only pole on the top sheet is at the Rayleigh slowness p_{R0} arising from the amplification factor $\mathbf{W}_F$. There are branch points at α_0^{-1}, β_0^{-1}, but the integrand is now well behaved and amenable to a reflectivity type integration (cf. Section 9.3.1), with a finite interval in p.

The displacement field corresponding to the remainder in (11.101) is best expressed in terms of the outgoing harmonics $\mathsf{T}_m^{(1)}$ and so we have

$$\bar{\mathbf{u}}_2(r, m, 0, \omega) = \tfrac{1}{2}\omega|\omega| \int_{-\infty}^{\infty} dp\, p\boldsymbol{\sigma}^T \times [\mathbf{I} - \mathbf{R}_F\mathbf{R}_D^{0L}]^{-1}\mathbf{R}_F\mathbf{R}_D^{0L}\mathbf{R}_F\mathbf{R}_D^{0L}\mathbf{W}_F^T\mathbf{T}_m^{(1)}(\omega pr). \tag{11.103}$$

Now all the poles of the original expression are still present and since we have extracted a regular term the residues have the same value as in (11.99). We have, however, significantly modified the nature of integrand in the p plane away from the poles.

At low frequencies there are only a few poles in the slowness range $\beta_L^{-1} < p < p_{R0}$ and then it is fairly easy to calculate the modal effects. We would like to preserve this simplicity at high frequencies by distorting the contour of integration to pick up only a few modes and supplement these modal contributions with a numerical contour integral. Thus

$$\bar{\mathbf{u}}_2(r, m, 0, \omega) = \tfrac{1}{2}\omega|\omega| \int_{E} dp\, p\boldsymbol{\sigma}^T[(\mathbf{I} - \mathbf{R}_F\mathbf{R}_D^{0L})^{-1} - 1]\mathbf{R}_F\mathbf{R}_D^{0L}\mathbf{W}_F^T\mathbf{T}_m^{(1)}(\omega pr) + \pi i\omega^2 \sum_{j=0}^{N(\omega)} p_j \operatorname*{Res}_{\omega, p=p_j} \{\mathbf{w}_0^T\mathbf{T}_m^{(1)}\}, \tag{11.104}$$

where we have absorbed an $\mathbf{R}_F\mathbf{R}_D^{0L}$ term into the expression in square brackets. Since we would like the contour E to be independent of frequency (above some threshold) the number of poles $N(\omega)$ will be a function of frequency. We are then left with the choice of contour E and we can be guided in this choice by making use of asymptotic results for large ω.

We consider *P-SV* propagation for angular order 0, and look at the vertical component of displacement. The contour of integration needs to lie close to the real axis for $p < \beta_L^{-1}$ so that we get a good representation of multiple *P* reflections. In order to get a limited number of poles we require the new contour E to cross the p axis in the slowness interval $\alpha_0^{-1} < p < \beta_0^{-1}$. For this region at high frequencies we can neglect the PP and PS components of $\mathbf{R}_D^{0L}$ and so, for a surface source, we find

$$\bar{u}_{2z}(r, 0, \omega) = \tfrac{1}{2}\omega|\omega| \int_{E} dp\, p\, (W_F)^{US}(R_D^{SS}\psi_D^0 + \psi_U^0) \times R_F^{SS}R_D^{SS}[(1 - R_F^{SS}R_D^{SS})^{-1} - 1]H_0^{(1)}(\omega pr). \tag{11.105}$$

To the same approximation the pole locations are given by $R_F^{SS}R_D^{SS} = 1$. The integrand in (11.105) has a saddle point when

$$r + \omega^{-1}\partial_p\{2\arg R_D^{SS} + \arg R_F^{SS}\} = 0, \qquad (11.106)$$

and if the contour E is taken over this saddle the main contribution will come from the neighbourhood of this slowness p_{SS}. In a smoothly varying medium, $\arg R_D^{SS} = \omega\tau_\beta(p)$ and so (11.106) becomes

$$r = 2X_\beta(p_{SS}) - \omega^{-1}\partial_p(\arg R_F^{SS}) \qquad (11.107)$$

where $2X_\beta(p_{SS})$ is just the geometrical range for an *SS* phase. When a saddle point lies midway between two poles.

$$[(1 - R_F^{SS}R_D^{SS})^{-1} - 1] \approx -\tfrac{1}{2} \qquad (11.108)$$

(cf. Felsen & Isihara 1979), and so the saddle point contribution to (11.105) will be just one-half of the size of the saddle point approximation for *SS* from (11.102) but reversed in sign.

In a more realistic Earth model the situation will be more complicated, but once again the cumulative effect of the higher multiple *S* reflections will resemble part of the last extracted multiple with surface wave terms.

A suitable numerical contour for E at large p values is an approximation to the steepest descents path across the saddle at p_{SS}. This defines a slowness contour which varies with distance and as r increases p_{SS} diminishes uncovering more poles. In order to link the contour E at small and large p values we have to take E off the top Riemann sheet and may uncover 'leaky-mode' poles whose residue contributions have to be added to the line integral along E. The leaky mode contributions help to refine the multiple reflection representation.

If it is desired to keep the number of 'surface wave' pole small it is necessary to change the expansion for $\boldsymbol{w}_0$ (11.74) to now separate second order surface reflections beyond some range r_0 (cf. fig 8.11). For $r > r_0$ we would then switch to a path E for the remainder term through the *SSS* saddle point defined by

$$r + \omega^{-1}\partial_p\{3\arg R_D^{SS} + 2\arg R_F^{SS}\} = 0 \qquad (11.109)$$

As the range increases further changes in the expansion have to be made to allow the separate representation of the various multiple *S* phases.

Appendix: Modal residue contributions

In Chapter 7 we have shown how the displacement response of a stratified medium can be represented in terms of displacement matrices $\boldsymbol{W}_{1S}$, $\boldsymbol{W}_{2S}$ which satisfy the free-surface and the lower radiation boundary conditions respectively. The pole singularities of the response are controlled by the inverse invariant

$$<\boldsymbol{W}_{1S}, \boldsymbol{W}_{2S}>^{-1} \qquad (11a.1)$$

and we now seek to find the residues at these poles. We will consider initially performing a slowness integral at fixed frequency ω, as in (7.77), and then the modal contribution is a sum of residue terms of the form cf. (7.82)

$$2\pi i p_j \operatorname*{Res}_{\omega, p=p_j} [\boldsymbol{w}_0^T(p, m, \omega)] \mathbf{T}_m^{(1)}(\omega p_j r), \tag{11a.2}$$

where p_j is the jth pole location for frequency ω.

For *SH* waves we may use standard results for the residue of a ratio of two analytic functions, but for coupled *P* and *SV* waves the situation is more complex. We may illustrate the process by considering the field

$$\boldsymbol{W}_0(p) = \boldsymbol{W}_{1S}(p, 0) \langle \boldsymbol{W}_{1S}, \boldsymbol{W}_{2S} \rangle^{-1} \boldsymbol{W}_{2S}^T(p, z_S), \tag{11a.3}$$

with simple poles when $\det\langle \boldsymbol{W}_{1S}, \boldsymbol{W}_{2S} \rangle$ vanishes. The displacement matrix $\boldsymbol{W}_{1S}$ is constructed from two linearly independent column vectors $\boldsymbol{w}_1, \boldsymbol{w}_2$ which satisfy the free-surface condition. $\boldsymbol{W}_{2S}$ has columns $\boldsymbol{w}_3, \boldsymbol{w}_4$ satisfying the radiation condition. Thus

$$\begin{aligned} \mathsf{H}^{-1}(p, 0) &= \langle \boldsymbol{W}_{1S}, \boldsymbol{W}_{2S} \rangle^{-1} \\ &= \frac{1}{\langle 42 \rangle \langle 31 \rangle - \langle 32 \rangle \langle 41 \rangle} \begin{bmatrix} \langle 42 \rangle & -\langle 32 \rangle \\ -\langle 41 \rangle & \langle 31 \rangle \end{bmatrix}, \end{aligned} \tag{11a.4}$$

where

$$\langle 31 \rangle = \langle \boldsymbol{w}_3, \boldsymbol{w}_1 \rangle, \tag{11a.5}$$

the composition of vectors introduced in (2.36).

The residue of $\boldsymbol{W}_0$ at a pole p_j is given by

$$\operatorname*{Res}_{\omega, p=p_j} [\boldsymbol{W}_0] = \boldsymbol{W}_{1S}(p_j, 0) \frac{1}{\partial_p (\det \mathbf{H})|_{p=p_j}} \mathsf{H}^A(p_j, 0) \boldsymbol{W}_{2S}^T(p_j, z_S), \tag{11a.6}$$

where H^A is the matrix appearing in (11a.5). We need therefore to construct a suitable expression for $\partial_p \det \mathbf{H}$. Now

$$\partial_z(\det \mathbf{H}) = \langle 42 \rangle g_{31} + \langle 31 \rangle g_{42} - \langle 32 \rangle g_{41} - \langle 41 \rangle g_{32}, \tag{11a.7}$$

where $\partial_z \langle 31 \rangle = g_{31}$. When we integrate over the entire interval in z we obtain

$$\det \mathsf{H} = \int_0^\infty d\zeta \{\langle 42 \rangle g_{31} + \langle 31 \rangle g_{42} - \langle 32 \rangle g_{41} - \langle 41 \rangle g_{32}\}, \tag{11a.8}$$

At the exact surface wave pole we have a displacement eigenvector $\boldsymbol{w}_{ej}(z)$ which satisfies both sets of boundary conditions i.e. vanishing tractions at $z = 0$ and the radiation condition into the lower half space. At the pole we can therefore choose $\boldsymbol{w}_1 = \boldsymbol{w}_{ej}$ and also $\boldsymbol{w}_3 = \gamma \boldsymbol{w}_{ej}$ for some constant γ. The vectors $\boldsymbol{w}_2$ and $\boldsymbol{w}_4$ will be distinct and will correspond to displacement solutions which satisfy only one of the boundary conditions. Since $\boldsymbol{w}_1$ and $\boldsymbol{w}_2$ satisfy the same boundary conditions

$$\langle 12 \rangle = \langle \boldsymbol{w}_1, \boldsymbol{w}_2 \rangle = 0, \tag{11a.9}$$

from (2.71). Similarly since $\boldsymbol{w}_3$ and $\boldsymbol{w}_4$ satisfy the radiation condition $\langle 34 \rangle = 0$. Thus *at the pole*

$$\langle 32 \rangle = \gamma \langle 12 \rangle = 0, \quad \langle 41 \rangle = \gamma^{-1} \langle 43 \rangle = 0, \tag{11a.10}$$

and since $\boldsymbol{w}_1$ and $\boldsymbol{w}_3$ are multiples $\langle 31 \rangle = 0$. There is therefore only one non-zero entry in H^A from $\langle 42 \rangle$.

In order to get an expression for $\partial_p \det \mathbf{H}$ at the pole p_j with fixed frequency ω, we construct $<\mathbf{W}_{1S}, \mathbf{W}_{2S}>$ with $\mathbf{W}_{1S}, \mathbf{W}_{2S}$ evaluated for slightly different slownesses. We take $\boldsymbol{w}_1, \boldsymbol{w}_2$ at the pole p_j and $\boldsymbol{w}_3, \boldsymbol{w}_4$ with slowness $p_j + \Delta p$ and then consider the limit $\Delta p \to 0$ for which $\boldsymbol{w}_3 \to \gamma \boldsymbol{w}_1$. From (2.48)

$$g_{31} = \Delta p\{(2p_j+\Delta p)\omega\rho\nu V_1 V_3 + U_1 S_3 + U_3 S_1 - (1-2\beta^2/\alpha^2)(V_1 P_3 + V_3 P_1)\}, \quad (11a.11)$$

where

$$\nu = 4\beta(1-\beta^2/\alpha^2). \quad (11a.12)$$

Now at the *pole* itself $\det \mathbf{H}$ vanishes and so

$$\frac{\partial}{\partial p} \det \mathbf{H}|_{p=p_j} = \lim_{\Delta p \to 0} \left\{ \frac{1}{\Delta p} \det \mathbf{H} \right\} \quad (11a.13)$$

$$= 2\gamma <42> \int_0^\infty d\zeta \{\omega p_j \rho \nu V_{ej}^2 + U_{ej} S_{ej} - (1-2\beta^2/\alpha^2) V_{ej} P_{ej}\}, \quad (11a.14)$$

where we have made use of the depth invariance of $<42>$, when both vectors are evaluated at the same frequency and slowness, and the vanishing of the other invariants. We may now recognise the integral in (11a.14) as $\omega p_j J_j$ which we have introduced in our variational treatment (11.74).

The residue contribution from (11a.6) is therefore

$$\operatorname*{Res}_{\omega, p=p_j} [\mathbf{W}_0] = \frac{1}{2\omega p_j J_j} \mathbf{W}_{1S}(p_j, 0) \begin{bmatrix} 1 & 0 \\ 0 & 0 \end{bmatrix} \mathbf{W}_{2S}^T(p_j, z_S), \quad (11a.15)$$

and so involves only the eigenvector entries. Also from (11.72) the group slowness for the mode

$$g_j = I_j / p_j J_j, \quad (11a.16)$$

where

$$I_j = \int_0^\infty d\zeta\, \rho(\zeta) \boldsymbol{w}_{ej}^T(\zeta) \boldsymbol{w}_{ej}(\zeta), \quad (11a.17)$$

and so we can achieve a convenient and compact representation of the residue. Thus

$$\operatorname*{Res}_{\omega, p=p_j} [\mathbf{W}_0] = \frac{g_j}{2\omega I_j} \boldsymbol{w}_{ej}(0) \boldsymbol{w}_{ej}^T(z_S), \quad (11a.18)$$

and this expression is valid for both the *P-SV* and *SH* wave cases, and also for full anisotropy.

A comparable development can be made at a fixed slowness p to find the residue at a pole ω_k in frequency. In this case we have to construct the frequency derivative of $\det \mathbf{H}$ and we find from (2.48) that

$$\frac{\partial}{\partial \omega} (\det \mathbf{H})|_{\omega=\omega_k} = 2\gamma <42> \{p^2 J_k - I_k\}, \quad (11a.19)$$

and so the residue is given by

$$\operatorname*{Res}_{p, \omega=\omega_k} [\mathbf{W}_0] = \frac{g_k}{2(p-g_k) I_k} \boldsymbol{w}_{ej}(0) \boldsymbol{w}_{ej}^T(z_S). \quad (11a.20)$$

Appendix: Table of Notation

Stress and Strain

$\mathbf{x}$	-	position vector
x_i	-	position coordinates
ξ	-	initial position vector
ξ_i	-	initial position coordinates
$\mathbf{u}$	-	displacement vector
$\mathbf{v}$	-	velocity vector
$\mathbf{f}$	-	acceleration vector
$\mathbf{g}$	-	external force vector
$\mathbf{n}$	-	normal vector
τ_{ij}	-	incremental stress tensor
σ_{ij}	-	stress tensor (including pre-stress)
e_{ij}	-	strain tensor
$\mathbf{t}$	-	traction vector
τ	-	traction vector at a surface
$\mathbf{n}$	-	normal vector
c_{ijkl}	-	elastic modulus tensor
C_{ijkl}	-	anelastic relaxation tensor
κ	-	bulk modulus
μ	-	shear modulus
ρ	-	density
λ	-	Lamé modulus
R_λ, R_μ	-	relaxation functions
A, C, F, L, N, H	-	moduli for transversely isotropic media

Waves and Rays

z	-	depth
r	-	horizontal distance
R	-	radius
t	-	time
ω	-	angular frequency
θ, ϕ	-	coordinate angles
α	-	*P* wavespeed
β	-	*S* wavespeed
p	-	horizontal slowness, phase slowness
$\wp$	-	angular slowness
g	-	group slowness
Δ	-	epicentral angle (for propagation in a sphere)
p	-	ray parameter (horizontal stratification) ($= dT/dX$),
q	-	auxiliary ray parameter
τ	-	intercept time, delay time
q_α	-	vertical slowness for *P* waves
q_β	-	vertical slowness for *S* waves
i	-	angle of incidence for *P* waves
j	-	angle of incidence for *S* waves
Q^{-1}	-	loss factor
ϑ	-	wavefront/phase function

Sources

R	-	distance from source
$\boldsymbol{\gamma}$	-	unit direction vector from source
γ_i	-	direction cosines
$\mathbf{h}$	-	source position
$\mathbf{R}$	-	receiver location
$\mathbf{S}$	-	source location
$\mathbf{n}$	-	normal vector
$\boldsymbol{\nu}$	-	slip vector
ϵ	-	force vector
$\mathcal{E}$	-	summary force vector
G_{ij}	-	Green's tensor
H_{ijp}	-	stress tensor derived from Green's tensor
M_{ij}	-	Moment tensor
$m_{ij}(\mathbf{x}, t)$	-	moment tensor density
$\mathbf{Q}, \mathbf{T}$	-	radiation vectors for *P*, *S*

Propagation terms

Q_α, Q_β	-	quality factors for *P* and *S* waves
$\mathbf{A}$	-	coefficient matrix
$\mathbf{b}$	-	stress-displacement vector
$\mathbf{B}$	-	fundamental matrix
$\mathbf{C}$	-	transformation matrix
$\mathbf{D}$	-	eigenmatrix of stress-displacement vectors
$\mathfrak{m}$	-	displacement partitions of eigenmatrix
$\mathfrak{n}$	-	traction partitions of eigenmatrix
D	-	stress-displacement matrix for gradients
$\mathsf{E}, \hat{\mathsf{E}}$	-	phase matrices for wavetype
$\mathbf{E}$	-	phase matrix
E	-	phase matrix for gradients
$\mathbf{F}$	-	source vector
$\mathbf{H}$	-	propagation matrix
H	-	partitions of propagation matrix
$\mathbf{L}, \mathsf{L}$	-	correction terms for gradient zones
$\mathcal{W}$	-	displacement matrix
T	-	traction matrix
$\mathbf{P}$	-	propagator matrix
P	-	partitions of propagator matrix
$\mathbf{R}_k^m, \mathbf{S}_k^m, \mathbf{T}_k^m$	-	vector surface harmonics
T_m	-	tensor surface harmonic
$\mathbf{Q}$	-	interface matrix
t	-	traction vector in $p - \omega$ domain
$\boldsymbol{\nu}$	-	wave vector
$\mathbf{w}$	-	displacement vector in $p - \omega$ domain
U, V, W	-	displacement components in $p - \omega$ domain
P, S, T	-	traction components in $p - \omega$ domain
$\boldsymbol{\nu}_{U,D}$	-	wave vector components
$\mathfrak{m}_{U,D}$	-	displacement matrices
$\mathfrak{n}_{U,D}$	-	traction matrices
$\mathbf{E}_{U,D}$	-	phase delay matrices
$\mathbf{S}$	-	source jump vector
$\mathbf{S}_W, \mathbf{S}_T$	-	components of source jump
Σ_U, Σ_D	-	upgoing, downgoing waves from source

Propagation terms (cont.)

$\epsilon_\alpha, \epsilon_\beta, \epsilon_H$	-	normalisation factors for wave energy
$\mathbf{R}_F$	-	free-surface reflection matrix
$R_F^{HH}, R_F^{PP}, \ldots$	-	free surface reflection coefficients
$\mathbf{W}_F$	-	free-surface amplification matrix
$\mathbf{R}_U, \mathbf{R}_D$	-	reflection matrices for incident upgoing, downgoing waves
$\mathbf{T}_U, \mathbf{T}_D$	-	transmission matrices for incident upgoing, downgoing waves
$\mathbf{r}_u, \mathbf{r}_d$	-	generalised reflection matrices for incident upgoing, downgoing wa
$\mathbf{t}_u, \mathbf{t}_d$	-	generalised transmission matrices for incident upgoing, downgoing
$P_{U,D}$	-	weighting coefficients for upgoing, downgoing *P* waves
$S_{U,D}$	-	weighting coefficients for upgoing, downgoing *SV* waves
$H_{U,D}$	-	weighting coefficients for upgoing, downgoing *SH* waves
φ	-	phase of *PP* reflection coefficient
ψ	-	phase of *SS* reflection coefficient
χ	-	phase of *HH* reflection coefficient
$\eta, \hat{\eta}$	-	generalised vertical slownesses
$\gamma_A, \gamma_P, \gamma_S, \ldots$	-	coupling coefficients
$\mathcal{Q}$	-	attenuation factor
$\mathcal{M}(t)$	-	effective source time function
$\mathbf{g}$	-	amplitude weight for generalised ray
H	-	Cagniard path in complex p plane

Mathematical

a, b, c	-	constants
$\mathcal{H}$	-	Hilbert transform operator
$\mathcal{L}, \mathcal{M}, \mathcal{N}$	-	differential operators
Ai,Bi	-	Airy functions
Aj,Ak,Bj,Bk	-	Airy function wave variables
Ej,Ek,Ej,Ek	-	Airy function wave variables
$H_m^{(1),(2)}(x)$	-	Hankel functions
$J_m(x)$	-	Bessel function
$\mathbf{R}_k^m, \mathbf{S}_k^m, \mathbf{T}_k^m$	-	vector surface harmonics (cylindrical)
$\mathbf{T}_m, \mathbf{T}_m^{(1)}$	-	tensor field of vector harmonics
$j_l(x), h_l^{(1),(2)}(x)$	-	spherical Bessel functions
$P_l(x), P_l^m(x)$	-	Legendre functions
$Q_l^{(1),(2)}$	-	travelling wave form of Legendre function
$\mathbf{P}_l^m, \mathbf{B}_l^m, \mathbf{C}_l^m$	-	vector surface harmonics (cylindrical)
$Y_l^m, \mathcal{Y}_{lm}$	-	surface harmonics on a sphere
Λ	-	analytic time function

References

Abo-Zena, A.M., 1979. Dispersion function computations for unlimited frequency values, *Geophys. J. R. Astr. Soc.*, **58**, 91–105.

Abramovici, F., 1968. Diagnostic diagrams and transfer functions for oceanic wave guides, *Bull. Seism. Soc. Am.*, **58**, 427–456.

Ahluwahlia, D.S. & Keller, J.B., 1977. Exact and asymptotic representations of the sound field in a stratified ocean, *Wave Propagation and Underwater Acoustics*, ed. J.B. Keller & J.S. Papadakis, Springer-Verlag, Berlin.

Aki, K., 1960. Study of earthquake mechanism by a method of phase equilisation applied to Rayleigh and Love waves, *J. Geophys. Res.*, **65**, 729–740.

Aki, K. & Chouet, L.B., 1975. Origin of coda waves: source, attenuation and scattering effects, *J. Geophys. Res.*, **80**, 3322–3342.

Aki, K. & Richards, P.G., 1980. *Quantitative Seismology*, (2 vols), W. H. Freeman & Co, San Francisco.

Aki, K., 1981. Attenuation and scattering of short period seismic waves in the lithosphere, *Identification of Seismic Sources*, ed. E. S. Husebye, Noordhof, Leiden.

Alterman, Z., Jarosch, H. & Pekeris, C. L., 1959. Oscillations of the Earth, *Proc. R. Soc. Lond.*, **252A**, 80–95.

Altman, C. & Cory, H., 1969. The generalised thin film optical method in electromagnetic wave propagation, *Radio Sci.*, **4**, 459–469.

Anderson, D.L., BenMenahem, A. & Archambeau, C.B., 1965. Attenuation of seismic energy in the upper mantle *J. Geophys. Res.*, **70**, 1441–1448.

Anderson, D.L. & Minster, J.B., 1979. The frequency dependence of Q in the Earth and implications for mantle rheology and Chandler wobble, *Geophys. J. R. Astr. Soc.*, **58**, 431–440.

Archambeau, C.B., 1968. General theory of elastodynamic source fields, *Revs. Geophys.*, **16**, 241–288.

Archuleta, R.J. & Spudich, P., 1981. An explanation for the large amplitude vertical accelerations generated by the 1979 Imperial Valley, California earthquake, *EOS*, **62**, 323.

Azimi, Sh., Kalinin, A.V., Kalinin, V.V. & Pivovarov, B.L., 1968. Impulse and transient characteristics of media with linear and quadratic absorption laws, *Izv. Physics of Solid Earth*, **2**, 88–93.

Backus, G.E. & Mulcahy, M., 1976a. Moment tensors and other phenomenological descriptions of seismic sources I - Continuous displacements, *Geophys. J. R. Astr. Soc.*, **46**, 341-3-62.

Backus, G.E. & Mulcahy, M., 1976b. Moment tensors and other phenomenological descriptions of seismic sources II - Discontinuous displacements, *Geophys. J. R. Astr. Soc.* , **47**, 301–330.

Bamford, D., Faber, S., Jacob, B., Kaminski, W., Nunn, K., Prodehl, C., Fuchs, K., King, R. & Willmore, P., 1976. A lithospheric profile in Britain - I. Preliminary results, *Geophys. J. R. Astr. Soc.*, **44**, 145–160.

Bedding, R.J. & Willis, J. R., 1980. The elastodynamic Green's tensor for a half-space with an embedded anisotropic layer, *Wave Motion*, **2**, 51–61.

BenMenahem, A. & Vered, M., 1973. Extension and interpretation of the Cagniard-Pekeris method for dislocation sources, *Bull. Seism. Soc. Am.*, **63**, 1611–1636.

Biswas, N.N. & Knopoff, L., 1970. Exact earth-flattening calculation for Love waves, *Bull. Seism. Soc. Am.*, **60**, 1123–1137.

Boltzman, L., 1876. *Pogg. Ann. Erganzungbd*, **7**, 624-.

Booth, D.C. & Crampin, S., 1981. Calculating synthetic seismograms in anisotropic media by the reflectivity technique: theory, *Inst. Geol. Sci.*, GSU Report No. 133.

Braile, L.W. & Smith, R.B., 1975. Guide to the interpretation of crustal refraction profiles, *Geophys. J. R. Astr. Soc.*, **40**, 145–176.

Brekhovskikh, L.M., 1960. *Waves in Layered Media*, Academic Press, New York.

Brune, J.N., 1965. The Sa phase from the Hindu Kush earthquake of July 6, 1972 *Pure and Applied Geophysics*, **62**, 81–95.

Budden, K. G., 1961 *Radio Waves in the Ionosphere*, Cambridge University Press.

Bunch, A.W H. & Kennett, B.L.N., 1980. The development of the oceanic crust on the Reykjanes Ridge at 59° 30′ N, *Geophys. J. R. Astr. Soc.*, **61**, 141–166.

Burdick, L.J. & Helmberger, D.V., 1978. The upper mantle P velocity structure of the western United States, *J. Geophys. Res.*, **83**, 1699–1712.

Burdick, L.J. & Orcutt, J.A., 1979. A comparison of the generalized ray and reflectivity methods of waveform synthesis, *Geophys. J. R. Astr. Soc.*, **58**, 261–278.

Burridge, R., Lapwood, E.R. & Knopoff, L., 1964. First motions from seismic sources near a free surface. *Bull. Seism. Soc. Am.*, **54**, 1889–1913.

Burridge, R. & Knopoff, L., 1964. Body force equivalents for seismic dislocations, *Bull. Seism. Soc. Am.*, **54**, 1875–1888.

Burridge, R., 1968. A new look at Lamb's problem, *J. Phys. Earth*, **16**, 169–172.

Cagniard, L., 1939. *Réflexion et Réfraction des Ondes Séismiques Progressives*, Gauthier-Villars, Paris.

Calcagnile, G., Panza, G. F., Schwab, F. & Kausel, E. G., 1976. On the computation of theoretical seismograms for multimode surface waves, *Geophys. J. R. Astr. Soc.*, **47**, 73–82.

Červený, V., 1974. Reflection and transmission coefficients for transition layers, *Studia Geoph. et Geod.*, **18**, 59–68.

Červený, V. & Ravindra, R., 1975. *The Theory of Seismic Head Waves*, University of Toronto Press.

Červený, V., Molotkov, I. A. & Pšenčík, I., 1978. *Ray Method in Seismology*, Charles University Press, Prague.

Chapman, C. H. & Phinney, R. A., 1972. Diffracted seismic signals *Methods in Computational Physics*, **12** ed. B. A. Bolt, Academic Press, New York.

Chapman, C. H., 1973. The Earth flattening approximation in body wave theory *Geophys. J. R. Astr. Soc.*, **35**, 55–70.

Chapman, C. H., 1974a. Generalized ray theory for an inhomogeneous medium *Geophys. J. R. Astr. Soc.*, **36**, 673–704.

Chapman, C. H., 1974b. The turning point of elastodynamic waves *Geophys. J. R. Astr. Soc.*, **39**, 613–621.

Chapman, C. H., 1976. Exact and approximate generalized ray theory in vertically inhomogeneous media, *Geophys. J. R. Astr. Soc.*, **46**, 201–234.

Chapman, C. H., 1978. A new method for computing synthetic seismograms, *Geophys. J. R. Astr. Soc.*, **54**, 431–518.

Chapman, C. H., 1981. Long period correction to body waves: Theory *Geophys. J. R. Astr. Soc.*, **64**, 321–372.

Chapman, C. H. & Woodhouse, J. H., 1981. Symmetry of the wave equation and excitation of body waves *Geophys. J. R. Astr. Soc.*, **65**, 777–782.

Choy, G. L., 1977. Theoretical seismograms of core phases calculated by frequency-dependent full wave theory, and their interpretation, *Geophys. J. R. Astr. Soc.*, **51**, 275–312.

Choy, G. L., Cormier, V. F., Kind, R., Müller, G. & Richards, P. G., 1980. A comparison of synthetic seismograms of core phases generated by the full wave theory and the reflectivity method, *Geophys. J. R. Astr. Soc.*, **61**, 21–40.

Cisternas, A., Betancourt, O. & Leiva, A., 1973. Body waves in a 'real Earth', Part I *Bull. Seism. Soc. Am.*, **63**, 145–156.

Cooley, J. W. & Tukey, S. W., 1965. An algorithm for the machine calculation of complex Fourier series, *Mathematics of Computation*, **19**, 297–301.

Cormier, V. F. & Richards, P. G., 1977. Full wave theory applied to a discontinuous velocity increase: the inner core boundary, *J. Geophys.*, **43**, 3–31.

Cormier, V. F., 1980. The synthesis of complete seismograms in an earth model composed of radially inhomogeneous layers, *Bull. Seism. Soc. Am.*, **70**, 691–716.

Cormier, V. F. & Choy, G. L., 1981. Theoretical body wave interactions with upper mantle structure *J. Geophys. Res.*, **86**, 1673–1678

Crampin, S., Evans, R. , Uçer, B., Doyle, M., Davis, J. P., Yegorkina, G. V. & Miller, A., 1980. Observations of dilatancy induced polarisation anomalies and earthquake prediction, *Nature*, **286**, 874–877.

Dahlen, F. A., 1972. Elastic dislocation theory for a self gravitating configuration with an initial stress field, *Geophys. J. R. Astr. Soc.*, **28**, 357–383.

Dey-Sarkar, S. K. & Chapman, C. H., 1978. A simple method for the computation of body-wave seismograms, *Bull. Seism. Soc. Am.*, **68**, 1577–1593.

Doornbos, D.J., 1981. The effect of a second-order velocity discontinuity on elastic waves near their turning point, *Geophys. J. R. Astr. Soc.*, **64**, 499–511.

Douglas, A., Hudson, J. A. & Blamey, C., 1973. A quantitative evaluation of seismic signals at teleseismic distances III - Computed P and Rayleigh wave seismograms *Geophys. J. R. Astr. Soc.*, **28**, 345–410.

Dunkin, J. W., 1965. Computations of modal solutions in layered elastic media at high frequencies *Bull. Seism. Soc. Am.*, **55**, 335–358.

Dziewonski, A. & Anderson, D. L., 1981. Preliminary Reference Earth Model, *Phys. Earth Plan. Int.*, **25**, 297–356.

Ewing, W. M., Jardetsky, W. S. & Press, F., 1957. *Elastic Waves in Layered Media*, McGraw-Hill, New York.

Faber, S. & Müller, G., 1980. Sp phases from the transition zone between the upper and lower mantle, *Bull. Seism. Soc. Am.*, **70**, 487–508.

Felsen, L. B. & Isihara, T., 1979. Hybrid ray-mode formulation of ducted propagation, *J. Acoust. Soc. Am.*, **65**, 595–607.

Filon, L.N.G., 1928. On a quadrature formula for trigonometric integrals, *Proc. R. Soc. Edin.*, **49**, 38–47.

Fletcher, J.B., Brady, A.G. & Hanks, T.C., 1980. Strong motion accelerograms of the

Oroville, California aftershocks: Data processing and the aftershock of 0350 August 6 1975, *Bull. Seism. Soc. Am.*, **70**, 243–267.

Fowler, C.M.R., 1976. Crustal structure of the Mid-Atlantic ridge crest at 37°N. *Geophys. J. R. Astr. Soc.*, **47**, 459–492.

Frantsuzova, V. I., Levshin, A. L. & Shkadinskaya, G. V., 1972. Higher modes of Rayleigh waves and upper mantle structure *Computational Seismology*, ed. V. I. Keilis-Borok, Consultants Bureau, New York.

Frazer, R. A., Duncan, W. J. & Collar, A. R., 1938. *Elementary Matrices*, Cambridge University Press.

Fryer, G.J., 1981. Compressional-shear wave coupling induced by velocity gradients in marine sediments, *J. Acoust. Soc. Am.*, **69**, 647–660.

Fryer, G.J., 1980. A slowness approach to the reflectivity method of seismogram synthesis, *Geophys. J. R. Astr. Soc.*, **63**, 747–758.

Fuchs, K., 1968. Das Reflexions- und Transmissionsvermogen eines geschichteten Mediums mit belieber Tiefen-Verteilung der elastischen Moduln und der Dichte fur schragen Einfall ebener Wellen, *Z. Geophys.*, **34**, 389-411.

Fuchs, K. & Müller, G., 1971. Computation of synthetic seismograms with the reflectivity method and comparison with observations, *Geophys. J. R. Astr. Soc.*, **23**, 417–433.

Fuchs, K., 1975. Synthetic seismograms of PS-reflections from transition zones computed with the reflectivity method, *J. Geophys.*, **41**, 445–462.

Gans, R., 1915. Fortpflanzung des Lichtes durch ein inhomogenes Medium, *Ann. Physik*, **47**, 709–732.

Gilbert, F., 1964. Propagation of transient leaking modes in a stratified elastic waveguide, *Revs. Geophys.*, **2**, 123–153.

Gilbert, F., 1966. The representation of seismic displacements in terms of travelling waves, *Geophys. J. R. Astr. Soc.*, **44**, 275–280.

Gilbert, F. & Backus, G. E., 1966. Propagator matrices in elastic wave and vibration problems, *Geophysics*, **31**, 326-332.

Gilbert, F., 1971. The excitation of the normal modes of the Earth by earthquake sources *Geophys. J. R. Astr. Soc.*, **22**, 223–226.

Gilbert, F. & Helmberger, D. V., 1972. Generalized ray theory for a layered sphere, *Geophys. J. R. Astr. Soc.*, **27**, 57–80.

Green, G., 1838. On the laws of reflexion and refraction of light at the common surface of two non-crystallized media, *Trans. Camb. Phil. Soc.*, **7**, 245–285.

Haddon, R.A.W. & Cleary, J. R., 1974. Evidence for scattering of PKP waves near the core-mantle boundary, *Phys. Earth Plan. Int.*, **8**, 211–234.

Harjes, H-P., 1981. Broad band seismometry - A unified approach towards a kinematic and dynamic interpretation of seismograms, *Identification of Seismic Sources*, ed. E. S. Husebye, Noordhof, Leiden.

Harkrider, D. G., 1964. Surface waves in multi-layered elastic media I: Rayleigh and Love waves from buried sources in a multilayered half space, *Bull. Seism. Soc. Am.*, **54**, 627–679.

Harvey, D. J., 1981. Seismogram synthesis using normal mode superposition: The locked mode approximation, *Geophys. J. R. Astr. Soc.*, **66**, 37–70.

Haskell, N. A., 1953. The dispersion of surface waves on multilayered media, *Bull. Seism. Soc. Am.*, **43**, 17–34.

Haskell, N. A., 1964. Radiation pattern of surface waves from point sources in a multi-layered medium, *Bull. Seism. Soc. Am.*, **54**, 377–393.

Healy, J.H., Mooney, W.D., Blank, H.R., Gettings, M.R., Kohler, W.M., Lamson, R.J. &

Leone, L.E., 1981. Saudi Arabian seismic deep-refraction profile: final report, *U.S. Geological Survey, Tech Report*.

Helmberger, D. V., 1968. The crust-mantle transition in the Bering Sea, *Bull. Seism. Soc. Am.*, **58**, 179–214.

Helmberger, D. V. & Wiggins, R. A., 1971. Upper mantle structure of the midwestern United States, *J. Geophys. Res.*, **76**, 3229–3245.

Helmberger, D. V., 1973. On the structure of the low velocity zone, *Geophys. J. R. Astr. Soc.*, **34**, 251–263.

Helmberger, D. V. & Malone, S. D., 1975. Modelling local earthquakes as shear dislocations in a layered half space, *J. Geophys. Res.*, **80**, 4881-4896.

Helmberger, D. V., 1977. Fine structure of an Aleutian crustal section, *Geophys. J. R. Astr. Soc.*, **48**, 81–90.

Helmberger, D. V. & Harkrider, D. G., 1978. Modelling earthquakes with generalised ray theory, *Modern Problems in Elastic Wave Propagation* ed. J. Miklowitz & J. D. Achenbach, 499–518, Wiley, New York.

Helmberger, D. V. & Engen, G., 1980. Modelling the long-period body waves from shallow earthquakes at regional ranges, *Bull. Seism. Soc. Am.*, **70**, 1699–1714.

Henry, M., Orcutt, J. A. & Parker, R. L., 1980. A new method for slant stacking refraction data *Geophys. Res. Lett.*, **7**, 1073–1076.

Hirn, A., Steimetz, L., Kind, R. & Fuchs, K., 1973. Long range profiles in Western Europe II: Fine structure of the lithosphere in France (Southern Bretagne), *Z. Geophys.*, **39**, 363–384.

Hoop, A. T. de, 1960. A modification of Cagniard's method for solving seismic pulse problems *Applied Science Research*, **38**, 349–356.

Hron, F., 1972. Numerical methods of ray generation in multilayered media, *Methods in Computational Physics*, **12**, ed. B. A. Bolt, Academic Press, New York.

Hudson, J. A., 1962. The total internal reflection of SH waves, *Geophys. J. R. Astr. Soc.*, **6**, 509–531.

Hudson, J. A., 1969a. A quantitative evaluation of seismic signals at teleseismic distances I - Radiation from a point source, *Geophys. J. R. Astr. Soc.*, **18**, 233–249.

Hudson, J. A., 1969b. A quantitative evaluation of seismic signals at teleseismic distances II - Body waves and surface waves from an extended source, *Geophys. J. R. Astr. Soc.*, **18**, 353–370.

Hudson, J. A., 1980. *The Excitation and Propagation of Elastic Waves*, Cambridge University Press.

Husebye, E. S., King, D. W. & Haddon, R.A.W., 1976. Precursors to PKIKP and seismic scattering near the core-mantle boundary, *J. Geophys. Res.*, **81**, 1870–1882.

Jackson, D. D. & Anderson, D. L., 1970. Physical mechanisms of seismic wave attenuation, *Rev. Geophys. Sp. Phys.*, **8**, 1–63.

Jeans, J. H., 1923. The propagation of earthquake waves, *Proc. R. Soc. Lond.*, **102A**, 554–574.

Jeffreys, H., 1928. The effect on Love waves of heterogeneity in the lower layer, *Mon. Not. R. Astr. Soc., Geophys. Suppl.*, **2**, 101–111.

Jeffreys, H. & Bullen, K. E., 1940. Seismological Tables,Brit. Ass., Gray-Milne Trust.

Jeffreys, H., 1957. Elastic waves in a continuously stratified medium. *Mon. Not. R. Astr. Soc., Geophys. Suppl.*, **7**, 332–337.

Jeffreys, H., 1958. A modification of Lomnitz's law of creep in rocks, *Geophys. J. R. Astr. Soc.*, **1**, 92–95.

Jeffreys, H., 1961. Small corrections in the theory of surface waves, *Geophys. J. R. Astr. Soc.*, **6**, 115–117.

Jobert, G., 1975. Propagator and Green matrices for body force and dislocation, *Geophys. J. R. Astr. Soc.*, **43**, 755–760.

Johnson, L. R. & Silva, W., 1981. The effect of unconsolidated sediments upon the ground motion during local earthquakes, *Bull. Seism. Soc. Am.*, **71**, 127–142.

Keilis-Borok, V. I., Neigauz, M. G. & Shkadinskaya, G. V., 1965. Application of the theory of eigenfunctions to the calculation of surface wave velocities, *Revs. Geophys.*, **3**, 105–109.

Kennett, B.L.N., 1974. Reflections, rays and reverberations, *Bull. Seism. Soc. Am.*, **64**, 1685–1696.

Kennett, B.L.N., 1975. The effects of attenuation on seismograms, *Bull. Seism. Soc. Am.*, **65**, 1643–1651.

Kennett, B.L.N. & Simons, R.S., 1976. An implosive precursor to the Columbia earthquake 1970 July 31, *Geophys. J. R. Astr. Soc.*, **44**, 471–482.

Kennett, B.L.N., 1977. Towards a more detailed seismic picture of the oceanic crust and mantle, *Marine Geophys. Res.*, **3**, 7–42.

Kennett, B.L.N. & Woodhouse, J. H., 1978. On high frequency spheroidal modes and the structure of the upper mantle, *Geophys. J. R. Astr. Soc.*, **55**, 333–350.

Kennett, B.L.N., Kerry, N. J. & Woodhouse, J. H., 1978. Symmetries in the reflection and transmission of elastic waves, *Geophys. J. R. Astr. Soc.* , **52**, 215–229.

Kennett, B.L.N. & Kerry, N. J., 1979. Seismic waves in a stratified half space, *Geophys. J. R. Astr. Soc.*, **57**, 557–583.

Kennett, B.L.N. & Nolet, G., 1979. The influence of upper mantle discontinuities on the toroidal free oscillations of the Earth, *Geophys. J. R. Astr. Soc.*, **56**, 283–308.

Kennett, B.L.N., 1979a. Theoretical reflection seismograms for an elastic medium, *Geophys. Prosp.*, **27**, 301–321.

Kennett, B.L.N., 1979b. The suppression of surface multiples on seismic records, *Geophys. Prosp.*, **27**, 584–600.

Kennett, B.L.N., 1980. Seismic waves in a stratified half space II - Theoretical seismograms, *Geophys. J. R. Astr. Soc.*, **61**, 1–10.

Kennett, B.L.N., 1981. Elastic Waves in Stratified Media, *Advances in Applied Mechanics*, **21**, ed. C-S Yih, Academic Press, New York.

Kennett, B.L.N. & Illingworth, M. R., 1981. Seismic waves in a stratified half space III - Piecewise smooth models, *Geophys. J. R. Astr. Soc.*, **66**, 633–675.

Kerry, N. J., 1981. The synthesis of seismic surface waves, *Geophys. J. R. Astr. Soc.*, **64**, 425–446.

Kind, R., 1976. Computation of reflection coefficients for layered media, *J. Geophys.*, **42**, 191–200.

Kind, R., 1978. The reflectivity method for a buried source, *J. Geophys.*, **45**, 450–462.

King, D. W., Haddon, R.A.W. & Husebye, E. S., 1975. Precursors to PP, *Phys. Earth Plan. Int.*, **10**, 103–127.

Knopoff, L., Schwab, F. & Kausel, E., 1973. Interpretation of Lg, *Geophys. J. R. Astr. Soc.*, **33**, 389–404.

Knott, C. G., 1899. Reflection and refraction of elastic waves, with seismological applications, *Phil. Mag. Lond.*, **48**, 64–97, 467–569.

Kosminskaya, I. P., 1971. *Deep Seismic Sounding of the Earth's Crust and Upper Mantle*, Consultants Bureau, New York.

Lamb, H., 1904. On the propagation of tremors over the surface of an elastic solid, *Phil. Trans. R. Soc. Lond.*, **203A**, 1–42.

Langer, R.E., 1937. On the connection formulas and the solution of the wave equation *Phys. Rev.*, **51**, 669–676.

Langston, C. & Helmberger, D. V., 1975. A procedure for modelling shallow dislocation sources, *Geophys. J. R. Astr. Soc.*, **42**, 117–130.

Lapwood, E. R., 1948. The disturbance due to a line source in a semi-infinite elastic medium, *Phil. Trans. Roy. Soc. Lond.*, **242A**, 63–100.

Lapwood, E. R. & Hudson, J. A., 1975. The passage of elastic waves through an anomalous region, III - Transmission of obliquely incident body waves, *Geophys. J. R. Astr. Soc.*, **40**, 255–268.

Lapwood, E. R. & Usami, T., 1981. *Free Oscillations of the Earth*, Cambridge University Press.

Levin, F. K., 1979. Seismic velocities in transversely isotropic media, *Geophysics*, **44**, 918–936.

Liu, H-P., Anderson, D. L. & Kanamori, H., 1976. Velocity dispersion due to anelasticity: implications for seismology and mantle composition, *Geophys. J. R. Astr. Soc.*, **47**, 41–58.

Love, A. E. H., 1903. The propagation of wave motion in an isotropic solid medium, *Proc. Lond. Maths. Soc.*, **1**, 291–316.

Love, A. E. H., 1911. *Some Problems of Geodynamics*, Cambridge University Press.

Lundquist, G. M. & Cormier, V. C., 1980. Constraints on the absorption band model of Q, *J. Geophys. Res.*, **85**, 5244–5256.

McKenzie, D. P. & Brune, J. N., 1972. Melting on fault planes during large earthquakes, *Geophys. J. R. Astr. Soc.*, **29**, 65–78.

Mellman, G. R. & Helmberger, D. V., 1978. A modified first-motion approximation for the synthesis of body wave seismograms, *Geophys. J. R. Astr. Soc.*, **54**, 129–140.

Mendiguren, J., 1977. Inversion of surface wave data in source mechanism studies, *J. Geophys. Res.*, **82**, 889–894.

Menke, W. H. & Richards, P. G., 1980. Crust-mantle whispering gallery phases: A deterministic model of teleseismic Pn wave propagation, *J. Geophys. Res.*, **85**, 5416–5422.

Molotkov, L. A., 1961. On the propagation of elastic waves in media consisting of thin parallel layers, *Problems in the theory of seismic wave propagation*, Nauka, Leningrad, 240-251 (in Russian).

Müller, G., 1970. Exact ray theory and its application to the reflection of elastic waves from vertically inhomogeneous media, *Geophys. J. R. Astr. Soc.*, **21**, 261–283.

Müller, G., 1973. Amplitude studies of core phases, *J. Geophys. Res.*, **78**, 3469–3490.

Müller, G. & Kind, R., 1976. Observed and computed seismogram sections for the whole earth, *Geophys. J. R. Astr. Soc.*, **44**, 699–716.

Nolet, G., 1977. The upper mantle under western Europe inferred from the dispersion of Rayleigh modes, *J. Geophys.*, **43**, 265–285.

Nolet, G. & Kennett, B., 1978. Normal-mode representations of multiple-ray reflections in a spherical earth, *Geophys. J. R. Astr. Soc.*, **53**, 219–226.

O'Brien, P. N. S. & Lucas, A. L., 1971. Velocity dispersion of seismic waves, *Geophys. Prosp.*, **19**, 1–26.

O'Neill, M. E. & Hill, D. P., 1979. Causal absorption: Its effect on synthetic seismograms computed by the reflectivity method, *Bull. Seism. Soc. Am.*, **69**, 17–26.

Orcutt, J. A., Kennett, B.L.N. & Dorman, L. M., 1976. Structure of the East Pacific Rise from an ocean bottom seismometer survey, *Geophys. J. R. Astr. Soc.*, **45**, 305–320.

Panza, G. F., Schwab, F. A. & Knopoff, L., 1972. Channel and crustal Rayleigh waves, *Geophys. J. R. Astr. Soc.*, **30**, 273–280.

Pao, Y-H. & Gajewski, R. R., 1977. The generalised ray theory and transient responses of layered elastic solids, *Physical Acoustics*, **13**, ed. W. Mason, Academic Press, New York.

Peacock, K. L. & Treitel, S., 1969. Predictive deconvolution - theory and practice, *Geophysics*, **34**, 155–169.

Pearce, R. G., 1980. Fault plane solutions using relative amplitude of P and surface reflections: further studies *Geophys. J. R. Astr. Soc.*, **60**, 459–473.

Pekeris, C. L., 1948. Theory of propagation of explosive sound in shallow water, *Geol. Soc. Am. Memoirs no. 27.*

Pekeris, C. L., 1955. The seismic surface pulse, *Proc. Nat. Acad. Sci. U.S.A.*, **41**, 469–480.

Poupinet, G. & Wright, C., 1972. The generation and propagation of shear-coupled PL waves, *Bull. Seism. Soc. Am.*, **62**, 1699–1710.

Raitt, R. W., 1969. Anisotropy of the upper mantle, *The Earth's Crust and Upper Mantle*, ed. P. J. Hart, American Geophysical Union, Washington, D.C.

Randall, M. J., 1966. Seismic radiation from a sudden phase transition, *J. Geophys. Res.*, **71**, 5297–5302.

Rayleigh, Lord, 1885. On waves propagated along the plane surface of an elastic solid *Proc. Lond. Maths. Soc.*, **17**, 4–11.

Richards, P. G., 1971. Elastic wave solutions in stratified media, *Geophysics*, **36**, 798–809.

Richards, P. G., 1973. Calculations of body waves, for caustics and tunnelling in core phases, *Geophys. J. R. Astr. Soc.*, **35**, 243–264.

Richards, P. G., 1974. Weakly coupled potentials for high frequency elastic waves in continuously stratified media, *Bull. Seism. Soc. Am.*, **64**, 1575–1588.

Richards, P. G., 1976. On the adequacy of plane wave reflection/transmission coefficients in the analysis of seismic body waves, *Bull. Seism. Soc. Am.*, **66**, 701–717.

Richards, P. G. & Frasier, C. W., 1976. Scattering of elastic waves from depth-dependent inhomogeneities, *Geophysics*, **41**, 441–458.

Saito, M., 1967. Excitation of free oscillations and surface waves by a point source in a vertically heterogeneous Earth, *J. Geophys. Res.*, **72**, 3689–3699.

Schilt, S., Oliver, J., Brown, L., Kaufman, S., Albaugh, D., Brewer, J., Cook, F., Jensen, L., Krumhansl, P., Long, G. & Steiner, D., 1979. The heterogeneity of the continental crust; results from deep crustal seismic reflection profiling using the vibroseis technique, *Revs. Geophys. Sp. Phys.*, **17**, 354–368.

Scholte, J. G. J., 1956. On seismic waves in a spherical earth, *Koninkl. Ned. Meteorol. Inst. Publ.*, **65**, 1–55.

Schwab, F. & Knopoff, L., 1972. Fast surface wave and free mode computations, *Methods in Computational Physics*, **11**, ed. B. A. Bolt, Academic Press, New York.

Seidl, D., 1980. The simulation problem for broad-band seismograms, *J. Geophys.*, **48**, 84–93.

Sezawa, K., 1935. Love waves generated from a source of a certain depth, *Bull. Earthquake Res. Inst. Tokyo*, **13**, 1–17.

Smith, M. L. & Dahlen, F. A., 1981. The period and Q of the Chandler Wobble, *Geophys. J. R. Astr. Soc.*, **64**, 223–281.

Spudich, P. & Orcutt, J. A., 1980. Petrology and porosity of an oceanic crustal site: results from waveform modelling of seismic refraction data, *J. Geophys. Res.*, **85**, 1409–1434.

Stephen, R. A., 1977. Synthetic seismograms for the case of the receiver within the reflectivity zone, *Geophys. J. R. Astr. Soc.*, **51**, 169–182.

Stevens, J. L., 1980. Seismic radiation from the sudden creation of a spherical cavity in an arbitrary prestressed medium, *Geophys. J. R. Astr. Soc.*, **61**, 303–328.

Stokes, G. G., 1849. Dynamical theory of diffraction, *Trans. Camb. Phil. Soc.*, **9**,.

Stoneley, R., 1924. Elastic waves at the surface of separation of two solids, *Proc. R. Soc. Lond.*, **106A**, 416–420.

Sykes, L. R., 1967. Mechanisms of earthquakes and the nature of faulting on mid-ocean ridges, *J. Geophys. Res.*, **72**, 2131–2153.

Takeuchi, H. & Saito, M., 1972. Seismic Surface Waves, *Methods of Computational Physics*, **11**, ed. B. A. Bolt, Academic Press, New York.

Telford, W. M., Geldart, L. P., Sheriff, R. E. & Keys, D. A., 1976. *Applied Geophysics*, Cambridge University Press.

Thomson, W. T., 1950. Transmission of elastic waves through a stratified solid medium, *J. Appl. Phys.*, **21**, 89–93.

Titchmarsh, E.C., 1937. *An Introduction to the Theory of Fourier Integrals*, Oxford University Press.

Tolstoy, I. & Usdin, E., 1953. Dispersive properties of stratified elastic and liquid media, *Geophysics*, **18**, 844–870

Tolstoy, I., 1955. Dispersion and simple harmonic sources in wave ducts, *J. Acoust. Soc. Am.*, **27**, 897–910.

Vered, M. & BenMenahem, A., 1974. Application of synthetic seismograms to the study of low magnitude earthquakes and crustal structure in the Northern Red Sea region, *Bull. Seism. Soc. Am.*, **64**, 1221–1237.

Walton, K., 1973. Seismic waves in pre-stressed media, *Geophys. J. R. Astr. Soc.*, **31**, 373–394.

Walton, K., 1974. The seismological effects of prestraining within the Earth, *Geophys. J. R. Astr. Soc.*, **36**, 651–677.

Wang, C.Y. & Herrmann, R.B., 1980. A numerical study of P-, SV- and SH-wave generation in a plane layered medium, *Bull. Seism. Soc. Am.*, **70**, 1015–1036.

Ward, S.N., 1978. Upper mantle reflected and converted phases, *Bull. Seism. Soc. Am.*, **68**, 133-154.

Ward, S.N., 1980. A technique for the recovery of the seismic moment tensor applied to the Oaxaca, Mexico, earthquake of November 1978, *Bull. Seism. Soc. Am.*, **70**, 717–734.

Werth, G.C., 1967. Method for calculating the amplitude of the refraction arrival, *Seismic Refracting Prospecting*, ed. A. W. Musgrave, Society of Exploration Geophysicists,Tulsa, Oklahoma.

White, R.S., 1979. Oceanic upper crustal structure from variable angle seismic reflection/refraction profiles, *Geophys. J. R. Astr. Soc.*, **57**, 683–722.

White, R.S. & Stephen, R.A., 1980. Compressional to shear wave conversion in the oceanic crust, *Geophys. J. R. Astr. Soc.*, **63**, 547–565.

Wiggins, R.A. & Helmberger, D.V., 1974. Synthetic seismogram computation by expansion in generalised rays, *Geophys. J. R. Astr. Soc.*, **37**, 73–90.

Wiggins, R.A., 1976. Body wave amplitude calculations - II, *Geophys. J. R. Astr. Soc.*, **46**, 1–10.

Woodhouse, J.H., 1974. Surface waves in a laterally varying layered structure, *Geophys. J. R. Astr. Soc.*, **37**, 461–490.

Woodhouse, J.H., 1976. On Rayleigh's Principle, *Geophys J. R. Astr. Soc.*, **46**, 11–22.

Woodhouse, J.H., 1978. Asymptotic results for elastodynamic propagator matrices in plane stratified and spherically stratified earth models, *Geophys. J. R. Astr. Soc.*, **54**, 263–280.

Woodhouse, J.H., 1981. Efficient and stable methods for performing seismic calculations in stratified media, *Physics of the Earth's Interior*, ed. A. Dziewonski & E. Boschi, Academic Press, New York.

Zoeppritz, K., 1919. Erdbebenwellen VIIIB: Uber Reflexion and Durchgang seismicher Wellen durch Unstetigkeitsflachen, *Göttinger Nachr.*, **1**, 66–84.
Zvolinskii, A.V., 1958. Reflected waves and head waves arising at a plane interface between two elastic media, *Izv. ANSSSR, Ser. Geofiz*, **1**, 3–16.

Index

www.ingramcontent.com/pod-product-compliance
Lightning Source LLC
Chambersburg PA
CBHW061227270326
41928CB00025B/3387
9781921536724